STRUCTURAL MATERIALS TECHNOLOGY

An NDT Conference

Edited by

PAUL E. HARTBOWER
PHILIP J. STOLARSKI, PE

Sponsored by

THE CALIFORNIA DEPARTMENT OF TRANSPORTATION
THE FEDERAL HIGHWAY ADMINISTRATION
CALIFORNIA TRANSPORTATION FOUNDATION

FEBRUARY 20–23, 1996
SAN DIEGO, CALIFORNIA

TECHNOMIC
PUBLISHING CO., INC.
LANCASTER · BASEL

Structural Materials Technology—An NDT Conference

a TECHNOMIC publication

Published in the Western Hemisphere by
Technomic Publishing Company, Inc.
851 New Holland Avenue, Box 3535
Lancaster, Pennsylvania 17604 U.S.A.

Distributed in the Rest of the World by
Technomic Publishing AG
Missionsstrasse 44
CH-4055 Basel, Switzerland

10 9 8 7 6 5 4 3 2 1

Main entry under title:
   Structural Materials Technology—An NDT Conference, 1996

A Technomic Publishing Company book
Bibliography: p.
Index p. 387

Library of Congress Catalog Card No. 94-62454
ISBN No. 1-56676-424-6

HOW TO ORDER THIS BOOK
BY PHONE: 800-233-9936 or 717-291-5609, 8AM–5PM Eastern Time
BY FAX: 717-295-4538
BY MAIL: Order Department
Technomic Publishing Company, Inc.
851 New Holland Avenue, Box 3535
Lancaster, PA 17604, U.S.A.
BY CREDIT CARD: American Express, VISA, MasterCard
BY WWW SITE: http://www.techpub.com

PERMISSION TO PHOTOCOPY–POLICY STATEMENT
Authorization to photocopy items for internal or personal use, or the internal or personal use of spe-
cific clients, is granted by Technomic Publishing Co., Inc. provided that the base fee of US $3.00 per
copy, plus US $ .25 per page is paid directly to Copyright Clearance Center, 222 Rosewood Drive,
Danvers, MA 01923, USA. For those organizations that have been granted a photocopy license by
CCC, a separate system of payment has been arranged. The fee code for users of the Transactional
Reporting Service is 1-56676/96 $5.00 + $ .25.

Contents

SESSION 2
C2–Bridge Inspection Using Radar

S2–Fracture Critical Inspections

C2A–NDE of Bridge Foundations

S2A–Fatigue Life and NDT

Preface

I have been honored to have had the opportunity to serve as Chairperson of the Steering Committee for this conference. It has been rewarding to have been able to contribute to the continuation of the conference begun in 1994 in Atlantic City, New Jersey as a forum for the exchange of information about new and existing technologies in NDT and continues to bring NDT practitioners together from the Federal, state, academia and private industry to exchange ideas on how to better manage the infrastructure. I know that with the dedicated individuals involved in this year's conference, the future NDT Conferences will be a success. Thank you for this opportunity to serve you.

NANCY MCMULLIN-BOBB
Steering Committee Chairperson
FHWA

Before the 1994 NDT Conference ended, I was handed the "baton" to continue what Bob Scancella and others in the New Jersey Department of Transportation developed and named the Structural Materials Technology and Nondestructive Testing Conference.

As I discovered this was no small task to undertake. No task can be accomplished by one person but by the collective energy of many. My personal gratitude goes out to the following people: Nancy McMullin-Bobb and the FHWA for their support of the conference; Bob Scancella, Cheryl Wilson and Mary Ellen Callahan from the New Jersey Department of Transportation who encouraged me to keep the vision of the conference in focus and on track; my colleagues at CALTRANS including Paul Hartbower and Rosme Aguilar who supported me with their work on the Technical Committee and Erol Kaslan for his excellent work as Session Moderator Chairman; and finally, from CALTRANS, Linda Petavine and Carol Howard for their assistance in the development and mailing of the conference brochure.

My very special appreciation goes to the California Transportation Foundation (CTF) and Heinz Heckeroff of the CTF. Without the leadership of Heinz and the CTF this conference could not have happened. Thank you Heinz for being the clear thinker I needed when things were getting complicated.

I also thank my wife Cindy for being empathetic and understanding during the many nights and weekends I was preoccupied with the conference planning.

I leave the future NDT Conference Committees with the words of one of our great Presidents, Theodore Roosevelt. This is what it takes to be committed to the conference.

"It is not the critic who counts: not the man who points out how the strong man stumbled or where the doer of deeds could have done them better. The credit belongs to the man who is actually in the arena; whose face is marred by dust and sweat and blood; who strives valiantly; who errs, and comes short again and again, because there is no effort without error and shortcoming; who does actually try to do the deed; who knows the great enthusiasm, the great devotion and spends himself in a worthy cause; who at the worst, if he fails at least fails while daring greatly.

Far better it is to dare mighty things, to win glorious triumphs even though checkered by failure, than to rank with those poor spirits who neither enjoy nor suffer much because they live in the gray twilight that knows neither victory nor defeat."

PHILIP J. STOLARSKI
Program Chairman
CALTRANS

Developing NDT Technologies for the Next Century

S. B. CHASE

It is my distinct pleasure to be addressing this group again. I was privileged to make an address at the first Structural Materials Technology Conference in Atlantic City in February, 1994. At that conference, Dr. Karl Frank in his keynote address made an honest assessment that there had not been much advancement over the previous twenty years in nondestructive evaluation technology as applied to highway bridges. At that conference I presented an outline of a program which FHWA was initiating to identify and respond to the needs of the bridge inspection community. I would like to take this opportunity to bring you up to date on the status of that program.

The FHWA research and development program in nondestructive evaluation which I will describe is currently targeted at developing specific tools to meet the needs of the bridge inspection and management community. The highest priority needs identified are the detection and evaluation of fatigue cracks in steel bridges, the rapid assessment of asphalt covered bridge decks, global bridge evaluation, the evaluation of unknown foundations, the evaluation of prestressing steel in prestressed concrete, and integration of nondestructive evaluation into bridge management systems.

Finding fatigue cracks in steel bridges

The most important research need identified for steel highway bridges was a better way of detecting fatigue cracks under paint. The FHWA responded by soliciting for the development of an improved fatigue crack detection system in 1993. The result was the **NUMAC.**

NUMAC, an improved fatigue crack detection system. The NUMAC (New Ultrasonic and Magnetic Analyzer for Cracks), a new fatigue crack detection system combining ultrasonic and magnetic inspection capabilities into a single instrument has been successfully developed, demonstrated and delivered to FHWA. This system consists of a backpack computer, a heads-up display and features one hand operation (essential for use on a bridge). This system will greatly improve our capabilities to detect and quantify fatigue cracks in steel bridges, even though they may be covered with paint

Steven B. Chase, Research Structural Engineer, Federal Highway Administration, Turner-Fairbank Highway Research Center, 6300 Georgetown Pike, McLean, VA 22101-2296

Even better ways of finding fatigue cracks

While the NUMAC is a significant advance in fatigue crack detection technology, it still has some limitations. It is a contact system requiring the inspector to place a transducer on the bridge within a foot or so of where a crack is suspected. This is a severe impediment to the rapid detection of fatigue cracks given the number of possible location which need to be inspected on a typical highway bridge. The ideal technology would be non-contact and allow the inspector to rapidly scan an area. We needed a more innovative solution. In 1995 we solicited for innovative proposals through the Department of Transportation's Small Business Innovative Research (SBIR) Program. We asked for proposals for innovative yet feasible methods of detecting fatigue cracks in steel highway bridges using remote sensing. We received twenty-eight proposals and after review, recommended two for funding.

Thermographic Imaging to Detect and Quantify Fatigue Cracks in Steel Highway Bridges
The first project was initiated in October of 1995. It is based upon the use of high resolution thermographic imaging systems which are commercially available. The method will use both active and passive thermographic methods. In the active mode, a special pattern of hot and cold regions are created on the steel bridge using a heat lamp. The thermographic imaging system presents the operator with an image of heat flow patterns. If a crack is present, a characteristic pattern should be observed. In the passive mode, the thermoelasic and thermoplastic heating associated with the stress concentrations at the ends of active fatigue cracks will be detected. The objective of this six month study is to prove the concept by demonstrating the ability to detect paint covered fatigue cracks.

Use of Moire Deflectometry for Remote Sensing of Fatigue Cracks. The second SBIR project, also initiated in October 1995, is remote fatigue crack detection using optical moire pattern analysis. This six month study, will demonstrate the feasibility of using lasers to project a pattern of closely spaced , parallel lines on the surface of a steel. Under stress, the areas around the fatigue crack will deform. Using standard video technology, it should be possible to detect the characteristic, out-of-plane deformations near the crack with the moire pattern produced by the closely spaced lines of laser light.

What do you do if you find a fatigue crack ?

Finding fatigue cracks in steel bridges will be facilitated by these new technologies. However, not every fatigue crack is critical. Another important objective of FHWA's NDE R&D program is to develop better tools for the assessment of fatigue cracks. We are attaching this problem on two fronts. First, we are developing improved acoustic emission (AE) technology to determine if a fatigue crack is growing and if so how quickly and under what conditions and second, we are developing better tools to measure and quantify the fatigue loading experienced by a bridge.

Acoustic Emission Monitoring

Acoustic emission is a technology the FHWA has pioneered and developed over the last 20 years. This technology is based upon the use of very sensitive piezoelectric sensors which convert the very high frequency vibrations (several hundred kilo-hertz) generated by a propagating fatigue crack into electrical signals which can be detected and analyzed. The FHWA Office of Technology Applications recently completed a study which evaluated the capabilities and utility of AE for fatigue crack detection and monitoring. The study demonstrated that AE can be useful for the evaluation of a fatigue crack but that the available systems are not engineered for the bridge monitoring application.

Acoustic Emission Monitor for Bridges. In response, FHWA solicited for a cooperative agreement in 1995. In this solicitation we explained our desire to work with industry to co-sponsor the development of an acoustic emission monitoring system which was specifically engineered and packaged meet the need of monitoring a fatigue crack on an in-service highway bridge. The result was a cooperative agreement where the FHWA and an industrial partner are sharing the cost to develop this system. The new system will be small. ruggedized, battery powered and could be left in place for unattended monitoring for up to one week. It is important to note that this acoustic emission system could also be used to determine the effectiveness of a fatigue crack retrofit. This portable AE system will be very useful for monitoring and evaluating fatigue cracks.

Measurement and Quantification of Fatigue Loading

Once a fatigue crack has been found it is often desired to estimate the rate of growth of the crack. Once formed. fatigue cracks will increase as a function of the stress intensity and the number of stress cycles. The preferred method of determining these parameters is to measure the random, variable amplitude stress cycles at or near the crack. the FHWA is developing better tools for these measurements.

Wireless Strain Measurement System

One of the impediments to the measurement of fatigue loading is the need to install a strain gage near the fatigue crack and then to monitor the random, variable amplitude strains for a period long enough to capture the loading spectrum. With traditional strain gages this has proven to be difficult because of the difficulty of access to the locations where fatigue cracks typically form and the need for long wire runs back to a data acquisition system. What was needed was a portable, rugged yet accurate system for measuring strains at inaccessible locations.

The Small Business Innovative Research Program was used to sponsor the development of a wireless strain measurement system. This highly innovative system consists of rugged, battery powered (solar cells optional), radio transponder modules. These modules feature the ability to accept up to four standard resistive stain gages with all power and signal conditioning provided by the transponder. The system features 16 bit analog to digital conversion and an effective 500 Hz sampling rate. Up to ten of these transponders can be used simultaneously. They can be configured to form local radio telemetry networks with extensive data error checking and multi-path redundancy for very stable and accurate wireless data transmission. A local transponder is attached to a PC for data acquisition. This system should greatly facilitate the field measurement of fatigue loads.

A Passive Fatigue Load Measurement Device. The wireless strain measurement system is an excellent tool, but it is somewhat expensive (about $10,000 per transponder) and has a limited battery life. A totally passive and inexpensive device for the measurement of fatigue loading was needed. Under a new contract, initiated in October of 1995, a low cost, passive fatigue load measurement device will be developed. It is based upon the use of two, precracked fatigue coupons which strain along with the bridge. The cracks in the two coupons, made of different materials, grow at different rates. Special foil gages are attached to the coupons to accurately measure the lengths of the cracks. The measurement is made by plugging in a crack length reader into the device . It is possible to measure the fatigue loading by measuring how much the cracks have grown.

Fatigue Loading Measurement using Electromagnetic Acoustic Transducers As a part of a congressionally mandated study with the Constructed Facilities Center, West Virginia University, a device that will measure the cumulative fatigue loading of a typical highway bridge using an innovative strain measurement technology is under development. This technology uses electromagnetic acoustic transducers, the generation and detection of high frequency stress waves in steel using electromagnetic fields, to measure the strain in steel members. The advantage of this system is that it attaches magnetically to the steel bridge and no surface preparation is required.

Bridge Deck Inspection

Quantitative and rapid inspection of bridge decks, especially if they are covered with bituminous concrete, has been identified as a high priority need by the bridge inspection and bridge management community. The FHWA has two projects underway to address this need.

Development of Dual-Band Infrared Thermography Imaging System for Bridge Deck Inspection. This project is adapting defense technology developed for buried land mine detection to the quantitative inspection of bridge decks. Looking at bridge decks using two different infrared wavelengths simultaneously will overcome some of the operational problems (primarily surface emissivity variations) that have been experienced with infrared thermography as applied to the detection and quantification of delaminations on bridge decks. A first phase evaluation on test slabs demonstrated that dual-band infrared thermography could detect delaminations in both bare concrete and asphalt-covered concrete and that surface emissivity variations could be compensated for by application of image processing techniques. The current phase is to develop a fully operational mobile IR imaging system and to evaluate dual-band infrared thermography on actual bridge decks.

Ground-Penetrating Radar Imaging for Bridge Deck Inspection. This project will develop an engineering prototype of a new generation ground-penetrating radar for bridge deck inspection. The system will use impulse radar, synthetic aperture techniques, and sophisticated signal processing and imaging algorithms to image an entire lane width of a bridge deck at one time. The goal is a system that will travel at traffic speeds, image a lane width of a bridge, and provide two and three dimensional images of the interior of the bridge deck. Using a small scale prototype, preliminary tests have been able to provide images of the interior of a reinforced-concrete test bed that show test voids and reinforcement.

Global Bridge Evaluation

Another area where the bridge community has indicated there is a need for better methods is in global bridge evaluation. This means looking at the overall performance or response of a bridge as a system, rather than as individual components. An example of global bridge evaluation is static load testing.

Global Bridge Measurement using Coherent Laser Radar. This project is an adaptation of a system developed for NASA. It is a portable laser scanning system which will quickly measure the deflected shape of a bridge with sub-millimeter accuracy. It will also measure the vibration of the bridge and has potential to facilitate the application of modal analysis as a bridge inspection tool. The system has been demonstrated in the field and will be delivered to FHWA in the Fall of 1995.

Global Bridge Monitoring with Wireless Transponders. This project is developing a wireless bridge monitoring system. It will consist of a number of sensor/transponder modules which will communicate via microwaves to a local controller. The system development emphasizes the use of off-the shelf components developed for the cellular telephone and automotive applications and consequently minimizes technical risk. There will be modules measuring strain, rotation, deformation and vibration. It is the goal the develop technology which will make it possible to instrument a bridge at a dozen locations for a cost of less than $5000. The implications for bridge management are obvious.

Bridge Deflection Measurement using Precision Differential Global Positioning System. An innovative approach to monitoring and measurement of large bridges is being developed under another contract initiated in October of 1995. This project will make the transition from a proven concept into a operational system prototype. The concept has already been proved by research sponsored by the Louisiana and Texas DOT's. The objectives are to produce a system which can be easily configured, installed and affordably operated by state and local authorities. The system will provide a cost effective means of performing structural deformation surveys. The resolution of the system (sub-centimeters) limits its application to large bridges.

Bridge Overload Measurement and Monitoring using TRIP Steel Sensors. A major contributing factor to the deterioration of the nations bridges are overloads. These overloads could be caused by heavy trucks or a seismic event. A very useful tool from a bridge management perspective and as a tool for helping to make the most efficient use of bridge inspection resources would be a passive device which could detect and measure the maximum load experienced by a bridge. Continuing work which was initially sponsored by the Georgia Department of Transportation, a contract has been initiated in November of 1995 to develop such a system. The system is based upon the use of transformation induced plasticity (TRIP) steel sensors. This steel undergoes a permanent change in crystal structure in proportion to peak strain. It changes from a non-magnetic to a magnetic steel. This change can be measured easily. This project will improve the design and development of these peak strain sensors and test their performance on instrumented bridges. These sensors could provide a reliable, inexpensive and easy to implement means for quantitative bridge assessment as a key element of a comprehensive bridge management system.

Unknown Foundations.

Bridge Substructure Evaluation using Forced Vibration Response. Substructure deterioration is a major reason for structural deficiency of bridges. There is also a pressing need to evaluate substructures for scour vulnerability and for post earthquake evaluation. An innovative approach for quantitative substructure evaluation will be tested under a new contract initiated in November 1995. This project will use measured structural movements due to induced vibrations to determine the condition of the substructure. The response will enable engineers to determine the presence of piles and to establish a base line response for subsequent evaluations. This technology could help evaluate the scour vulnerability of the approximately 100,000 bridges with unknown foundations.

Evaluation of Prestressed Concrete Bridges

Prestressed concrete bridges have been characterized as bridges constructed of high strength steel cables which have been totally encased in concrete so as to be uninspectable. The global bridge evaluation technologies described above can be used to evaluate such bridges but there is also a need for more local inspection tools.

Impact Echo System for Detection of Voids in Post-tensioning Ducts. A project was recently completed which developed a device for the detection of voids in the grout on post-tensioned bridges. The long term reliability and safety of these bridges depends upon the integrity of the post-tensioning system. This device uses the impact echo principle of striking a concrete surface with a known energy pulse and measuring the local response using a piezoelectric transducer. The frequency and energy content of the response can be used to detect voids in grouted ducts. The system is smaller and more suited for use on vertical and irregular surfaces than another impact echo system designed primarily for bridge deck evaluation. The system is currently being evaluated at the Virginia Transportation Research Council.

Magnetic-Based System for NDE of Prestressing Steel in Prestressed Concrete. This project will develop a portable and versatile inspection tool for the detection and quantification of corrosion and strand breakage in prestressed concrete. The concept builds upon some of the technologies used in other projects. It is similar to the Magnetic Flux Leakage Inspection System for Bridge Cables and it also uses wireless communications. The system is built around a magnetic scanning head. The scanning head includes a strong permanent magnet, a Hall-effect sensor array (detects changes in magnetic field), a position encoder, and a wireless communications unit. This portable, self-contained system will be used to scan prestressed concrete girders and beam and telemetry the information back to a portable PC for signal processing and analysis. The results can be displayed as an image for rapid anomaly identification. It might also be useful for inspecting decks, columns and abutments.

Special Inspection needs. There are a few projects underway which are targeted at some very special inspection needs which have been identified by the bridge community.

Magnetic Flux Leakage Inspection System for Bridge Cables. This project is the continuation of the development of a specialized inspection system for bridge cables. A prototype system using an array of shunt able, permanent magnets has been designed and fabricated. This system detects the changes in the strong magnetic field set-up by the magnets if a broken or corroded cable is present. There are commercially available systems which inspect smaller cables using similar technology, but typical bridge stay cables are too large. The system is currently undergoing test and evaluation at the Volpe Transportation Systems Center.

Cable Stay Force Measurement Using Laser Vibrometers. Dynamic analysis will also be the basis for a new approach to the quantitative measurement of the forces in stay cables. This innovative approach will use non-contact laser vibrometers (commercially available) to provide a rapid, low cost, yet accurate method for force measurement. The forces in the stay cables are an excellent indicator of overall structural health for these type of bridges. The coherent laser radar system being developed in a separate project could also be used to perform dynamic cable stay measurements.

Integration of NDE with Bridge Management Systems

What is the best way to use new nondestructive evaluation technologies? The objective of the FHWA NDE research and development program is to address specific needs identified by the bridge inspection and bridge management community. The technologies being developed will provide new, and heretofore unavailable, capabilities for the quantitative evaluation of bridges and bridge components. For example, it will become feasible to quantitatively measure and track bridge deterioration and to develop new deterioration models. Condition states of bridge elements, the basis of modern bridge management systems, could be defined in terms of quantitative measurements rather than subjective visual indications of distress. There is a need to study the best way to use these new technologies within the context of an overall bridge management system. FHWA will be sponsoring two studies to begin to investigate these issues.

Integration of Quantitative Nondestructive Evaluation Methods into Bridge Management Systems. This first study will investigate how to develop a unified quantitative methodology for the integration of nondestructive bridge evaluation into bridge management systems. The study will establish relevant measures of damage for bridge components, it will establish formal links between the results of NDE measurements and condition states, and it will develop a methodology for NDE assisted bridge inspections. The methodology and procedure will be demonstrated in field inspections on at least twelve highway bridges in six states. Deliverables from this contract will include complete damage descriptions of all commonly recognized elements, a complete basis for the translation of NDE measurements to condition states and guidance in the application of NDE assisted inspections for highway bridges. This will be an ambitious step toward the development of bridge management for the next century.

The previous project is working toward a shift in the bridge management paradigm, and as such is somewhat future directed. These is a pressing immediate need, identified by the bridge management community, for a better method for asphalt covered reinforced concrete bridge deck condition state assessment. In response, FHWA has initiated a contract in November 1995 to look specifically at this problem.

Improved Bridge Deck Condition State Descriptions Using Quantitative Nondestructive Evaluation Methods. This study will take the results of previous research sponsored under the Strategic Highway Research Program in measuring and predicting the condition of concrete bridges decks. This study will develop methodology to improve bridge deck deterioration models, to develop a methodology to incorporate preventative treatments into bridge management systems, and to develop criteria to use nondestructive testing for these purposes.

The wide array of projects and technologies we are developing are targeted to meet the needs identified by this community. As I mentioned in my address in 1994, our success will be determined by how well we meet your needs. To be successful we cannot stop with simply producing a prototype system and then claim success. We need to thoroughly test and evaluation the technology in real world conditions with actual users. Recognizing this, the FHWA Office of Engineering Research and development was reorganized in 1995 to include a new engineering support division. This division's mission is to take the products of the research I have outlined and to put them through a rigorous test and evaluation program. This will involve end users and other FHWA engineers. Some of you might already have been contacted to help with program.

Laboratory Support Contract for Test and Evaluation of New Technologies. As can be seen from the wide array of new technologies being studied and under development the testing and evaluation of these systems will be a big job. Recognizing this, the FHWA awarded contract in October 1995 to provide technical, professional and logistical support to the Special Projects and Engineering Division for the testing and evaluation of these new systems.

Conclusion.

I have summarized FHWA's current research and development program developing new nondestructive evaluation technologies. We are beginning to take delivery of some of the first systems and will be looking for assistance and input from this audience as we thoroughly test and evaluate these systems. FHWA has benefited greatly by the willingness of this community to assist us in identifying what your needs are. This conference is an excellent opportunity to share your experience and to provide additional input to FHWA. Your active participation in this conference will help us develop the NDT technologies needed for the next century. I hope that twenty years from now, the keynote speaker at the 2016 Structures Materials Technology and Nondestructive Testing Conference can report that things have changed dramatically.

Nondestructive Testing and Assessment of Bridges

J. W. FISHER

ABSTRACT

Fatigue and fracture as well as loss of section due to corrosion are time-dependent performance characteristics of concern that have the potential to jeopardize the integrity of bridge structures. During the past 25 years these conditions have developed in a number of bridges resulting in loss of service, costly repairs and concern with the safety of these structures. Fortunately, few have caused loss of life as a result of collapse. A decade ago, the collapse of the Mianus River Bridge in Connecticut was one such structure. Both corrosion and fatigue cracking played a role in that structure's performance.

This paper will review the experience with such time-dependent damage over the 25 year period. It will examine the experience with crack and corrosion detection. Changes in structural behavior as a result of corrosion, deterioration of the deck and other adverse characteristics can result in unanticipated fatigue cracking as the analysis model and assumptions used in design are no longer valid. Often these changes in behavior are the cause of damage and crack development.

Damage assessment requires a knowledge of the live load stresses while the structure is in service so that reasonable estimates of safe service life can be made for critical structural details and inspection techniques and NDE tools can be rationally focused. Hence, methods of monitoring structural response and evaluating its live load stress range history are as important as crack detection and crack growth.

IMPACT OF CORROSION

Extensive corrosion damage is often observed in girders, floor beams, and stringers of steel girder bridges and in the truss members and their floor systems of truss bridges. The loss of material often results in complete corrosion penetration components.

Crevices or pockets exist at the edge of the angles used to connect end connections and flange angles to the girder web. Crevices have a tendency to corrode at a far faster rate than the adjacent flat metal areas. An example is the significant role that crevice corrosion played in the

John W. Fisher, NSF Engineering Research Center on Advanced Technology for Large Structural Systems, Lehigh University, 117 ATLSS Drive, Bethlehem, PA 18015-4729

development of the collapse of the Mianus River Bridge carrying I-95 in 1983 (Demers and Fisher, 1989; National Transportation Safety Board, 1983). Figure 1 shows the pin and remaining hanger that had supported the suspended span. A crack undetected prior to the collapse resulted in a segment of the pin supporting the hanger to fracture, as illustrated in Fig. 2. Final collapse immediately followed. Corrosion and the geometric changes it produced at the pin connection played a major role in the development of the cracked pin. This can be seen from a cross-section through the center of the pin, washers and girder web which is given in Fig. 3.

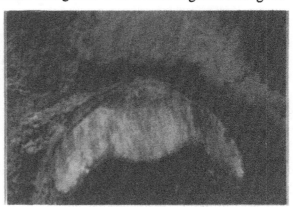

Fig. 1 Remaining Pin and Hanger Assembly   Fig. 2 Crack that Formed at End of Pin
        After Mianus River Bridge Collapse

The corrosion packout between the spacer washers and the hanger forced the outside hanger toward the end of the pin causing yielding and crack limitation in the pin and hangers (Demers and Fisher, 1989).

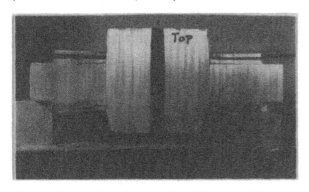

Although close visual inspection would likely provide an indication of geometric change and the existence of corrosion packout, other inspection tools and sensors which can identify active corrosion cells as well as the crack initiation and the unusual displacements that were observed are needed.

Fig. 3 Cross-Section of Pin, Web and Washers
        Showing Corrosion Packout

Another example of corrosion related cracking is the fixity that occurs at pin connection as a result of debris and contaminated water causing corrosion which can cause some degree of joint fixity. This can result in cyclic stresses that are up to an order of magnitude greater than the design stress range. Figure 4 shows the hanger of a deck truss that developed fatigue cracks in the riveted built-up sections (Demers and Fisher, 1989). Strain measurements demonstrated that the pin fixity caused bending stress as well as sudden joint release. This resulted in stress range conditions well above the fatigue resistance of riveted joints. Figure 5 shows a closeup view of the fatigue crack that initiated from a rivet hole in the web plate. The crack was arrested with a drilled hole.

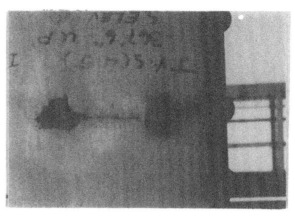

Fig. 4 Crack in Hanger Web Plate
Near Pinned Connection

Fig. 5 Close-up View of Fatigue Crack
From Rivet Hole

LOW FATIGUE STRENGTH DETAILS

The possibility of fatigue cracks forming at the ends of welded cover plates was demonstrated at the AASHO Road Test in the 1960's (Fisher and Viest, 1964). Multibeam bridges subjected to relatively high stress range cycles, 83 MPa, under controlled truck traffic experienced cracking after 500,000 vehicle crossings. In general, few cracked details were known to exist until a cracked beam was discovered in Span 11 of the Yellow Mill Pond in 1970. Between 1970 and 1981, the Yellow Mill Pond multibeam structures located at Bridgeport, Connecticut, developed numerous fatigue cracks at the ends of cover plates (Fisher, 1984). These cracks resulted from the large volume of truck traffic and the unanticipated low fatigue resistance of the large cover-plated beam members (Category E') (Fisher, Hausammann, Sullivan and Pense, 1979).

The stress range at these cover-plated details was measured and found to vary between 4 MPa and 35 MPa, which exceeded the constant amplitude fatigue limit. Hence, fatigue cracks formed because of the large number of stress cycles (more than 35×10^6).

Similar cover-plated beams on a ballasted deck girder railroad bridge are shown in Fig. 7. This structure was placed in service in the 1940's, and no cracks have formed to date. Measurements demonstrated that the effective stress range was about 10 MPa, however, the total accumulated stress cycles was only 20×10^6. Hence, no cracking would be predicted.

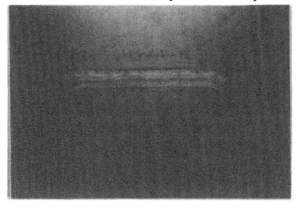

Fig. 6 Crack at End of Cover-Plated Weld Toe

Fig. 7 Cover-Plated Beams on Ballasted
Deck Girder Railroad Bridge

 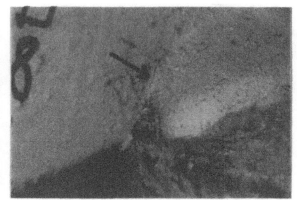

Fig. 8 Lateral Gusset Welded to Floor Beam   Fig.9 Small Weld Toe Crack on Flange Tip

It is important to detect very small weld toe cracks at fillet or groove welds - terminations of low fatigue strength details. An example of such a detail is the lateral gusset plate welded to the bottom of the floor beam flange seen in Fig. 8 (Demers and Fisher, 1989). This structure, located on a heavily traveled interstate (ADTT ~ 3500), was observed to develop small cracks at the ends of longitudinal welds along the floor beam flange tip, as can be seen in Fig. 9. Early detection allowed these cracks to be arrested and the detail retrofitted by peening the weld toe. Field measurements indicated the effective stress range in the floor beam flange was about 7 MPa with a maximum value of about 14 MPA.

LARGE INITIAL DEFECTS

Large initial defects and cracks make up the next category of cracked members and components. In several cases, the defect resulted from poor quality welds that were produced before nondestructive test methods were well developed. Many of these cracks occurred because the groove-welded component was considered a secondary member or attachment, and no weld quality criteria were established, nor were nondestructive test requirements imposed on the welds. Splices in longitudinal stiffeners frequently fall into this category (Fisher, 1984).

The fitness for use of some of these details and crack-like conditions can often be established from examination of the ultrasonic inspection tests, the calibration of these results to actual defects taken in core samples, the measurement of service stresses, and a crack growth assessment using fracture mechanics. Figure 10 shows a core with its edge polished and etched to establish the groove weld connection and the lack of fusion defect near midthickness. A view of the polished and etched surface at 400X in Fig. 11 shows the embedded slag and small crack extending from the lack of fusion condition. This structure had been in service for about 25 years on I-95, and the initial flaw was stable and had experienced no crack extension. The ultrasonic test defect rating for the lack of fusion condition was -7Db.

The core was saw cut and the defect exposed by cooling in liquid nitrogen and fracturing the net section to expose the crack surface. Figure 12 shows the exposed defect which has a crack-like width 2a = 3.8 min. It can be seen that the defect width is nearly uniform across the segment.

An examination of the crack tips in 9 core samples taken from the structure yielded no evidence of fatigue crack extension with the maximum defect $2a \cong 5$ mm. Measurements of the stress range at the flange groove weld demonstrated that the maximum stress range did not exceed 29 MPa.

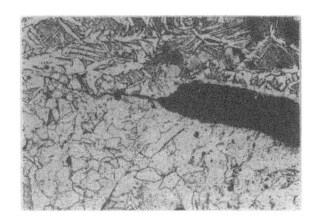

Fig. 10 Core removed from Groove Weld
at UT Indication of - 7Db.

Fig. 11 Small Weld Metal Crack at Root
Slag @ 400 X

An analysis of the defects can be made using the relationship

$$\Delta K = S_r \sqrt{\Pi a} \qquad (1)$$

This relationship yields ΔK = 2.6 MPa $\sqrt{m}$ for maximum defect and stress range which is below the fatigue crack growth threshold Δk_{th} ~ 3 MPa $\sqrt{m}$. Hence, no crack extension has occurred in these joints, and none will be expected in the future. Weld metal toughness was also found to provide good resistance to crack instability.

Figure 13 shows the ultrasonic test result dB levels from a number of sources as a function of the measured defect width, 2a. Significant scatter exists with defects between 2 and 13 mm which provided dB levels from +4 to -10. The bulk of the defects in this dB range are bounded between 1.5 and 5 mm. It is clear that the dB rating is not a suitable vehicle to determine the fitness for service of the joints. Defect size needs more accurate assessment normal to the cyclic stress in the groove weld. Characterization of the defect shape is required for a rational assessment.

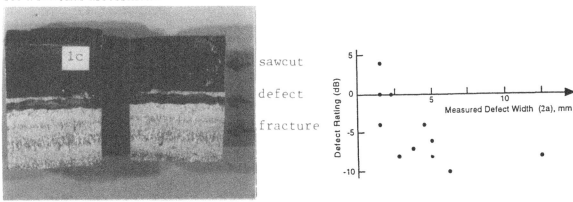

Fig. 12 Exposed Defect Surface

Fig. 13 Defect Rating vs. Measured
Defect Width

SUMMARY

The two primary time-dependent performance mechanisms of concern for aging steel structures are loss of section due to corrosion and fatigue and fracture. This paper has examined the performance of several structures that have developed distress from one or both of these characteristics.

Damage assessment requires information from NDE tools, models of damage accumulation, the characteristics of the material, particularly fracture resistance, and an evaluation of the actual service stress ranges that the structure experiences. This latter information is particularly needed when the structure's behavior changes, i.e. pinned members become fixed. The aging of our infrastructure insures ever increasing numbers of stress cycles and damage accumulation. Reasonable estimates of service life require better information on internal defects and weld toe cracks so that the damage assessment can be made.

REFERENCES

Demers, C. E. and J. W. Fisher. April 1989. "A Survey of Localized Cracking in Steel Bridges 1981 to 1988," Vol. 1, FHWA-RD-89-106.

Fisher, J. W. 1984. "Fatigue and Fracture in Steel Bridges, Case Studies," John Wiley and Sons.

Fisher, J. W., H. Hausammann, M. D. Sullivan and A. W. Pense. June 1979. "Defection and Repair of Fatigue Damage in Welded Highway Bridges," NCHRP Report 206, Transportation Research Board.

Fisher, J. W. and I. M. Viest. 1964. "Fatigue Life of Bridge Beams Subjected to Controlled Truck Traffic," Preliminary Publication, 7th Congress, IABSE.

National Transportation Safety Board. June 28, 1983. "Highway Accident Report: Collapse of a Suspended Span of Interstate Route 95 Highway Bridge Over the Mianus River, Greenwich, Connecticut," Report No. NTSB/HAR-84/03.

Session 1

C1—ULTRASONIC TESTING TECHNIQUES FOR CONCRETE

S1—ULTRASONIC TESTING OF STEEL

Ultrasonic Techniques for the Bonding of Rebar in Concrete Structures

A. A. AFSHARI, D. G. FRAZER AND R. C. CREESE

ABSTRACT

A through ultrasonic technique was used to evaluate the bonding and a semi-through ultrasonic technique was used to find the location of unbonded metal reinforcement rods in a concrete specimen by measuring and comparing the maximum amplitude response of received signals. In addition, a practical system for implementing the semi-through ultrasonic technique using a rotary sender transducer was developed and tested to reduce measurement times.

1. INTRODUCTION

Many types of discontinuities in homogeneous structures have been able to be identified by analyzing how the ultrasound passes through that structure. Discontinuities in non-homogeneous composite materials, such as reinforced concrete, however, have proven to be much more difficult to analyze using the same techniques. Since the integrity of reinforced concrete structures critically depends upon the bonding between the concrete and reinforcement rod, it would be advantageous to develop new ultrasound techniques that are capable of evaluating the interface between reinforcement rod and concrete surfaces.

In the past it has been shown that pulse-echo ultra-sonic techniques may not be well suited for identifying discontinuities in concrete because of the frequency constraints imposed on the applied signals, and the introduction of increased noise due to dispersion, absorption and scattering (Carlton et al.,1986). A through transmission technique has recently been described, however, which gives information concerning how well a metal reinforcement rod is bonded with concrete (Gaydecki et al., 1992).

The first objective of this study was to verify the use of a through transmission technique to determine the overall quality of bonding between concrete and individual reinforcement rods. The second objective was to examine a new semi-through transmission method for locating the coordinates of discontinuities in a structure caused by unbonded regions between cement and reinforcement rods, and finally this new technique was

Ali Akbar Afshari, Research Assistant, Industrial and Management System Engineering. Dept.
David G. Frazer, Adjunct Prof. Physiology Dept.
Robert C. Creese, Professor, Industrial and Management System Engineering Dept.
West Virginia University, Morgantown WV 26506

implemented and evaluated using a moveable transducer to decrease testing time. Results showed the semi-through technique can be used to locate unbonded regions between the cement and reinforcement rods.

2.   EXPERIMENTAL SYSTEMS

The pea gravel concrete specimen used in this study is illustrated in Figure 1. Once cast, the specimen was allowed to cure at room temperature for 28 days. Dimensions of the concrete were 86 cm in length, 51 cm in width and 20 cm thick.

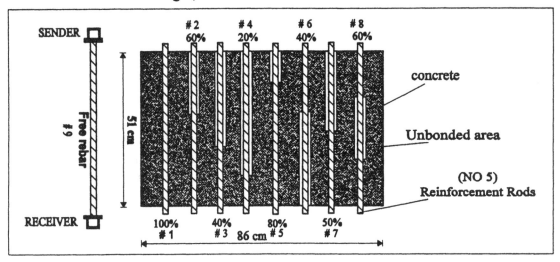

Figure 1- Diagram of a Concrete Specimen with Eight Imbedded Reinforcement Rods

Eight reinforcement rods 65 cm long were positioned in the concrete at 10 cm intervals at a depth of 7.5 cm. One was fully bonded and the seven remaining rods had a range of bonding lengths indicated in Figure 1. A ninth, 65 cm long, unbonded reinforcement rod was also prepared for comparison. Three ultrasonic methods were used to test the reinforced concrete specimen. The first used a through transmission technique in which transducers were attached to both ends of each of the nine reinforcement rods, including the free rod, to determine if the differences in bonding between the reinforcement rods and concrete could be detected. In this part of the study, a pair of transducers having a diameter of 2.54 cm and a central frequency of 250 kHz were used as shown in Figure 2.

The second technique used semi-through ultrasonic transmission to locate the region along each reinforcement rod which was not bonded to the concrete (Figure. 3-A) . Using this method a 7.5 cm diameter, 200 kHz ultrasonic sender was positioned on a two dimensional matrix drawn on the concrete surface. This matrix had 18 columns drawn normal to the reinforcement rods which were separated by 2.5 cm and 11 rows parallel to the rods that were separated by 1.25 cm. The 6th row was always chosen to be directly above the rod being tested.

The third method used the semi-through transmission technique performed with a 250 kHz sender mounted on a roller mechanism to accelerate the testing procedure (Figure 3-B).

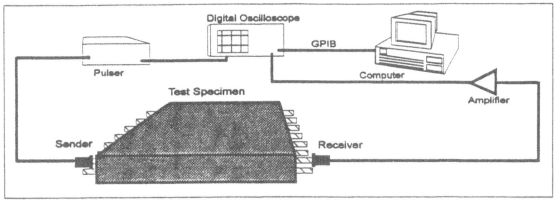

Figure 2- Configuration of Equipment to Measure Bonding Between Reinforcement Rod and Concrete Using an Ultrasonic Through Transmission Technique.

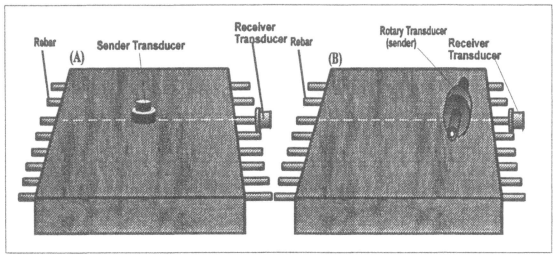

Figure 3- The Position of Transducer During Testing with Semi-through Transmission Technique Using the (A) Normal and the (B) Roller Sender Transducers

Transducers were coupled to either reinforcement rods or the concrete surface with petroleum jelly in all measurements, except those using the roller mechanism. The roller was coupled to the concrete surface through a thin layer of water. The sender was excited at its resonant frequency with a pulser (RITEC BP-9600) which generated a 400 V pulse every 10 msec. Pulse width was 2-μ sec for the 2.54 cm sender and 1.6-μsec for the 7.5 cm sender. The receiver was attached to an amplifier (RITEC 4000) whose output was recorded on a digital oscilloscope (Tektronix 2232). The oscilloscope was triggered by the pulser. Data collected on the oscilloscope was transferred to a computer (Zenith 386) through a GPIB interface. Individual digital records were analyzed and the results plotted using MATLAB software on a second computer (IBM 486).

3. EXPERIMENTATION

In the first set of experiments the through transmission technique was used to test unbonded, fully bonded, and partially unbonded reinforcement rods that are illustrated in Figure 1. During these tests the sender and receiver transducers were coupled to opposite ends of each reinforcement rod. The recorded time response of each received waveform for the nine individual reinforcement rods is shown in Figure 4. Table 1 lists the maximum amplitude response and percentage bonding for the same rods.

In the second set of experiments, illustrated in Figure 3-B, the semi-through transmission technique was used to retest the embedded reinforcement rods. The receiver transducer was attached to one end of a rod, and the sender was positioned at points of a rectangular matrix drawn on the concrete surface. In concrete the attenuation coefficient is greater than in steel (Krautkrämer *et al.*,1983). In order to minimize the effects of geometric attenuation, a different sender was chosen for these experiments whose diameter (D) was calculated by assuming its near-field distance (N) in concrete should be greater than the 7.5 cm ultrasonic path length between the sender and reinforcement rod and by assuming that the ultrasonic wavelength (λ) should be greater than the maximum gravel size (λ ≥ d=1.5 cm) according to relationship:

$$D \geq \sqrt{4N\lambda + \lambda^2} = 7.45 \, (cm) \qquad (1)$$

A sender transducer having a diameter of 7.5 cm and nominal frequency of 200-kHz was chosen for these experiments by assuming that the ultrasonic velocity in concrete was 3500 m/sec.

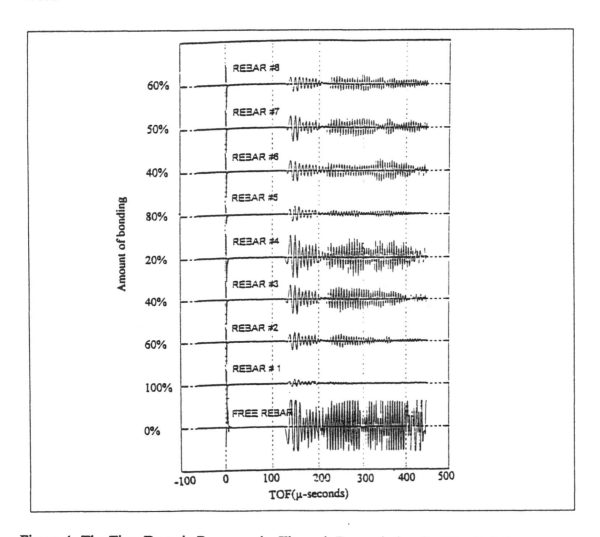

Figure 4- The Time Domain Response by Through Transmission for Nine Reinforcement Rods with Different Amounts of Bonding Shown in Figure 1.

The maximum amplitude response and percent bonding for each of the nine reinforcement rods illustrated in Figure 1 is shown in Table 1. A typical receiver amplitude response for a reinforcement rod (rod #7 in Figure 1) is illustrated in three dimensional form in Figure 5-A. This figure shows that a region of low amplitude responses corresponds to the location of the unbonded region of the reinforcement rod. The unbonding of a rod, which is normally caused by corrosion, was simulated by creating a small air gap between the reinforcement rod and concrete. This discontinuity due to the air gap increased the reflection coefficient between concrete and steel from 0.0017 % to 29%.

Table 1- Summary of Through Transmission Testing of Reinforcement Rods

Experiment #	% of Bonding	Gain (dB)	Amplitude (v)
1-full bonded	100% bonded	32	0.64 (8.9 %)
2-close to right edge	60% bonded	32	1.17 (20 %)
3-close to right edge	40% bonded	32	1.92 (34 %)
4-close to right edge	20% bonding	32	4.37 (76 %)
5-close to right edge	80% bonding	32	0.85 (15 %)
6-close to left	40% bonded	32	1.49 (26 %)
7- close to right	50% bonding	32	1.92 (34 %)
8- middle	60% bonded	32	1.28 (22%)
9-free in air	0% bonding	28	5.72 (100%)

When the semi-through transmission technique was implemented using the roller sender transducer, the path length was increased and ultrasonic wave intensity was decreased because of a decreased contact area between the transducer and concrete, but the total testing time was significantly reduced. Figure 5-B shows the results using the semi-through transmission technique when the roller sender was moved along the concrete and stopped at 6 locations directly above rod #7.

The results of both semi-through transmission techniques illustrate that the bonding between the concrete and reinforcement rod is responsible for determining how much ultrasonic energy is transferred through the concrete to the reinforcement rod. With diminished bonding between the concrete and rod, there is a decreased ultrasonic signal amplitude detected at the receiver.

4.   SUMMARY AND CONCLUSION

New ultrasonic methods for testing reinforced concrete structures were examined in this study. First, a through transmission technique described by others (Gaydecki *et al.*, 1992) was used to determine the quality of bonding between concrete and metal reinforcement rods. Second, a new semi-though transmission technique was designed and evaluated which was capable of locating the coordinates of discontinuities within a concrete structure caused by unbonded regions between the concrete and metal reinforcement rods. Finally, the semi-through technique was successfully modified and tested using a movable transducer in order to decrease the time necessary to evaluate a reinforced concrete structure. In the future it is expected that the semi-through ultrasonic technique will be used successfully in the field to locate and determine the length of bonding flaws in concrete structures, such as those made from prestressed concrete, which allow access to the metal reinforcement rods.

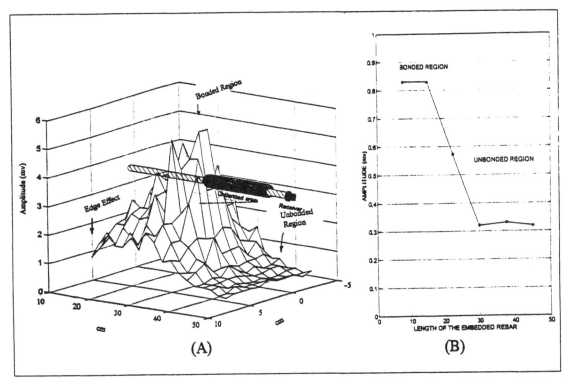

Figure 5- Display of Maximum Amplitude Response Signal in Locating of Unbonded Region in Concrete Using Semi-through Transmission Technique (A).Display of Amplitude Response Signal Versus The Length of Reinforcement Rod Using Roller Transducer (B).

5.   ACKNOWLEDGMENT

The research was performed by the Constructed Facilities Center, WVU, under the direction of Dr. Hota GangaRao. This project was sponsored by West Virginia Dept. of Transportation and National Science Foundation under contract No. WVDOH-T6994-CIDDMOC and also founded by West Virginia University.

6.   REFERENCES

1-Carlton H R and Muratore J.F., 1986. Ultrasonic Evaluation of Concrete IEEE Ultrason. Symp. 2 1017-20

2-Gaydecki,P. A., F M Burdekin, W Damaj, D. G. John, 1992. Digital Deconvolution of Ultrasonic Signals Influenced by the Presence of Longitudinally Aligned Steel Cables in Pre-stressed Concrete Meas. Sci. Technol. Vol 4, 909-917

3-Krautkrämer J. and Krautkrämer H., 1983 Ultrasonic Testing of Materials Berlin Springe

Design, Optimization, and Management of Cathodic Protection of Rebars in Bridges

R. SRINIVASAN, P. GOPALAN, P. R. ZARRIELLO, C. J. MYLES-TOCHKO,
J. H. MEYER AND D. E. TEAGLE

ABSTRACT

The future of cathodic protection (CP) as applied to rebars in concrete bridges is strongly dependent upon one important factor, namely the design of CP systems. The design, in turn, determines the effectiveness of CP in minimizing corrosion and the cost of implementation and maintenance. The major drawback of contemporary designs has been excessive flow of CP current in some parts of the bridge, and little or no current at others. The primary objective of our work is to achieve a uniform distribution of current over the entire structure at all times. Our long-term goal is to design "expert" CP systems controlled by a variety of current and environmental sensors and a dedicated microprocessor. Our short-term goals include: (1) development of numerical techniques that can be used in CP system designs; (2) remote sensing and mapping the distribution of CP currents in bridges; and (3) correlating the effect of micro-climatic changes to the distribution of CP currents. In this work, a Maryland state highway bridge which is presently under CP has been studied using a variety of sensors. Based on that study, a comprehensive approach to design and management of cathodic protection that maintains uniform protection at all times and in all locations in bridges is presented.

INTRODUCTION

Cathodic protection is a popular technique that is commonly used to minimize corrosion of metals in a wide variety of structures including, bridges, and parking lots. This technique is based on the principles of electrodekinetics, which can be briefly described as follows. In the absence of any polarization, a metal in contact with concrete or an electrolyte will remain at its corrosion potential (E_{cor}). At this potential, the metal surface sustains at least two reactions occurring at equal rates: a metal dissolution (or anodic metal oxidation) reaction, and a cathodic conjugate reaction, such as oxygen reduction or hydrogen evolution. If the metal is electrically polarized to potentials positive to E_{cor}, the metal dissolution reaction will be accelerated, while the cathodic conjugate reaction will be decelerated. The converse is true when the metal is polarized negative to E_{cor}. Thus, when the metal is polarized away from E_{cor} to a positive or negative value, a net anodic or a net cathodic current, respectively, will flow across the metal/electrolyte interface. A metal is said to be under cathodic protection (CP) when it is polarized sufficiently negative to E_{cor} to reduce the metal dissolution rate by three orders of magnitude or more.

R. Srinivasan, P. Gopalan, P. R. Zarriello, C. J. Myles-Tochko, J. H. Meyer and D. E. Teagle, Applied Physics Laboratory, The Johns Hopkins University, Laurel, MD 20723-6099

Under most conditions, a polarization of about 300 mV is sufficient to achieve cathodic protection. Excessive cathodic polarization should, however, be avoided to prevent onset of hydrogen evolution reaction, and to reduce the possibility of hydrogen embrittlement of the metal. Furthermore, cathodic polarization, like corrosion, is a surface process. Therefore, to achieve uniform protection at all locations on a given surface, it is imperative that the cathodic current density is uniform at all locations. Any nonuniformity in the current flow, especially with values less than some critical minimum, can cause localized variations in the metal dissolution rate. This can result in the structure corroding more severely in some places than the others. In a bridge, for example, if the CP current is nonuniformly distributed, those parts of the bridge that do not receive the current will continue to corrode, while those that do receive CP current will be well-protected from corrosion.

The installation cost of cathodic protection systems on a bridge is reportedly in the range of 15% of the cost of the construction of the bridge. In addition, the recurring expenses due to maintenance and management of CP over its lifetime can increase its cost by several folds. Therefore, in the future, the use of CP techniques is likely to be determined as much by cost considerations, as by its technical merits. There is a wide variety of options for reducing design and installation costs, as well as maintenance and management costs. The purpose of this work is to explore these options through the use of modern techniques that were not available or utilized in present CP system designs.

During the past 15 years federal and state highway administrations have been testing and assessing the merits of the CP technique. As a part of that initiative, over 300 bridges in North America are now under CP [1]. In most cases, the voltage source is a rectifier, and the ground-bed is either a palladium-coated titanium mesh or a conducting polymer mixed with concrete, which overlays the rebars. In general, the electrical connections between the rebars and the rectifier is made at one or two remote locations on the bridge. Similarly, the electrical connections between the ground-bed and the rectifier is also made at one or two remote locations. Thus, in most cases, the ground-bed is distributed evenly with respect to the rebars; however, the electrical contact points are highly localized. Most of these bridges are located in a wide variety of geographical locations, from Washington to Maryland, and Florida to New York. Thus, intended or not, they are exposed to a wide variety of environmental and climatic conditions.

If the past 15 years of history were considered as the first step toward exploring CP as a viable technique to protect bridges from corrosion, then the results from those studies are somewhat mixed at best. During the past 10 years, there has been a profusion of reports on the status of CP in several state and federal bridges, (e.g., see References 1-4). These reports suggest that cathodic protection has had mixed success and appears to be more effective on some bridges [1,2], as compared to others [3,4]. The causes of failures have been attributed to a variety of reasons, ranging from inadvertent shutdown of the rectifiers to improper electrical connections. Studies correlating climatic and environmental factors to the effectiveness of CP have yet to appear in any published literature.

There are several obvious steps one can take to improve the CP technique. For example, the CP system can now be designed more efficiently than in the past, with greater emphasis on uniform distribution of current and potential throughout the bridge. Using numerical techniques such as the finite element method (FEM), one can predict the geometric arrangement of ground-beds and the ideal locations for the electrical contacts vis-a-vis the geometry of the bridge and the rebars. Variations in the electrical conductivity of the concrete overlayers, the presence of coated and uncoated (bare) rebars, and spatial variations in the level of exposure to moisture and salt can all be easily incorporated into the CP design. Furthermore, after implementation, the validity of the FEM results can be verified through mapping of the CP current over the entire bridge. CP current mapping can be done using magnetic sensors.

The main objective of this work is to demonstrate the possibility of improving the CP technique, and making it fully functional and completely automated. As a first step toward that goal, we show the use of FEM technique to predict current and potential distribution in concrete, and magnetic sensors to map currents, in prototypical models as well as over the entire deck of a CP-protected, in-service concrete bridge.

CP CURRENT MAPPING AND NUMERICAL SIMULATION

Computer modeling of current and voltage distribution and mapping of the current distribution are the two most important steps used in the present work. Therefore, a brief introduction to finite element analysis and magnetometer sensors is in order. In this work, the tool used to model the current and voltage distribution in concrete is the finite element method (FEM). Magnetometer sensors are used for mapping the current. These sensors are lightweight instruments and can be powered by batteries. They are totally isolated and insulated from their surroundings, and can be operated while buried in concrete, immersed in water, or left under any condition that surrounds bridges, able to measure A. C. and D. C. currents, are vector instruments, that measure both amplitude and flow direction of the current. Their spatial resolution decreases with the increase in the liftoff distance between the sensors and the rebars; when the sensor is 6 inches away from the rebar, it will measure the current that is present over a 6-inch length of the rebar. Note that magnetic sensors are not reference half-cells; they are not used to measure polarization potentials, and are not limited by resistive or iR drop across the concrete resistance. These properties of the sensors are quite useful in mapping localized distribution of the CP current.

The FEM analysis used in numerical simulations can be conducted both in two and in three dimensions, using 486-type computers. A wide range of dimensions, from the large size of a bridge deck to the fine thin coating on rebars can be incorporated simultaneously in FEM models. Linear and nonlinear properties of electrical conductivities (as in the case of metal/electrolyte interface) can also be simultaneously incorporated in them. Similarly, different kinds of materials properties can all be included in the models. The FEM analysis can be used to simulate potentiostatic (constant potential) or galvanostatic (constant current) conditions that are present in CP systems. FEM analysis can be used to predict potential, current and magnetic field distributions in bridges.

CURRENT DISTRIBUTION MAPPINGS ON A MODEL CONCRETE BLOCK

As the first step toward using computer simulation and current mapping on in-service bridges, the concept of optimizing CP system designs was verified on a small, steel-reinforced, 12"x12"x6" block of concrete . The block (Figure 1) contained two layers of six #5 rebars (0.5" diameter), three on top and three on the bottom. These rebars were bare; each was 14 inches long, with 12 inches buried inside the concrete, and 2 inches (1 inch on each side) projecting outside the concrete. The three rebars in the top layer (Rebars 1, 2 and 3) were shorted to each other. The three rebars in the bottom layer (Rebars 4, 5 and 6)were also shorted to each other. The rebars on the top layer was cathodically polarized, while the rebars

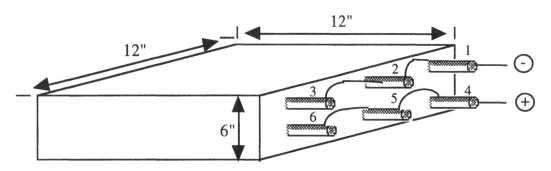

Figure 1: Schematic representation of the model concrete block.

in the bottom layer were used as the ground return, under two different configurations described below. The objective was to identify the effect of the electrical contact configuration on the distribution of the CP current.

ASYMMETRIC ELECTRICAL CONTACTS

In an asymmetric electrical configuration, Rebar 1 in the top layer was connected to the negative terminal of the voltage source. Rebar 4 in the bottom layer was connected to the positive terminal. The resulting currents determined by the FEM analysis is shown in Figures 2a and 2b. Figure 2a shows the CP current that is present on one of the rebars in the block. The current between the rebar and the concrete is shown in Figure 2b. The magnitude of the current is maximum at that end of the rebar which has the electrical contact, and minimum at the end far away from it.

The current map, obtained using a magnetic sensor on the actual concrete block is shown in Figure 3, which also shows that the current is unevenly distributed over the block. Clearly, both the numerical simulation and the experimental data show that although the rebars are good electrical conductors (relative to concrete), the choice of asymmetric electrical contact to them can cause a large flow of electrical current in one end of the pipe, and relatively much less at the other end. What is most surprising is that nonuniformity can occur even when the size of the concrete block is as small as 12"x12"x6". The only way that the observed asymmetry can be explained is that: (1) the current is limited by the resistance due to concrete; (2) currents tend to go through the path of least resistance, and (3) the path close to the contact

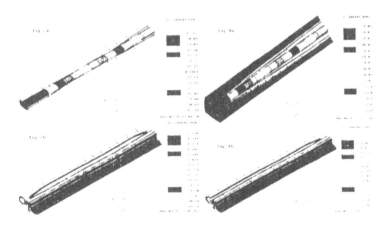

Figures 2 & 4: Current distribution on: (a) one of the rebars; and (b) from rebar to concrete, as obtained from the FEM analysis, for asymmetric (Fig. 2), and symmetric contacts (Fig. 4).

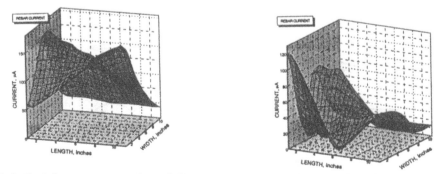

Figures 3 & 5: Magnetometer-based CP current map made over the concrete block in Figure 1 under asymmetric contact conditions (Fig. 3) and symmetric contact conditions (Fig. 5).

points, perhaps, is least resistive. In order to test this hypothesis, we repeated the experiment using a symmetric electrical connection, which is described in the next section.

SYMMETRIC ELECTRICAL CONTACTS

In a symmetric electrical configuration, Rebars 1, 2 an 3 were all connected on both ends to the negative end of the power source. Similarly, Rebars 4, 5 and 6 were all connected on both ends to the positive end. As in asymmetric case, the current distribution was determined by numerical simulation, and direct mapping was done with a magnetometer. The results of the FEM analysis are shown in Figures 4a and 4b. The current mapped with the sensor is shown in Figure 5. Both the simulation and direct mapping show a fairly uniform distribution of current in the rebars and block.

CP CURRENT MAPPING ON A 93x133-FOOT CONCRETE BRIDGE

Figure 6a shows a steel-reinforced concrete bridge located in Maryland. The schematic of the bridge is shown in Figure 6b. The bridge is 93 feet long in the east-west direction and 133 feet wide in the north-south direction. The bridge deck is cathodically protected by a single rectifier. The deck has two layers of uncoated rebars, one on top, and other on the bottom, with concrete in between. All rebars are shorted to one another. A palladium-coated titanium mesh, spread over the entire bridge, is placed over the top layer of the rebars, and acts as the ground return. A latex-concrete mix covers the titanium mesh. A 133-foot-long conducting bar, which runs along the north-south axis, placed at about 46 feet from the west end of the bridge, is connected to the titanium mesh. A point contact made to the conducting bar at about 60 feet from the north end of the bridge, is connected to the positive terminal of the rectifier. The negative terminal of the rectifier is connected to the rebars at two locations along the west end of the bridge. Thus, the bridge is a textbook combination of a uniformly distributed ground return laid over a uniformly distributed rebars, with the non-textbook condition of remote electrical connections.

CP currents were mapped from the top of bridge deck. For this purpose, the deck was divided into a matrix of several parallel and perpendicular lines at intervals of 10 feet. At each intersection of these lines, the currents flowing along the east-west axis and the north-south axis were measured using magnetometer sensors. The resulting current maps are shown in

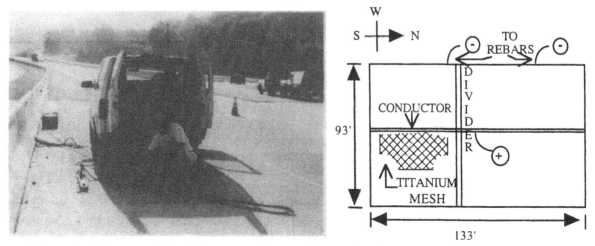

Figure 6a: Picture of the CP protected, steel-reinforced, in-service concrete bridge.

Figure 6b: Schematics of the electrical contacts from the rectifier to the rebars and ground bed.

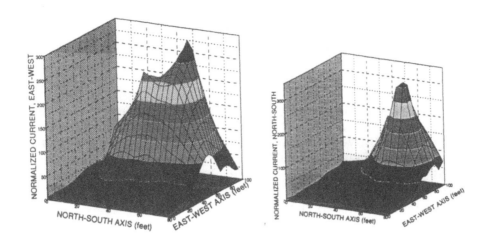

Figure 7: CP current distribution along: (a) the east-west axis; and (b) the north-south axis of the bridge in Figure 6. The current was mapped using a magnetometer.

Figures 7a and 7b. Note that these two figures show only an 80x80-foot part of the 93x133-foot area of the bridge deck; the amplitude of the CP current in rest of the deck is less than 0.1 µA. The CP currents are concentrated only in the northwest part of the bridge. Furthermore, while the current in the north-south direction covers approximately a 40x40-foot area, the current in the east-west direction covers even less. The reason for the concentration of all the currents in the north-west section of the bridge is due to the locations of the electrical contacts. While the geometry of the ground return, namely a mesh spread over the entire deck, appears intuitively correct, it did not help achieve uniform current distribution. The CP current mapped with magnetometer sensors over the entire deck of the bridge shows that more than 60% the area did not receive any current.

SUMMARY

The key to the success in cathodic protection is the ability to achieve uniform distribution of current over the entire metal/electrolyte interface. This appears easy to achieve under the controlled conditions of a laboratory, using relatively simple electrode geometry. However, this is not necessarily the case under field conditions, as can be seen from several published reports on the corrosion status of over 300 bridges in North America that are currently under CP. The reason for the failure of CP in these bridges and - by extension, future bridges - should be critically evaluated in terms of: (1) the uniformity of distribution of the CP current; (2) the design of not only the ground bed vs. rebar geometry, but also the locations of electrical contacts; and (3) the effect of the environmental and climatic factors on the distribution of CP current.

We have demonstrated here that even the use of a uniformly distributed ground bed spread over the entire deck does not ensure uniform distribution of CP current; we used a magnetometer to map the CP current on an in-service bridge in Maryland. In this case, the nonuniformity in the current distribution appears to be due to improper choice of locations for the electrical contacts. We have also shown here that it is possible to identify the ideal geometry of the ground bed and the electrical contact points through the use of a numerical simulation technique known as the finite element method.

In order to ensure proper cathodic protection, the design of a CP system should not only include electrode geometric parameters, but also the spatial and temporal effects of micro-environmental and micro-climatic factors that affect cathodic reaction. In other words, temperature, humidity, wetness, oxygen and chloride concentration, and pH should all be included as a part of the design, maintenance, and management of CP systems. Thus, the future designs of cathodic protection ought to be based on a sensors-based feedback systems that use microprocessor-controlled rectifiers, which we have called "expert" CP systems.

REFERENCES

1. Broomfield, J. P. 1994. "Life Prediction of Corrodible Structures," Ed.: R. N. Parkins, NACE International, Houston, Texas, 1994, p.632.
2. Jurach, P. J. 1981. "An Evaluation of Effectiveness of Cathodic Protection of Seven Bridge Decks," FHWA, Sacramento, CA, California Division, Report #624180.
3. Telford, E. R. 1986. "Cathodic Protection Check-out: SR-24, Yakima River Bridge," FHWA, Olympia, WA, Washington Division, Report #WA-RD-87.2.1.
4. Telford, E. R. 1986. "Cathodic Protection and Corrosion Monitoring Check-out: -I-405-SR 522 Interchange, Woodinville, Washington," FHWA, Olympia, WA, Washington Division, Report #WA-RD-87.1.1.

Measuring Dielectric Properties of Portland Cement Concrete: New Methods

I. L. AL-QADI, R. MOSTAFA, W. SU AND S. M. RIAD

ABSTRACT

Electrical characterization of Portland cement concrete (PCC) is necessary for reliable operation of electromagnetic nondestructive evaluation (NDE) techniques. Examples of such techniques include ground penetration radar (GPR). To better understand the dielectric properties of PCC, a study was conducted to evaluate PCC complex permittivity and magnetic permeability over a wide band of frequencies (100 KHz-10 GHz) using both time domain and frequency domain techniques. To achieve this, three measuring devices were designed and built: a parallel plate capacitor, a coaxial line fixture and a TEM horn antenna. The parallel plate capacitor covers the low radio frequencies (RF) characterization (0.1- 40 MHz). The coaxial setup covers the frequency spectrum of 100-1000 MHz while the antenna covers the microwave frequencies (1-10 GHz). The PCC specimen under test assumes different geometry for different setups: a prism for capacitor setup, a cylinder for the coaxial setup, and a rectangular slab for the antenna setup. This paper presents the design of the three setups.

INTRODUCTION

Nondestructive testing (NDT) may be an important tool in assessing the performance and characteristics of constructed facilities specially for the purpose of preserving existing structures through maintenance and rehabilitation activities. With time, Portland cement concrete (PCC) structures deteriorate due to freeze-thaw damage, reinforcing steel corrosion and aggregate/matrix separation caused by chemical reactions. Inspections based on visual observations fail to reveal internal deterioration until it reaches an advanced stage and damage is visible on the surface. Thus to avoid costly repair and maintenance work at advanced stage of damage, it is necessary to assess the condition of the structures quantitatively on a regular basis. Nondestructive testing techniques can be employed in situations which require no

Imad L. Al-Qadi, The Via Department of Civil Engineering, Virginia Polytechnic and State University, Blacksburg, VA 24061-0105

Raquibul Mostafa, Wansheng Su, and Sedki M. Riad, The Bradley Department of Electrical Engineering, Virginia Polytechnic and State University, Blacksburg, VA 24061-0111

The Center for Infrastructure Assessment and Management, Virginia Polytechnic and State University, Blacksburg, VA 24061-0105

perturbation to the structures. These techniques are quicker and cover larger area than the existing testing techniques. Among the current testing techniques are direct contact measurements such as chloride content and corrosion current of reinforced PCC structures, but they lack the desired accuracy and reliability.

Some nondestructive techniques are usually based on electromagnetic (EM) characterization of test material. Electromagnetic characterization of PCC can be utilized to reveal information about its composite, which in turn can be used to reveal information about its mechanical response properties. The EM properties of interest in this regard can be conductivity, permittivity and permeability. These EM properties are related to the composite properties of the aggregate, aggregate size, water to cement (w/c) ratio, chloride content, internal flaws, and reinforcing steel. The mechanisms of deterioration in PCC result in changes in local and bulk dielectric properties of the material. Thus, early damage detection may be possible.

The EM properties of PCC and the effect of deterioration on them are currently investigated in an ongoing project at Virginia Tech. This involves wideband (0.1 MHz- 10 GHz) dielectric characterization of PCC specimens employing both time domain and frequency domain techniques. The complete frequency spectrum is covered by three different EM configurations: parallel plate capacitor (0.1-40 MHz), coaxial transmission line (100-1000 MHz) and TEM horn antenna (1-10 GHz). The development of the parallel plate capacitor setup has been reported in (Al-Qadi et al., 1994a), while the coaxial fixture and measurement techniques have been reported in (Al-Qadi et al., 1994b,1995).

This paper presents an overall description of wideband characterization. The parallel plate capacitor fixture is investigated solely in frequency domain while both frequency domain and time domain measurements were employed in the coaxial and antenna setups. For the coaxial and antenna configurations, the emphasis was on time domain measurements because of the simplicity of measurement technique, accuracy and efficiency concerning time required to perform measurements. These aspects are important when characterization involves a significant number of large test specimens.

THEORETICAL BACKGROUND

PLANAR TRANSMISSION LINE (PARALLEL PLATE CAPACITOR)

The measurement techniques are based on planar transmission line principles. The measuring device assumes a plane transmission line in the form of a parallel plate capacitor configuration. The specimen under test forms the dielectric media for some part between the plates. A lumped capacitor model can be used to represent the test specimen.

A signal flow diagram of the measurement system is shown in figure 1. Γ_M and Γ_L are the complex reflection coefficients at the capacitor terminals and at the specimen load, respectively. The interface between the specimen and the measuring instrument is modeled by a scattering parameter model. S_{11}, S_{12}, S_{21} and S_{22} are the complex S-parameters of the interface network, which correspond to the complex signal gains of the signal between the network ports. The interface network models the propagation of the measurement signal across the connector, the capacitor terminals, and the capacitor through the center.

In terms of impedance on both sides of a discontinuity, the complex reflection coefficient can be expressed as

$$\Gamma = \frac{Z_L - Z_0}{Z_L + Z_0} \tag{1}$$

where Z_L is the load impedance and Z_0 is the characteristic impedance of the transmission line.

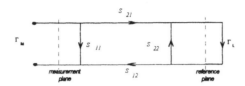

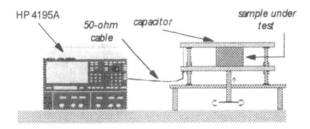

Figure 1. Signal flow graph representation and a schematic of the parallel plate capacitor setup.

The reflection coefficient seen at the capacitor terminals can be expressed as

$$\Gamma_M = S_{11} + \frac{S_{21}.S_{12}.\Gamma_L}{1 - S_{22}.\Gamma_L} \tag{2}$$

The mathematical development needs to incorporate the calibration of the measuring device. Microwave measurement principles identify the error sources involved in the measurement process and try to minimize them by standard calibration techniques. The reflection from load terminal, Γ_L, gets modified by the interface or embedding network and thus necessitates de-embedding process for accurate material characterization. This leads to calibrating for obtaining the S-parameters of the interface network. Three known standard terminations are adequate to determine the three unknowns, S_{11}, S_{22} and $S_{21}.S_{12}$. For open, short and matched standard terminations, the reflection coefficients at the load plane are the following:

$$\Gamma_{Lopen} = 1 \tag{3a}$$
$$\Gamma_{Lshort} = -1 \tag{3b}$$
$$\Gamma_{Lload} = 0 \tag{3c}$$

With the substitution of the above Γ_Ls in equation 2, one can come up with the following set of equations for determining the S-parameters of the interface network:

$$S_{11} = \Gamma_{Mload} \tag{4}$$

$$S_{22} = \frac{\Gamma_{Mopen} + \Gamma_{Mshort} - 2\Gamma_{Mload}}{\Gamma_{Mshort} - \Gamma_{Mopen}} \tag{5}$$

$$S_{21}.S_{12} = (\Gamma_{Mopen} - S_{11}).(1 - S_{22}) \tag{6}$$

With a knowledge of the above S-parameters as a function of frequency, the reflection at the load plane, Γ_L, can be determined from the measured Γ_M from the following equation:

$$\Gamma_L = \frac{\Gamma_M - S_{11}}{S_{21}.S_{12} + S_{22}(\Gamma_M - S_{11})} \tag{7}$$

The calibrated impedance of the specimen load is therefore given by

$$Z_L = \frac{1 + \Gamma_L}{1 - \Gamma_L} \tag{8}$$

The load impedance thus derived can represent the capacitor model in terms of dimensions and permittivity of the specimen.

An admittance model can be assumed for the capacitor in terms of complex capacitance, C^*, given by

$$Y_L = \frac{1}{Z_L} = j2\pi f(C^* - C_0) = G_L + jB_L \tag{9}$$

where Y_L is the complex load admittance, G_L and B_L correspond to load conductance and susceptance, respectively, f is the frequency, and $j = \sqrt{-1}$. C_0 is the capacitance of an equivalent specimen of air. In equation 9, the difference (C^*-C_0) is used since C_0 is included as part of the open circuit calibration.

With the assumption of uniform electric field distribution at the center of the capacitor and homogeneous specimen, the complex capacitance is given by

$$C^* = \varepsilon_0 \varepsilon_r^* \frac{S}{d} = \varepsilon_0 (\varepsilon_r' - j\varepsilon_r'') \frac{S}{d} \tag{10}$$

where ε_0 is the free space permittivity (= 8.854×10^{-12} F/m), ε_r^* is the complex relative permittivity of the specimen, ε_r' and ε_r'' are the real relative dielectric constant and the relative dielectric loss factor, respectively.

The capacitance of the air-filled capacitor is given as follows:

$$C_0 = \varepsilon_0 \frac{S}{d} \qquad (11)$$

The dielectric constant and the loss factor can be expressed as

$$\varepsilon_r' = 1 + \frac{B_L.d}{2\pi f S \varepsilon_0} \qquad (12a)$$

$$\varepsilon_r'' = \frac{G_L.d}{2\pi f S \varepsilon_0} \qquad (12b)$$

The loss tangent can be defined as

$$\tan \delta = \frac{\varepsilon_r''}{\varepsilon_r'} \qquad (13)$$

COAXIAL TRANSMISSION LINE

Frequency Domain Measurement

The measurement techniques involved in this case involve coaxial transmission line geometry to contain the specimen. The tested specimen is assumed to be placed between two reference or measurement planes in the case of two-port measurement and at one reference plane in one-port measurement. The measurement techniques assume a scattering parameter or S-parameter model for the specimen under test. The four parameters of this model, S_{11}, S_{21}, S_{12} and S_{22} are determined after a calibration process by a network analyzer as a function of frequency. Among the four parameters, S_{11} and S_{21} account for reflection from and transmission through the specimen and indicate the electrical properties of it. Reflection and transmission at a discontinuity is dependent upon the properties of the material providing the discontinuity. The information about electrical properties of a material such as permittivity and permeability are embedded in the S-parameters.

Assume the specimen has a length d, an impedance Z and a propagation constant γ_d. Two sides of the specimen are taken as two reference planes. On both sides of the specimen, the adjacent media has characteristic impedance of Z_o as shown for a coaxial fixture in figure 2. The S-parameter model for the specimen is shown in the same figure.

The complex reflection coefficient, ρ, of the specimen can be expressed in terms of impedance as given in equation 1. In terms of relative permittivity, ε_r, and permeability, μ_r,

$$\rho = \frac{\sqrt{\mu_r/\varepsilon_r} - 1}{\sqrt{\mu_r/\varepsilon_r} + 1} \qquad (14)$$

A parameter τ, incorporating the propagation constant can be expressed as

$$\tau = e^{-j\omega\sqrt{\mu_r\varepsilon_r}\frac{d}{c}} \qquad (15)$$

where radian frequency, ω, $=2\pi f$ and c is the velocity of light in free space.

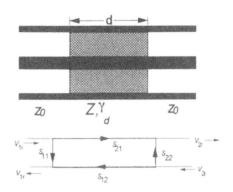

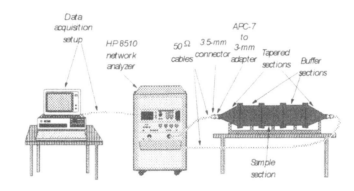

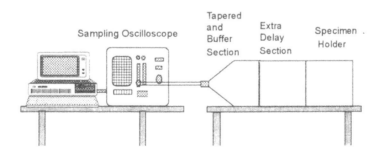

Figure 2. Signal flow graph representation and a schematic for the coaxial transmission line setup (frequency domain and time domain, respectively).

The S-parameters can be related to the reflection coefficient, ρ, and the parameter τ. In terms of ρ and τ, S-parameter can be written as

$$S_{11} = \frac{\rho(1 - \tau^2)}{1 - \rho^2 \tau^2} \tag{16a}$$

$$S_{21} = \frac{(1 - \rho^2)\tau}{1 - \rho^2 \tau^2} \tag{16b}$$

From equations 14 and 15, it can be noted that ρ and τ are function of ε_r and μ_r; thus two transcendental equations can be formed whereby ε_r and μ_r can be solved iteratively. To overcome this tedious process of solving for ε_r and μ_r, the method proposed in (Nicolson and Ross, 1970) is followed:

$$\varepsilon_r = \frac{1-\rho}{1+\rho} \frac{j\ln(1/\tau).c}{2\pi fd} \tag{17a}$$

$$\mu_r = \frac{1+\rho}{1-\rho} \frac{j\ln(1/\tau).c}{2\pi fd} \tag{17b}$$

Thus, over a particular range of frequency, a complete set of frequency dependent S_{11} and S_{21} will lead to determination of ε_r and μ_r.

Time Domain Measurement

Measurements are performed in time domain by employing time domain reflectometry or transmission (TDR or TDT) principles. This involves injecting an electrical signal from a signal generator to the test fixture and observing the reflected and transmitted waveforms in a sampling oscilloscope. From these time domain signals S_{11} and S_{21} of the specimen can be evaluated and the electrical properties can be determined. S_{11} and S_{21} can be related to the time waveforms by the following expressions:

$$S_{11} = \frac{F(V_r(t))}{F(V_i(t))} \tag{18a}$$

$$S_{21} = \frac{F(V_t(t))}{F(V_i(t))} \tag{18b}$$

where, V_i, V_r, V_t are incident, reflected and transmitted waves, respectively and F denotes Fourier Transform.

Under the condition of free-space permeability for PCC ($\mu_r \cong 1$), only the reflected signal is sufficient to determine ε_r. This involves a time record of incident waveform and the reflected waveform. The measurement technique uses the principle of gating the first reflection from the specimen and utilizing this gated waveform to determine the complex reflection coefficient in frequency domain. The reflection coefficient in this condition equals S_{11} of the specimen (Miller,1986). Conversion from time domain to frequency domain is done by Fourier Transform of step signals (Sharaawi and Riad, 1986). From equation (14),

$$S_{11} = \rho = \frac{1-\sqrt{\varepsilon_r}}{1+\sqrt{\varepsilon_r}} \tag{19}$$

Thus, ε_r can be determined from S_{11} by the following equation

$$\varepsilon_r = \left(\frac{1-S_{11}}{1+S_{11}}\right)^2 \tag{20}$$

OPEN SPACE RADIATION

The principle of two-port measurement is applied in this case. A scattering parameter model or S-parameter model is assumed for the specimen. The specimen, a 60×60×13 cm PCC slab, because of its geometry and overall homogeneity, provides a symmetrical S-parameter model where $S_{11} = S_{22}$ and $S_{21} = S_{12}$.

Assume that the length of the specimen is d and it has impedance Z and propagation constant γ_d. The two sides of the specimen are taken as the two reference planes. The adjacent media has characteristic impedance Z_0. Figure 3 shows the schematic arrangement of the measurement principle.

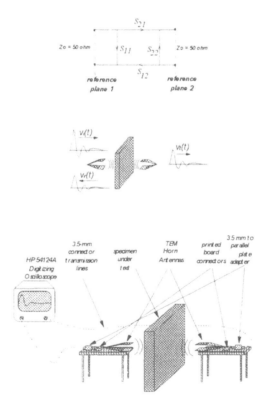

Figure 3. Schematic of measurement principle and experimental setup for open space antenna.

Portland cement concrete is nonmagnetic in nature and can be assumed to have the same permeability as that of free-space, i.e. $\mu_r = 1$. Thus, the only unknown is ε_r and either a measurement of S_{11} or S_{21} suffices to determine ε_r. If S_{21} is measured, then in terms of ε_r it can be written as (from equation 16[b])

$$S_{21} = \frac{4\sqrt{\varepsilon_r}\,\tau}{\left(1 + \sqrt{\varepsilon_r}\right)^2 - \left(1 - \sqrt{\varepsilon_r}\right)^2 \tau^2} \tag{21}$$

The expression above can be rearranged to form the following transcendental equation:

$$\left[x + \frac{1}{x}\right]\sinh(\gamma x) + 2\cosh(\gamma x) - \frac{2}{S_{21}} = 0 \qquad (22)$$

where $x = \sqrt{\varepsilon_r}$ and $\gamma = j\omega\frac{d}{c}$.

Equation 22 can be solved by numerical techniques to yield ε_r as a function of frequency.

DESIGN OF FIXTURES AND EXPERIMENTAL SET-UPS

CAPACITOR FIXTURE

A parallel plate capacitor configuration has been designed to investigate the low RF behavior of the PCC specimens. A schematic of the designed fixture is shown in figure 1. Each plate is about 35×35 cm and made of stress-relieved steel. The upper plate is fixed to the base by means of four supporting rods which pass through the lower plate, one at each corner, while the lower plate is mounted on a threaded rod at the center, which allows it to move vertically. The lower plate is maintained in a horizontal position by means of 5-cm long sleeves on the supporting rods. This setup allows a range of 5-13 cm displacement between the two plates. The dimensions of the test specimen are 15×15×7.5 cm.

The experimental setup consists of the capacitor fixture, a network analyzer and connecting cable and adapters. The impedance of the capacitor is measured using an HP 4195A Network Analyzer. The setup is calibrated to account for errors due to connections as well as errors due to the signal propagation across the capacitor plates. The complex permitivitty then can be calculated using the equations developed in the theory section.

COAXIAL FIXTURE

The coaxial fixture is designed with dimensions that can accommodate large PCC specimens. The principal mode of propagation is the TEM mode. The material used for the conductors is brass. The coaxial fixture is a symmetrical device in design and consists of the following sections:
- Two pieces of Tapered Section.
- Two pieces of Buffer Section.
- One piece of Specimen Holder.

The test specimen is cylindrical with a brass rod at the center, as a center conductor. The outer diameter of the specimen is 15 cm and its length is 15 cm; the diameter of the brass rod is 2.5 cm. The fixture begins at one end with the tapered section. This section combines with the buffer section to form one end for input signal. The buffer or isolation section is followed by the specimen holder. The other side of the fixture is symmetrical about the specimen holder. The whole fixture when assembled takes the form of a large cylindrical object with two conical ends.

A two-port measurement technique is employed. An HP 8510 network analyzer is used for calibration and determining the S-parameters of the PCC specimen (figure 2). An S-parameter model is assumed for the specimen holder section containing the specimen. After the calibration process, the network analyzer yields the *S*-parameters of the specimen which can then be processed to obtain the required electrical properties. Thru, Reflection and Line (TRL) method was used for calibration (Engen and Hoer, 1979). Thru standard was realized by directly connecting the two buffer sections. A circular brass plate was placed at the end of the buffer section to provide the short or Reflection standard. The Line standard was achieved by using an empty specimen holder and this provided the reference impedance for the measurement system. A program (Hewlett-Packard, 1988) was used to perform the calibration and provided corrected measurement data.

For the time domain measurements, a sampling oscilloscope (Tektronix 11801A) was used. An extra buffer section was added to provide an additional time delay (figure 2). This extra delay facilitates the gating process by maintaining the waveform smooth around the reflection from the specimen.

The measurements involve two steps. The first is to acquire a reference waveform, which can be obtained by placing a metallic plate at the end of the delay section. The metal plate, also made from brass, acts as a short circuit and reflects back the input waveform. The second waveform is obtained by connecting the specimen holder containing the specimen, to the delay section.

TEM HORN ANTENNA

Among many antenna configurations available, TEM horn antenna is chosen in this research for its faithful broadband operation. As time domain measurements are involved, the requirements on an antenna are different from ordinary frequency domain antennae. To preserve the shape of a time waveform, an antenna operating in time domain should have a narrow impulse response, and this in turn implies very broadband operation. The design of the TEM horn antenna used in this research follows the design guidelines developed at NIST (Ondrejka et al.,1992). The dimensions of the antenna were kept close to the ones of the antenna from NIST. This is done to maintain the same level of performance. The antenna basically consists of three sections:

- •TEM horn
- •Parallel plate extension.
- •Phase shifter.

The PCC slab is illuminated by one of the antennae acting as the transmitter. The transmitted wave is picked up by the receiving antenna on the other side of the PCC slab. Two different waveforms, received by the receiving antenna, are acquired by the sampling oscilloscope: one without the PCC slab and the other, with the PCC slab. The propagation media between the slab and the antennae is air and is assumed to be homogeneous and isotropic. The experimental setup is shown in figure 3.

The transmitted and reflected signals or time signatures can be observed on the sampling oscilloscope. A digital sampling oscilloscope, Tektronix 11801A is used in the setup. The discontinuities between air and the slab in the signal path are manifested as amplitude changes in the time waveforms. A suitable time window is chosen for the waveforms so only regions of interest, i.e. changes indicating the discontinuities, can be observed and analyzed.

TYPICAL RESULTS

PCC specimens of different geometry were cast for dielectric characterization. Frequency domain technique was used at low RF while both frequency domain and time domain techniques were used for the higher RF and microwave frequency range. Figure 4 shows dielectric characterization of PCC specimens over broad frequency spectrum (100 KHz-10 GHz) by three different EM configurations. The dielectric constant ε_r' starts at a high value and decreases gradually to a constant value. The loss component ε_r'' exhibits high values in the low frequency range and decreases with increasing frequency. The loss is mainly due to the water contained in the PCC pores; polarization losses form water molecules contribute to this loss. At high frequencies, this loss decreases.

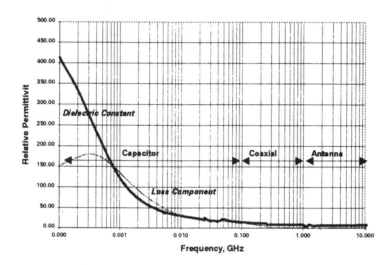

Figure 4. Broadband characterization of PCC.

CONCLUSIONS

Wideband dielectric characterization of PCC was studied using EM measurement techniques. Three different EM configurations were used to cover the total spectrum (100KHz-10GHz). Determination of ε_r was based on impedance measurements in all three cases. Both frequency domain and time domain techniques were employed in the measurements. For coaxial and antenna setup the main emphasis was on time domain measurement. Frequency domain measurements involved full calibration schemes: for capacitor setup, it was one-port calibration and for coaxial, it was two-port calibration. For capacitor and coaxial setups, ε_r was determined from analytical solutions of the relevant equations whereas numerical techniques were used for antenna method. Permittivity plot over the broad frequency spectrum, as obtained from three separate measurement methods, exhibited gradual change with small discontinuities at the transition regions between the different techniques. The results revealed the frequency dependent dielectric behavior of PCC.

REFERENCES

Al-Qadi, I. L., O. A. Hazim, W. Su, N. Al-Akhras, and S . M. Riad. 1994a. "Variation of Dielectric Properties of Portland Cement Concrete during Curing Time over Low RF Frequencies," *Transportation Research Board*, Paper No. 940479, 73rd Annual Meeting, Washington, DC.

Al-Qadi, I. L., W. Su, S. M. Riad, R. Mostafa, and O. A. Hazim. 1994b. "Coaxial Fixture Development to Characterize Portland Cement Concrete," Proceedings of the Symposium and Workshop on Time Domain Reflectometry in Environment, Infrastructure and Mining Applications, Evanston, Illinois. SP 19-94: 443-452.

Al-Qadi, I. L., S. M. Riad, R. Mostafa, and W. Su. 1995. "Design and Evaluation of a Coaxial Transmission Line Fixture to Characterize Portland Cement Concrete," Proceeding of the 6th International Conference on Structural Faults and Repairs, London, UK, :337-347.

Engen, G. F. and C. A. Hoer. 1979. "Thru-Reflection-Line: An Improved Technique for Calibrating the Dual Six-port Automatic Network Analyzer," *IEEE Transaction on Microwave Theory and Techniques*, MTT-27(12):987-993.

Hewlett-Packard. 1988. "HP 8510-B Network Analyzer Operating and Programming Manual," Santa Rosa, California.

Miller, E. K. 1986. "Time Domain Measurements in Electromagnetic," Van Norstrand Reinhold Co., New York, NY.

Nicolson, A. M. and G. F. Ross. 1970. "Measurement of Intrinsic Properties of Materials by Time Domain Techniques," *IEEE Transactions on Instrumentation and Measurement*, IM-19(4) :377-382.

Ondrejka, A. R., J. M. Ladbury, and H. W. Medley. 1991. "TEM Horn Antenna Design Guide", NIST Technical Report, Boulder, CO.

Shaarawi, A. M. and S. M. Riad. 1986. "Computing the Complete FFT of a Step-Like Waveform," *IEEE Transactions on Instrumentation and Measurement*, IM-35(1):91-92.

ACKNOWLEDGEMENT

This research has been sponsored by the National Science Foundation, grant No. MSS-9212318. The writers would like to acknowledge the support of the Structural, Systems and Construction Process Program at NSF, especially Ken P. Chong. The writers also appreciate the help of Rami Haddad, Denson Graham, and Brett Farmer. Thanks go to other members of the Virginia Tech Center for Infrastructure Assessment and Management, Gary Brown, John Duke, and Richard Weyers.

Ultrasonic Inspection of Bridge Pin and Hanger Assemblies

R. D. GESSEL AND R. A. WALTHER

ABSTRACT

Bridge pin and hanger assemblies are typically considered as critical elements whose failure may result in partial or complete collapse of the structure. As a result these elements are subjected to a more in-depth examination and testing. Pin and hanger inspection programs typically employ ultrasonic methods to assure that cracks and other discontinuities affecting pin/hanger performance are identified. However, the use of ultrasonic methods can prove difficult considering the complex geometry of typical pin elements which might include keyways, center bore holes, changes in pin diameter, threads, and cotter pin holes. Wear grooves and acoustic coupling further complicate ultrasonic pin testing. These factors and their influence on the success of the pin/hanger inspection are discussed.

Ultimately, the success of a pin/hanger inspection program is dependent upon establishment of a thorough ultrasonic test procedure specific to the bridge under consideration. The key elements of an effective inspection program are highlighted in this paper. The use of straight and various angled beam transducers will be discussed. The influence of signal frequency and transducer diameter on flaw identification will be reviewed. Construction of a model reference pin/hanger can aid in calibration of the ultrasonic inspection procedure.

INTRODUCTION

Bridges in various parts of the world have suffered catastrophic collapse following the failure of fracture critical pin and hanger link connections. After the Silver Bridge collapse in 1967, the United States issued a federal mandate requiring periodic inspection of bridge structures. Inspection programs utilizing ultrasonic techniques have been found to be thorough and economical. The key components of an effective ultrasonic pin/hanger examination are discussed in this paper.

Pin and hanger link connections used in bridges permit unrestrained movement from longitudinal expansion or contraction of adjacent spans. Typical connections consist of upper and lower connecting pins linked by hangers, as shown in Fig. 1. The upper pin penetrates the web of

Robert D. Gessel, Wiss, Janney, Elstner Associates, Inc., 83 South King Street, Suite 820, Seattle, Washington 98104

Richard A. Walther, Wiss, Janney, Elstner Associates, Inc., 330 Pfingsten Road, Northbrook, Illinois 60062

Suspended Girder   Cantilevered Girder

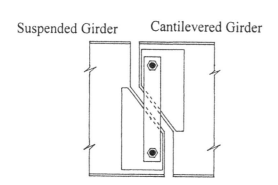

Fig. 1 - Typical pin and hanger connection

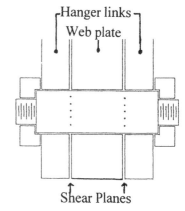

Fig. 2 - Shear planes of pin/hanger assembly

the cantilevered bridge girder, while the lower pin penetrates the suspended girder web. Such connections, where failure of a component may result in partial or total collapse of the structure, are classified as fracture critical.

Failure of the pin or hanger occurs when imposed forces exceed the resistance of the component. Factors influencing pin or hanger failure include but are not limited to cross-sectional area loss due to corrosion or wear, fatigue cracking, material or fabrication discontinuities, inadequate maintenance, unanticipated loadings resulting from "frozen" connections, and construction practices.

EXAMINATION PLAN AND PROCEDURES

The primary objective of an effective pin and hanger inspection program is to identify cracks or other defects that may eventually result in failure of either component. Failure to detect cracks at an early stage may have severe consequences; however misinterpretation of non-relevant indications may result in costly, unnecessary overhaul of pin and hanger link assemblies. Therefore, establishment of specific inspection parameters and evaluation criteria is essential to the successful examination of the pin and hanger assembly. ASTM A388, *"Standard for Ultrasonic Examination of Heavy Steel Forging"* is sometimes referenced with respect to the inspection of pins. However, procedures specific to the bridge under consideration which address the unique aspects of the pin and hanger connection are preferred.

As represented in Fig. 2, a typical pin will experience loading along planes oriented transverse to the pin axis and located at the shear plane between the hanger and girder web plate. Consequently, this area is of prime interest for ultrasonic examination. For hangers, the location of minimum cross section is the primary area of interest. For an effective ultrasonic examination of the critical area, sufficient access for the ultrasonic beam must be established. Examination procedures must adequately avoid or compensate for obstacles that may impede the detection of potential cracks or other discontinuities.

Pin or hanger geometry will control the selection of transducer size, frequency, beam angle, and scanning pattern. Typical obstacles in pins that may be encountered include threaded ends, stepped diameter changes, nut retention pin holes, and center bore holes. In many cases, pin ends provide the only practical access for the transducer introducing the examining ultrasonic wave.

In some instances the center bore hole may permit access for inspection using a specially designed angled transducer to scan from inside the bore hole.

Preliminary procedures are established by calculating beam spread and refraction angles to determine the region of the pin volume that may be accessed by the ultrasonic beam. An accurate graphic representation of the axial cross sectional geometry may be used to demonstrate the theoretical path of the ultrasonic beam within the pin. The benefits or consequences of beam spread from various transducer size and frequency combinations should be considered. Avoiding confusing reflectors and/or early returning signals resulting from mode conversion is nearly as important as reaching the critical shear plane at the outer circumference of the pin with the scanning ultrasonic beam.

Transducers do not always perform in practice exactly as they do in theory. Actual frequency and element size may vary somewhat from the nominal size/frequency, and material acoustic characteristics may also have a minor effect on beam angle and distance calibrations. It is therefore important to consider effects of minor variations while carefully calibrating test equipment to a verifiable standard.

Prior to inspection, defects or discontinuities which may potentially be encountered should be determined. The capabilities of various transducers may then be evaluated with respect to the potential reflectors in order to determine an appropriate test set-up. Larger (1 in. diameter) transducers exhibit the greatest penetration potential and produce smaller angles of divergence than smaller transducers with the same frequency. In some instances however the larger divergence angle produced by a smaller transducer provides the angle that is necessary to scan the critical area of the pin circumference. Good resolution and sensitivity characteristics are necessary for all transducer selections. Higher frequency transducers using low angled (6 to 10°) shoes are sometimes required to access the critical inspection area while avoiding interference from pin boundary surfaces. However, penetration capability with higher frequencies will be limited. In most cases transducer sensitivity for the various sizes, angles or frequencies which might be used are sufficient to detect a 0.125 in. notch in a reference standard. Incidence angle, beam divergence, access and range capabilities are significant factors that should be considered in the inspection procedures.

A test piece should be fabricated that may be used to qualify the ultrasonic examination procedures and calibrate for inspection. The test piece is used to prove the calibration of the function of the ultrasonic beam and to provide a suitable reference standard. A full scale replica of the pin or hanger with fabricated discontinuities at critical sections is the ideal reference standard. Size, weight, and fabrication costs often render such reference standards impractical. However, a reasonable facsimile of a portion of the pin or hanger fabricated from a 1 in. thick mild steel plate or bar stock will suffice under most circumstances. An example of a pin reference standard is shown in Fig. 3. A notch can be cut at the distant shear plane to function as a reflector

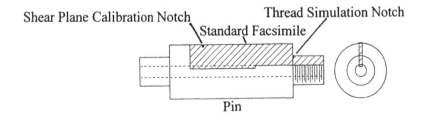

Fig. 3 - Sample pin and standard

for the sensitivity level calibration. Notches, grooves, or holes may be used to represent intrusive threads, diameter or length changes, or other obstructions to the ultrasonic beam. Reference standards that adequately demonstrate characteristics of the ultrasonic beam may also be adapted for other pins with similar geometries. Such use may require restricted transducer movement or cutting additional reference notches. Reference standards for bridge pins and hangers are not intended to be used to establish acceptance criteria on the basis of direct comparison to the indication level from the reference notch. Inspection parameters including transducers and angles used, as well as reference standard dimensions and notch depth should be recorded for reference in evaluation or future examinations.

Successful ultrasonic examination is heavily dependent upon the qualifications of the inspector, and the type and quality of equipment used. Instrument and personnel qualifications should be in accordance with the procedure developed and desired results.

FIELD INSPECTION OF PINS

In actual ultrasonic examination, compensation may be necessary to adjust for potential acoustic differences between the pin and reference standard. Reflections from the pin geometry oriented transverse to the pin axis at the end distant from ultrasonic beam entry may be used to establish the approximate attenuation ratio between the two materials. Sensitivity level is then adjusted by calculating the relative ratio. Reflections from geometric surfaces may also be used as "markers" to more accurately establish the location of the actual shear plane. Scanning levels of 10 to 14 db above the notch reference indication are usually appropriate.

Once approximate attenuation and sensitivity ratios, as well as shear plane locations are determined, it is often advantageous during inspection of the critical shear plane area to reduce the range of the flaw detector screen. The instrument display may be adjusted to correspond more closely to the width of the critical area. The trace may be delayed to display primarily the critical area of the pin. Instrument gating of the critical inspection area may also enhance the inspection. The on-screen separation of potential indications may be appreciable for pin lengths that exceed 20 inches or where geometric reflections are near potential shear plan indications. In addition the five or ten inch screen range is more readily interpreted for location of indications. Care must be used to assure proper delay of the flaw detector signal trace to coincide with the critical area of the pin. Tolerance must be provided for minor variations in length or relative shift of the shear plane. It is preferred to include a geometric reflector within the display range so that the quality of the ultrasonic beam transmission may be monitored. An initial scan traversing the full radius and length of the pin may be used to detect gross discontinuities, and to prequalify the quality of ultrasound transmission for subsequent examination of the critical area.

Preparation of the transducer contact surface at the pin end is important to maintain relative uniformity of the ultrasonic wave energy entering the pin. Surface conditions that permit intimate contact with the transducer are acceptable. Poorly adhered, non-uniform, or heavy build-up of paint may adversely affect the quality of the ultrasonic wave transmission and should be removed. Surfaces suffering significant corrosion, pitting or dents from hammer impact may require grinding to adequately transmit the examining ultrasonic wave. Care must be exercised when grinding is used for surface preparation to minimize deviations from flatness, and to maintain a surface which is normal to the pin axis.

The scanning plan for each pin will be successful when designed to systematically access the full circumference of the shear plane area. The complexity of the scan is dependent on the pin configuration. When using an angle beam transducer it is imperative to orient the beam toward the circumference of the pin. It may be necessary to scan the pin from both ends in order to achieve satisfactory coverage of the pin circumference at the shear plane. A scanning pattern using a sequential oscillation and movement of the transducer can be adapted to most pins when using an axial technique. Supplementary manipulation of the transducer may be necessary to overcome or minimize the effects of geometric obstructions.

Interpretation of indications and ultimate evaluation of the pin may be dependent upon logical deduction in addition to comparison with the fabricated reference standard. False or non-relevant indications may be recognized by the skilled operator and sometimes may be eliminated with the change of transducer frequency or angle. Cracking is expected to occur within the shear plane. Unfortunately, non-relevant indications such as wear grooves or acoustic coupling may also occur at or near the shear plane of the connection.

Wear grooves are sometimes detected at the shear plane of the pin and may be difficult to distinguish from potential cracks. Such grooves are often somewhat rounded and generally continue around the circumference of the pin. Reflections tend to be more pronounced with increasing incidence angle of the transducer as the ultrasonic beam becomes more perpendicular to the reflecting surface of the groove. Wear grooves may be more difficult or impossible to detect, especially in early development, with incidence angles of zero to 10 degrees. Angles of 15 to 20 degrees are believed more likely to detect minor wear grooves.

Indications attributed to load induced acoustic coupling have been observed in some pin and hanger link assemblies. Acoustic coupling occurs under concentrated loading and produces indications that correspond to distances between the transducer entry point and various plate components of the connection. It is theorized that the high bearing pressures allow sound transmission between the connected parts. Acoustic coupling produces low level signals that often occur in the shear plane with sufficient signal intensity to warrant careful evaluation. Angle of incidence from zero to approximately 10 degrees apparently does not significantly alter detection of these reflections. The apparent origin of such indications seem to occur where the girder web and associated reinforcing plates bear against the faying surface of the pin. In this example, reflections may be produced from any combination of one, two, three, or none of the supported girder web plate members. The signal magnitude may vary from plate to plate, and at different locations around the circumference of the pin depending on the fit and pressure of bearing surfaces. Scanning from the opposite end of the pin may yield "mirrored" or varied indications. Indications of this nature disappear from the screen of the flaw detector when the transducer is moved to scan the area of the pin circumference opposite the indication producing area. Acoustic coupling indications are typically observed on the bearing side of the pin, that is the lower pin circumference for upper pins, and at the upper circumference for lower pins when related to web plates. Conversely acoustic coupling indications that apparently originate from plates in hanger links are typically observed in the upper circumference for upper pins and in the lower circumference for lower pins. These indications are not present at the non-bearing surface or tension side of the pin where potential fatigue cracks would be expected to originate. Further investigation is recommended to verify the acoustic coupling phenomenon.

A systematic method of recording indication locations with both circumferential and axial orientation should be established prior to the examination. Pre-printed inspection forms are recommended. Records that may be used to compare similar pins may contribute to the evaluation of the pins with the repetition of similar indications. Recordings of the signal trace may be stored for documentation by some instruments and are required by some jurisdictions that oversee bridge inspection. The trace recordings are of greatest value when the specific transducer location is also recorded.

Division of pins into quadrants or pie sections are both practical methods of recording circumferential orientation. Axial orientation may be recorded as distance from the scanning transducer, from the shear plane, or other appropriate reference. Signal magnitude may also have significance and should be recorded.

HANGER LINKS

Hanger links may also be fracture critical, and may be examined using ultrasonic methods when the hanger geometry permits. Potential cracks may be expected to occur adjacent to the pin penetration where hanger cross section is minimized and stresses maximized. Hanger links that are built up from multiple plates and riveted together generally cannot be successfully examined using methods described herein.

Ultrasonic examination of the critical hanger area is most often accomplished by scanning from the edge surface of the hanger link, opposite the pin hole. A shear wave is directed toward the area of potential failure at the pin hole boundary adjacent to the long edges of the hanger. The angle of incidence selected will be partially dependent on access to the scanning surface and hanger configuration. Angles of 45° to 60° are usually appropriate. A crack or notch will create a corner trap reflector with the pin hole boundary. Scanning from two directions is preferred, but often is not practical due to rounded or obstructed edges of hanger links. Grinding is often necessary for the scanning contact surface, especially when the perimeter of the link has been flame cut.

A notched reference standard cut to replicate the critical hanger link area provides a suitable reference standard after horizontal linearity of the flaw detector has been established using an IIW block or equivalent. Indications from the reflection of the pin hole demonstrate quality of the wave and may provide a trace deflection "marker" for orientation while scanning. The reference standard is useful to establish a practical scanning level, and to orient the operator to characteristic response pattern from the ultrasonic beam in the hanger. It is not applicable as a reference for acceptance criteria.

All indications detected at the minimal section areas of the hanger are highly suspect and must be carefully evaluated. Potential for non-relevant indications is minimal as only the initial leg from the shear wave transducer is utilized with this inspection. Sound path distance to the critical area is shorter than to boundary surfaces that will produce ultrasonic reflections. Further investigation of questionable indications using alternate angles or techniques may be necessary to accurately evaluate the condition of the hanger link.

CONCLUSION

The examination of fracture critical pin and hanger connections by ultrasonic methods is an important and often complex process necessary to the maintenance of numerous bridges across the country. A quality evaluation is possible with the application of a carefully designed inspection procedure. Well qualified personnel and carefully selected quality equipment are essential to the success of the examination. Questions remain regarding detection and accurate interpretation of cracks, wear grooves, and acoustic coupling phenomena. Continued research and discussion amongst those responsible to perform examinations of these critical components may enhance general knowledge and quality of future inspections.

Ultrasonic Evaluation of Steel Bridge Girders over Long Ranges

C. WOODWARD, K. R. WHITE, A. PARASHIS AND V. CARRICA

ABSTRACT

This paper describes a preliminary study on the use of ultrasonic Lamb waves for detecting fatigue cracks and other types of defects in steel girders at long distances. Lamb waves are capable of propagating long distances, making them ideal for detecting defects in large portions of bridge girders with the transducer placed at an easily accessible location. In this study, practical field testing methods and procedures for detecting fatigue cracks were developed in the laboratory by testing small steel members and a large W section. These methods were then used in a field test of a steel girder bridge to determine if fatigue cracks could be detected in the area of a connection to the web at long ranges. Results from the laboratory tests indicate that cracks can be detected at long ranges using Lamb waves. Field test results indicated that Lamb waves can be used to detect fatigue cracks in the area of a connection to the girder web.

INTRODUCTION

Currently, most bridge inspection is carried out using visual methods. Visual methods are often inefficient when inspecting steel bridge girders for the presence of fatigue cracks. The inspector must examine each connection to the girder web carefully for the presence of fatigue cracks. To complete this inspection, lift equipment or snooper trucks are often required for the inspector to gain access to the areas requiring inspection. Additionally, traffic lanes may have to be closed so that this equipment can be used.

There exists a critical need for a nondestructive testing system which will eliminate some of the above mentioned problems.

C.Woodward, K.R.White, and A. Parashis, CAGE Dept., Box 3CE, New Mexico State University, Las Cruces, NM 88003.
V.Carrica, McCrossan Const. Co., P.O. Box 1717, Las Cruces, NM 88004

This system must be capable of detecting fatigue cracks in the girder web, particularly in areas of a connection to the web where fatigue cracks most often occur. The system should also be capable of evaluating the entire length of the girder or a substantial portion of it from an easily accessible location. Ideally, this location would be a pier or abutment so that the need for lift equipment or snooper trucks is minimized.

ULTRASONIC LAMB WAVES

Lamb waves are a particular type of ultrasonic waves which have the capability of propagating over long distances in plates (Alleyne and Cawley, 1992). Additionally, these waves have been shown to be sensitive to cracks of widely varying size and geometry (di Novi, 1963; Worlton, 1957). These long range properties combined with their sensitivity to cracks makes Lamb waves ideal for use in the long range inspection of steel bridge members (Woodward et al., 1996).

Lamb waves are generated when ultrasonic energy travels between two closely spaced parallel surfaces such as the faces of a plate. Thus, the webs and flanges of bridge beams and other structural components composed of plates may be inspected using Lamb waves. The type of Lamb wave generated and its mode depend on the frequency used and the thickness of the plate (Bray, Stanley, 1989).

LABORATORY TESTS

Initially, a series of laboratory tests were conducted to verify the ability of Lamb waves to detect cracks at long ranges and to detect cracks in the area of welded connections. A W21X62 beam 43 feet long was used to develop a testing procedure for long range crack detection. In this series of tests, a 1.5 inch long cut, used to simulate a fatigue crack, was placed in the web of the beam 33 feet from one end (Figure 1). Four frequencies (0.1, 0.2, 0.4, and 0.5 MHz) were used to determine their sensitivity to the simulated crack.

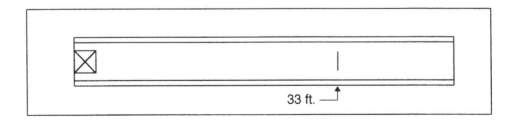

Figure 1. W21X62 beam with transducer at left end and cut at 33 ft.

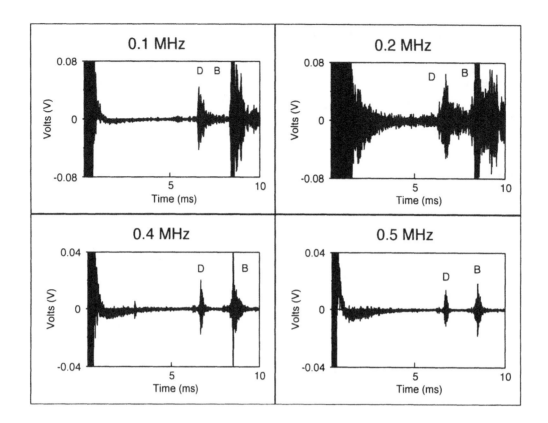

Figure 2. Defect waveforms for frequencies of 0.1, 0.2, 0.4
         and 0.5 MHz (B=backwall, D=defect).

Figure 2 shows the signals obtained from these tests. The
echoes from the cut are clearly visible for all frequencies
used indicating the presence of the defect.
    The same steel beam was also used to study Lamb wave
propagation and crack detection at longer ranges. In this
study a high power Lamb wave was generated at one end of the
beam and allowed to propagate back and forth along the beam as
it was reflected off each end of the beam. In this manner,
ranges considerably longer than the 43 foot length of the beam
could be evaluated. These results showed that the crack could
be detected at ranges of over 150 feet from the transducer
location.
    Cracks in bridge beams and girders most often originate
in the area of a connection to the member's web. Detection of
cracks in this area is difficult because Lamb waves are
reflected off the connection as well as the crack. At long
ranges these two echoes overlap and appear as one, making
crack detection difficult. This problem was addressed in the
second series of laboratory tests where a small 3/8 in. thick
plate was welded to a 3/8 in. x 12 in. plate 8 feet long to
simulate a diaphragm or stiffener connection. Two specimens
were prepared with one having a cut placed near this welded
connection, while the other was left undamaged for use as a
reference. Tests on the cut specimen indicated that the
echoes from the crack and the weld overlapped and were

difficult to separate. However, it was found that a significant and repeatable change in length occurred in the echo from an area containing a cut when compared to areas without a cut.

FIELD TESTS

In view of the very positive results obtained from the laboratory tests, arrangements were made with the New Mexico State Highway and Transportation Department to test the method on a bridge containing small unrepaired fatigue cracks. The objective of these field tests was to determine the feasibility of using this method under field conditions. To accomplish this it was first necessary to determine if fatigue cracks could be detected in the area of a welded connection to the girder web with the transducer and associated equipment located at the abutment. Secondly, it was necessary to determine if the testing equipment utilized in the laboratory tests could be used in a field situation.

The I40 - Edith street bridge in Albuquerque, NM was selected for use in this study. The girders are W30X99 sections with welded steel diaphragm connections to the web. These connection areas contain small fatigue cracks of several sizes and geometries. Two different girders with cracks of different sizes and geometries were selected for evaluation. Another girder with an uncracked welded diaphragm connection was selected to serve as a reference.

The first crack evaluated started at the top of the diaphragm connection and propagated down the web approximately 3/4 in. Its propagation path then turned approximately 45 degrees and propagated another 1 in. Figure 3 shows the echoes from this cracked area compared to the uncracked reference. The echo from the connection area containing the crack is approximately twice as long as the echo from the uncracked reference clearly indicating the presence of a defect.

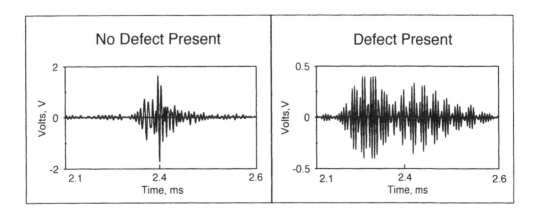

Figure 3. Waveforms from uncracked and cracked diaphragm connection.

The second crack evaluated started at the top of the diaphragm and propagated down approximately 3/8 in. It then turned and propagated along the beam's axis for approximately 1 in. This type of crack produces very little surface area perpendicular to the propagation direction of the Lamb wave. Thus, very little energy will be reflected from a crack with this orientation making detection difficult. The echo from this crack is approximately 33% longer than the echo from the uncracked reference, again clearly indicating the presence of the defect.

SUMMARY AND CONCLUSIONS

Results of the lab tests on a 43 foot girder showed that Lamb waves can be used to detect cracks at long ranges. Additional tests on a plate with a welded connection demonstrated that it is possible to detect a defect in the area of a connection at long ranges.

Field tests were conducted to verify the laboratory results and to evaluate the equipment used under field conditions. The results of these tests clearly indicated that small fatigue cracks in the area of a connection to a girder web may be detected using commercially available equipment.

The results of this study demonstrate the feasibility of using Lamb waves to detect fatigue cracks in steel bridge girders at long ranges. Additional studies are needed to further clarify optimal operating parameters such as frequency on the detectability of cracks of a variety of geometries and sizes. Other equipment types should also be evaluated to identify those most suitable for this type of crack detection. The detectability of cracks at ranges other than those evaluated in this study should also be addressed. Whenever possible, these studies should be conducted on bridges in the field so that realistic evaluations as to their effectiveness can be made.

ACKNOWLEDGMENTS

The authors would like to thank the New Mexico State Highway and Transportation Department and the Alliance for Transportation Research for their support of this project. The access to the bridge used for the field tests that the New Mexico State Highway and Transportation Department provided is greatly appreciated.

REFERENCES

Alleyne, D. N. and P. Cawley. 1992. "Optimization of Lamb Wave Inspection Techniques," *NDT&E International*, 25(1):11-22.

Bray, D.E. and R.K. Stanley. 1989. <u>Nondestructive Evaluation, A Tool for Design, Manufacturing, and Service</u>, McGraw-Hill, New York, NY.

di Novi, R. A. 1963. "Lamb Waves: Their Use in Nondestructive Testing," ANL-6630, Argonne National Laboratory, Argonne, IL.

Woodward, C., K.R. White, A. Parashis, and V. Carrica. 1996. "Long Range Bridge Girder Evaluation Using Lamb Waves," to be published <u>in Review of Progress in Quantitative NDE</u>, Vol. 15, ed by, D.O. Thompson and D.E. Chimeni, Plenum Press, New York, NY.

Worlton, D.C., 1957, "Ultrasonic Testing With Lamb Waves," *Materials Evaluation*, 15(4):218-222.

A Computerized Imaging System for Ultrasonic Inspection of Steel Bridge Structures

I. N. KOMSKY AND J. D. ACHENBACH

ABSTRACT

Non-destructive ultrasonic inspection is an effective technique to ensure structural integrity of steel bridges. Quantitative information on flaw size, location and orientation is required for cost-effective maintenance of bridge components by implementation of the "retirement-for-cause" concept. This is especially true for such fracture-critical bridge structures as pinned connections, fasteners and welded joints.

One of the major problems in presenting NDT data is how to visualize the results of an inspection. Using comprehensive ultrasonic images of the inspected areas bridge managers and engineers can make rational decisions on steps needed to correct the problem.

A two-step technique for ultrasonic inspection of pins in pin-hanger connections is discussed in this paper. First the ultrasonic technique is applied in conjunction with a portable ultrasonic flaw detector. This test serves to detect an irregularity of any kind in the inspection zone. The time required for the test will be small and the equipment used should be portable and inexpensive. If the pin condition requires a more detailed inspection, a sophisticated data acquisition and imaging system will be used. This system can provide information on flaw configuration, and distinguish cracks from wear groves in the load bearing locations.

The ultrasonic inspection technique is used in conjunction with a portable imaging device for real-time data acquisition, signal and image processing, and documentation of the inspection results. Ultrasonic images have been obtained for fatigue cracks implanted into pins. The sizes, geometries, and locations of the implanted cracks in the sample are based on statistical data from previous NDI of similar pins. All implanted flaws are fully documented with drawing and verification of location by both mechanical and ultrasonic measurements to provide a base for calibration and subsequent training of operators.

FORMATS FOR ULTRASONIC DATA PRESENTATION

To present visually the results of an ultrasonic inspection three basic presentation formats may be used: the A-scan, B-scan, or C-scan (Fig. 1).

An A-scan is the most common type of presentation used in commercial ultrasonic inspection. This scan displays signals on a horizontal base line which indicates elapsed time (Fig. 1A). Reflections are shown as deflections from the base line. The horizontal positions of the deflections indicate location of the reflector in the sample. The amplitudes represent the intensities of the reflected beams that can be related to flaw size, sample attenuation, and

Igor N. Komsky and Jan D. Achenbach, Center for Quality Engineering and Failure Prevention, Northwestern University, Evanston, IL 60208

other factors. The trace is usually set up to show a reflection from the top and the bottom of the part and reflections due to flaws will appear in between.

In a pin-hanger connection a pin face is the only part of the pin that is not covered by other structures. Therefore an ultrasonic inspection has to be performed with the ultrasonic transducer placed on the pin face. For this type of ultrasonic measurement an A-scan represents information on the pin condition underneath one point on the pin face (or at least for a local area that is restricted by the width of the ultrasonic beam). Stored in the memory of a modern ultrasonic flaw detector, a set of A-scans generally does not contain enough information that can be easily used by operators for evaluation of the pin condition.

A B-scan utilizes the same signal as an A-scan, but then presents the information along a line as an ultrasonic transducer is scanned along the part (for pin inspection the transducer is scanned along the circumference of the pin face). The display results in a cross-sectional view of the part along the direction of ultrasonic wave propagation (Fig. 1B). The location, length, spatial orientation (for planar defects), and depth (for volumetric defects) of the flaws can be determined, but no information on width or configuration of the flaw can be obtained.

A C-scan presentation involves scanning the transducer over an area. Results are displayed like a plane view looking down on the part (Fig. 1C). The display is similar to a radiograph of the part. A two dimensional scanner records the movement of the transducer and a scale image is created on a screen of the inspection system. With a C-scan all echoes within a pre-selected gate, or alternatively the echoes with a maximum amplitude between front and back echoes are recorded. The location and configuration of the defects are displayed. A three-dimensional presentation can be produced if it is built up from a number of two-dimensional recordings. A C-scan is the most complete type of ultrasonic data presentation. However, this kind of ultrasonic scanning is very time-consuming and requires expensive instrumentation. At the time of field inspection a C-scan should be performed only if the initial inspection procedure would indicate a real need for the application of such a sophisticated technique.

It is proposed to first apply a B-scan imaging technique as a method for initial inspection of pins in pin-hanger connections. This ultrasonic inspection can be used not only for flaw detection, but also to characterize quantitatively its length and location. An advantage of a B-scan is that the complete circumference of a pin can be inspected in one line of scanning. All ultrasonic data will be displayed on the screen of an ultrasonic flaw detector as an image of the pin along its circumference. This format of presentation makes data analysis and pin evaluation more accurate and reliable.

As a second step a C-scan image can be acquired for the areas with detected flaws or for the whole pin. The results of the C-scan should be compared with information obtained from the B-scan to verify location and configuration of the detected flaws.

The procedure for B- and C-scan imaging has been applied to the pins with machined slots, implanted crack, and wear grooves.

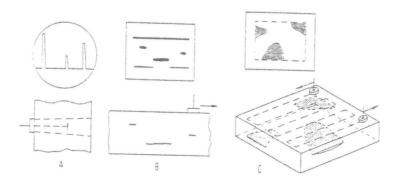

Figure 1. A-, B-, and C-scan images and principle of scanning

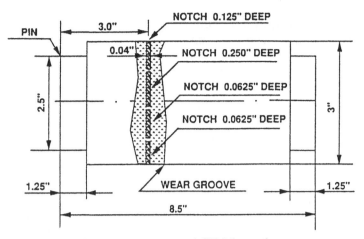

Figure 2. Pin with wear groove and EDM notches.

EXPERIMENTAL CONFIGURATION

A portable ultrasonic flaw detector EPOCH II (Panametrics, Inc.) was used in the pulse-echo mode in conjunction with contact transducers (AEROTECH MSW-QC 2.25MHz and 5MHz, 0.25 inch, Krautkramer-Branson, Inc., and VIDEOSCAN, 2.25 MHz and 5MHz, 0.5 inch, Panametrics, Inc.). Longitudinal or shear wave techniques were selected for each particular application based on pin size and configuration. The transducers were placed on one of the pin faces using special transducer holders with pre-selected angles of incidence. The ultrasonic signals reflected from the crack and the back face of the pin were digitized by the ultrasonic flaw detector and the data was subsequently acquired by a personal computer to generate B-scan images. An ULTRA IMAGE IV system (Ultra Image International, Inc.) was used to generate C-scan images.

PIN WITH WEAR GROOVES AND EDM NOTCHES.

A 3" diameter pin with a wear groove ranging from 0.030" to 0.125" in depth and from 0.125" to 0.5" in width was removed from a bridge and used for the tests. Four EDM notches with depths of 0.0625 inch (two notches), 0.125 inch, and 0.25 inch, were machined along the circumference of the pin at the bottom of the actual wear groove. A sketch of the pin with wear groove and EDM notches is shown in Fig. 2.

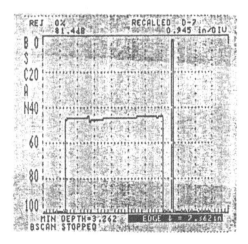

Figure 3. B-scan of the pin with wear groove and EDM notches.

The signals reflected from the EDM notches and from the back shoulder of the pin were detected and measured to generate a B-scan image. Figure 3 shows the B-scan as an image of the pin along its circumference. The four notches can be seen as lines approximately at 45% of the overall depth of the scan. The location of the notches (distance of 3.262" from the pin face) has been calculated using the known velocity of ultrasonic waves. The location of the back shoulder as the deepest reflector has also been calculated (7.362"). At one point, contact with the pin surface was lost resulting in the "zero depth" indication on the image.

Reflections of ultrasonic waves from wear grooves result in false crack indications and possibly unnecessary replacement of the pin. Hence, a discrimination of ultrasonic responses of wear grooves and cracks is of great importance for the reliability of ultrasonic inspection. As can be seen from Fig. 3 the wear groove did not provide a detectable response. Therefore the technique can be used in the future to detect cracks.

PIN WITH IMPLANTED FATIGUE CRACKS.

Due to substantial differences in the response functions of EDM notches and actual fatigue cracks of the same size, the detection and characterization of EDM notches can be considered only as the first step toward inspection of structures with fatigue cracks. Experiments using real flaws provide better results for calibration of sensitivity and resolution of the inspection system.

A novel specimen developed by FlawTech Inc. was used to optimize the ultrasonic technique for pin inspection. In this specimen a process of heating/cooling while under tension is used to initiate a thermal fatigue crack in a previously excavated piece. The number of cycles is controlled to obtain the desired crack " roughness." The thermal fatigue crack is made to initiate in the base material, thus allowing UT energy to impinge directly on the flaw face. After taking careful physical measurements to establish the flaw location, the implant is seal-welded back in place. The remaining weld groove is then filled with a standard welding procedure and machined to the required geometric condition.

Two fatigue cracks of depths 0.125 inch and 0.250 inch, and lengths 0.750 inch and 1.750 inch, respectively, were implanted into a pin normal to the side surface. The calibration specimen with implanted cracks is shown in Fig. 4.

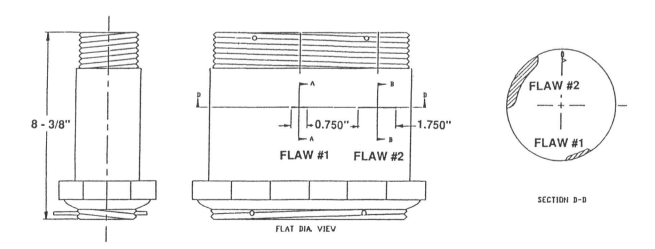

Figure 4. Pin with implanted fatigue cracks.

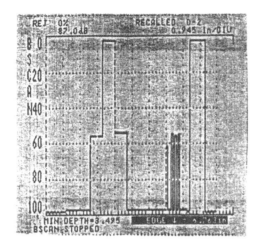

Figure 5. B-scan image of the pin with transducer placed on the closest pin face.

Two sets of data were acquired and analyzed. The ultrasonic transducer was first placed on the pin face which is closest to the flaw location. As shown on the acquired B-scan image (Fig. 5) both implanted cracks were successfully detected and imaged. However, the response from the smaller crack was not stable. To improve the quality of the imaging the transducer was placed on the face which is farthest from the flaws. In this case the responses from all cracks were stable (Fig. 6). On each of the images there are two areas with "zero depth" indications. One of the areas coincides with the longer crack, dividing the crack image into two parts. The origins of these indications are "shadows" from the cotter pin located just below the pin faces. In order to inspect the areas underneath these pins a special technique should be developed.

For more detailed information on the crack configurations a C-scan image has been acquired (Fig. 7). This image makes it possible to determine not only crack length but also the crack depth and even the crack shape. Hence the pin condition can be monitored more accurately. It can be seen that the C-scan image (Fig. 7) is in a good agreement with the B-scan images (Figs. 5, 6).

For rough surfaces of the pin faces application of the self-compensating technique, previously developed by the authors (Komsky and Achenbach, 1994), is recommended. This technique uses reference signals that can provide an adequate scale for crack depth measurements by B-scan imaging. The technique also make it possible to acquire amplitude C-scan images independently of the surface condition of a pin face.

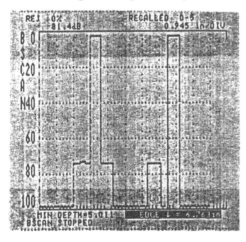

Figure 6. B-scan image of the pin with transducer placed on the farthest pin face.

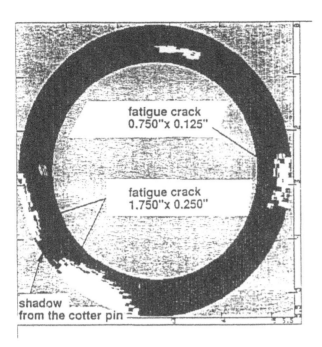

Figure 7. C-scan image of the pin with implanted fatigue cracks.

CONCLUSION

Application of different modalities of ultrasonic imaging to the inspection of pins has been discussed. A two-step procedure based on a combination of B-scan and C-scan imaging has been proposed. The imaging system has been tested on pins with artificial flaws and implanted fatigue cracks. The tests demonstrate high sensitivity of the imaging system to the detection of fatigue cracks, and a good consistency between the scanning techniques. An application of the self-compensating technique is recommended for pins with painted or corroded surfaces.

ACKNOWLEDGMENT

This work was supported by the Infrastructure Technology Institute of Northwestern University. The authors would like to thank the Wisconsin Department of Transportation and the Illinois Department of Transportation for supplying the pins used in this work.

REFERENCES

Komsky, I. N. and J. D. Achenbach. 1994. "A Modified Self-Compensating Ultrasonic Technique for Flaw Characterization in Steel Bridge Structures," *Review of Progress in Quantitative Nondestructive Evaluation*, 13B: 2107-2113, Plenum Press, New York, N.Y.

Krautkramer, J. and H. Krautkramer. 1990. Ultrasonic Testing of Materials, Springer-Verlag, Germany.

Session 2

C2—BRIDGE INSPECTION USING RADAR

S2—FRACTURE CRITICAL INSPECTIONS

C2A—NDE OF BRIDGE FOUNDATIONS

S2A—FATIGUE LIFE AND NDT

S2B—FATIGUE LIFE AND NDT

S2C—NDE APPLICATION

Underwater Non-Destructive Testing for Bridge Substructure Evaluation

M. J. GARLICH

ABSTRACT

The past several years have seen increased inspection of bridge substructures below water. In some cases, traditional visual and tactile inspection may not give sufficient data to accurately evaluate the substructure's condition. Underwater non-destructive test methods can be used to gather data to aid in such efforts. To assist state Departments of Transportation in utilizing these advanced inspection techniques, a portion of Demonstration Project 98, "Underwater Evaluation and Repair of Bridge Components", is devoted to presenting available equipment and techniques in underwater non-destructive testing applicable to bridges.

INTRODUCTION

Among changes included in the National Bridge Inspection Standards adopted in 1988, was a requirement for underwater bridge inspections to be conducted on a regular basis. While in most cases a bridge's submerged substructure can be satisfactorily examined by use of visual and tactile inspections supplemented with probing, sounding, and selected element cleaning, there are some instances where more detailed information on the condition of the construction materials is needed. As examples, the interior degradation of timber piles subject to marine borer or bacteriological decay, the corrosion losses on steel sheets in a sheet pile wall, and the extent of honeycomb concrete in a pier are all conditions which may require use of non-destructive testing (NDT) to obtain needed information for analysis. With increased frequency of underwater inspections, and concern for long term structural performance, added interest has been generated by the bridge community in underwater NDT techniques.

As a result of this interest, the Federal Highway Administrative has included a session on underwater NDT in its Demonstration Project 98, "Underwater Evaluation and Repair of Bridge Components", recently made available. The session presents an overview of NDT methods for timber, concrete, and steel elements including pulse-velocity testing, ultrasonic thickness and weld testing, magnetic particle testing, and partially destructive tests such as coring. Also discussed are developing techniques for underwater testing including radar

Michael J. Garlich, Collins Engineers, Inc., 165 North Canal Street, Suite 975, Chicago, Illinois 60606

evaluation of concrete. Several pieces of underwater test equipment are brought into the class and used in demonstrations. While not intended to train underwater NDT technicians, this training gives an overview of available and developing technology along with applications and limitations on its use. This information should enable engineers to utilize these techniques to obtain added detail information for use in evaluating below water structure elements.

GENERAL CONSIDERATIONS

Underwater NDT is generally performed in a manner similar to testing above water. Accommodation must be made for the environment's effect both on the equipment and technician, however. Much underwater testing is performed using standard equipment mounted in custom made waterproof housings. For some testing, such as ultrasonic thickness or weld testing, or pulse velocity testing, it is common to waterproof only the transducers and cables. These are then taken below water and manipulated by a technician-diver while the instrument is kept topside and adjusted and read by a second technician. Clearly this can present coordination difficulties, particularly in weld testing, and careful planning and continuous voice communication are essential for successful results. In addition, long instrument cables may be required causing problems with signal strength, or entanglement in piles. Self-contained instruments are also available, or can be made, which allow all test functions to be completed by the diver. Common examples include thickness testing using gauges such as those made by Krautkramer and Cygnus, and magnetic particle testing where both the particle containing gel and testing yoke are specifically manufactured for underwater usage. Data obtained by the diver can either be recorded on an underwater slate, or relayed by voice communication to a topside notetaker.

For bridge inspections, underwater data must often be obtained in conditions of near zero visibility, in high currents, and in cold waters, with heavily gloved hands. This makes placing and manipulating instruments and transducers, as well as reading or recording data, very difficult. For example, the testing of concrete may require the diver to place two transducers on the face of a structural element and apply pressure to obtain good transfer of ultrasonic pulses between the transducers and substrate. As he applies pressure however, his body, suspended in a fluid, is pushed away from the structure. To make the inspector's task easier and data gathering more reliable and rapid, underwater test equipment must be simple to operate with large knobs or lever actuated switches and output displays in large illuminated numbers. Any calibration or job specific adjustments should be able to be made prior to entering the water. Preferably, the test instrument should be able to be used with one hand, allowing the other hand to help hold onto the structure. The ability of the diver to hold his position while taking a test reading becomes extremely difficult in river currents. In other instances the visibility may be so poor that it is impossible to read data output, even from brightly illuminated displays. With the exception of weld testing, commonly used underwater NDT techniques may be considered relatively straight forward, and generally constitute data gathering, with data analysis and interpretation performed by structural or materials engineers. This suggests that for bridges, the underwater NDT testing can be performed by diver-technicians not fully certified to nationally accepted standards as Level I or II NDT inspectors. The diver-technician should; however, have received instruction in the theory and test procedure for the specific tests they are to run, and should follow detailed written testing procedures set up for each project. While use of fully qualified NDT diver-technicians is desirable, their availability in many areas of the country is quite limited, and in light of the types of tests performed and structures on which they are performed, reduced inspector NDT qualifications should not, in themselves, reduce the value of the data obtained.

STEEL TESTING

Testing of steel structural members most often involves thickness testing using ultrasonic gauges, though weld testing is also performed. Ultrasonic thickness testing of steel is well suited to testing pipe and H-piles as well as sheet piles, all of which occur often in bridge substructures. Where a member's surface can only be reached from one side, such as for sheet piles, pipe piles and H-pile webs, it is the only method to assess thickness, other than drilling a hole to allow a direct measurement.

A major problem in securing accurate underwater steel thickness data is the surface condition of the substrate required to yield reliable thickness readings. At best, splash zone and below water surfaces are corroded, often they are heavily corroded and pitted. Hand tool cleaning may not produce a surface suitable for obtaining good readings, thus requiring powered tools such as grinders or needle guns. A hard tightly adhering layer of corrosion products, smooth and shinny even after hard hand cleaning, may be present and is difficult to detect, though its removal is required. Pitting at the back face of the test areas can also adversely effect the measured thickness reading. This face is usually not accessible and hence the effect of back surface pitting can not be evaluated in reviewing test results. For underwater readings, several adjacent readings should be obtained at each test location and compared to expected reasonable values to help eliminate any inconsistent readings. The use of thickness testing equipment which utilizes separate transmitting and receiving transducers is recommended to help reduce the adverse effects of surface pitting. In critical areas, the drilling of small holes to measure thickness at a few location can add to the confidence level of the more readily obtained ultrasonic data.

Underwater weld testing is a well developed technology and widely used in the offshore industry. The need for such testing in bridge substructures is rare, and thus will only be briefly addressed here. Aside from visual examination, magnetic particle and ultrasonics are most often used underwater. In magnetic particle testing, the particles are carried in a water resistant gel which is placed on the test area. A specially made waterproof yoke is then used to apply the magnetic field. Ultrasonic inspection uses either a test instrument in a waterproof case or long waterproof transducers leading to an instrument above water. For either test, extensive surface cleaning at the weld areas is normally required in order to run the test. Good visibility is required to conduct these tests, even if the diver need only manipulate a transducer in a predetermined pattern. In one case in the midwest, testing pile splice welds was abandoned because near zero visibility and very cold water prohibited the diver from even being certain that he had found the welds.

CONCRETE TESTING

Submerged concrete structures generally perform well. Deterioration is most often found in splash zones as a result of reinforcing steel corrosion, and is easily detected by visual or tactile inspections. In some instances; however, internal defects generated by reactive aggregates or poor construction may be suspected. There may also be questions about concrete cover over existing reinforcing steel, or the depth of cracking identified by visual inspection. These are not readily detected by traditional testing techniques.

Nondestructive testing techniques commonly used underwater include pulse velocity testing, rebound hammer testing, and R-Meter testing. Manufacturers supply custom waterproof housings or special waterproof transducers for use of their equipment below water. Tests are performed as they would be above water. However, cleaning areas of marine growth will normally be required at test locations and recording the location of the test on the structure may require considerable effort in setting horizontal guide lines or hanging vertical down lines from which the diver can locate the test area. Other NDT methods adaptable to underwater testing include impact-echo and ground penetrating radar. Though to date, use of these methods underwater has been limited, both should be considered viable techniques.

Ground penetrating radar and pulse velocity test methods were recently used to examine the structural integrity of a concrete tunnel bulkhead. The bulkhead, located at a depth of 150 feet, seals a flooded water supply tunnel planned to be dewatered to allow further construction. The bulkhead's construction, placed by a previous contractor, was of unknown thickness and concrete strength. Radar was used to establish the bulkhead thickness and overall integrity, while pulse velocity tests were used to estimate concrete strength.

Before selecting the specific test methods and procedures to be used, a visual examination of the concrete bulkhead's exposed face was made utilizing a remotely operated vehicle equipped with a video camera. This investigation determined the general configuration of the bulkhead and the concrete's surface, and provided data to evaluate diving access in order to carry out the nondestructive testing program.

The thickness survey of the bulkhead was performed with a Geophysical Survey Systems Incorporated System 4 ground penetrating radar unit (GPR), modified for underwater use. A 500 Mhz antenna was installed in a watertight case that used marine penetrators to connect the antenna to the signal cable. A 450-foot signal cable was used to connect the antenna to the processing unit and the graphic display recorder located on the surface. The signal cable, which contains a kevlar stress member, was used to lower the submersible radar antenna approximately 150-feet down the entry shaft to the tunnel. A diver then carried the antenna an additional 150-feet into the tunnel to the concrete bulkhead that was to be investigated.

The diver held the GPR antenna perpendicular to the bulkhead surface and scanned the antenna across the bulkhead. Multiple passes were made in order to cover the bulkhead. As the antenna was moved across the bulkhead, it transmitted electromagnetic pulses through the concrete. When those pulses encountered a change in material property, such as an air or water filled void within or behind the bulkhead, they were reflected back to the antenna. The graphic display recorder presented a continuous profile of these subsurface features along the antenna travel path. The top of the radar record represents the surface on which the antenna is placed, and subsurface features, such as the back of the bulkhead, appear as linear features below the surface. The vertical axis on the radar record represents the time for the electromagnetic wave to travel from the antenna to a subsurface reflector, such as the back of the bulkhead, and back to the antenna. To convert this time scale to a depth or thickness scale requires knowledge of the speed at which an electromagnetic wave travels through the medium, concrete in this case. The electromagnetic wave velocity values for concrete typically vary from 5 nanoseconds/foot (dry concrete) to 7 nanoseconds/foot depending on factors such as the density and amount of water in the concrete. Conservatively, the calculation of a depth or thickness scale for this data was based on a velocity of 7 nanoseconds/foot. This value was selected based on published information.

Pulse velocity testing, utilizing the James Electronics, Inc. "V-Meter" was also carried out. The equipment used has a pulse frequency of 54 kHz with a repetition frequency of ten pulses per second. Similar equipment is available from other manufacturers. Because of concerns for signal degradation through the long cables which would have been needed to separate the transducers underwater from the signal unit located on the surface, a waterproof housing was built and the whole unit was taken to the bulkhead face by the diver. As only one side of the bulkhead was accessible, testing was performed using surface, or indirect, transmission techniques with both transducers located at the same surface. This method is less desirable than through transmission but was the only technique available for this application. A further limitation on the method imposed by the deep diving conditions with limited time to perform the testing, was that the distance between transducers was limited to approximately 5 feet. However, the various readings obtained during the testing indicated consistent concrete quality, and allowed an estimate of compressive strength to be made. Use of the two tests together, enabled reasonable conclusions to be reached on the bulkhead's thickness and overall structural integrity.

In order to assure that the custom underwater housings built for this equipment would not fail or leak at the 150 foot water depths at which they were to be used, tests were made using a recompression chamber. The housing was submerged in a tub of water placed within the recompression chamber, and then the chamber was pressurized to replicate the effect of a 150 foot water column. Passing the test not only reduced concerns for damaging expensive equipment, but also lessened the potential of having to abort an expensive diving operation due to test equipment malfunction.

TIMBER TESTING

Timber elements in water are subject to internal damage due to marine borer activity found in ocean environments, or due to decay. Timber located in fresh water is not subject to borer attack, but is subject to bacteriological deterioration of the wood structure. This type of deterioration reduces both wood strength and modulus of elasticity. These types of deterioration cannot be detected by external examination and thus other techniques such as partially destructive tests or NDT are required.

Nondestructive testing of timber above water has been used for several years. The NDT of timber piling below water received considerable efforts following the collapse of the Pokomoke River Bridge. Research work conducted under Dr. Aggour at the University of Maryland for the Maryland Department of Transportation and the Federal Highway Administration (FHWA) resulted in the publication in 1989 of the FHWA Publication No.

FHWA -1P-89-017, "Inspection of Bridge Timber Piling Operations and Analysis Manual". The testing technique developed through this work uses through transmission of pulses perpendicular to the axis of the pile.

An example of the use of the FHWA nondestructive testing technique for underwater bridge elements involved pile testing for a group of New Jersey bridges founded on timber pile bents. Each substructure first received a complete underwater visual and tactile inspection. The inspection team, headed by a professional engineer-diver, then selected locations and carried out NDT examination as an immediate follow-up activity. Approximately ten percent of the total piles inspected were subjected to NDT. Piles that exhibited signs of deterioration, based on initial visual and probing inspection results, were selected for testing. Thus, the weakest element which could initiate structural failure was being investigated.

Pulse velocity readings were made at three vertical locations along each pile as follows: areas to be tested were first thoroughly cleaned of marine growth and silt; the diver placed the transducers opposite one another across the diameter of the pile and when a steady and clear reading was obtained by the operator on the surface, the data was recorded and the diver notified; a second reading was then taken at the same level on a line perpendicular to the first reading; the diver would then move to the next predetermined location. Theoretically, the water between the transducer and the pile should provide adequate coupling for good sound transmission into the pile. However, use of a silicone or similar coupling material aids in securing consistent data. A second diver measured the pile diameter corresponding to each reading location. With this data, the velocity of the sound waves was calculated. Taking two readings at a given pile elevation is a recommended minimum to provide coverage of the pile cross section. Additional readings can be made, located at regular intervals, around the pile to further increase the area of pile actually traversed by the ultrasonic pulses.

Because the pulse velocity travel time in timber is influenced by water-filled internal voids, such as may be created by marine borers or decay, small diameter cores were taken in selected areas as an aid in determining the presence of decay voids or borers. In addition, a sample of pile was obtained from an abandoned (mis-driven) pile and subjected to laboratory testing for pulse velocity and compressive strength. The results of this additional data help substantiate the NDT test data and subsequent computed pile residual strengths. One of the major advantages of using NDT in such a test program is that large numbers of tests that can be performed in a short time. This is very beneficial in difficult diving conditions, and is quite cost effective.

SUMMARY

Several NDT techniques are available for detailed examination of underwater bridge elements. Though not used as part of routine inspection, data obtained through such NDT efforts are needed for in-depth evaluation of some structures. While their use increases inspection cost, careful planning can minimize these costs. Information on underwater NDT applicable to bridges has been included in the FHWA Demonstration Project 98 so that bridge owners can better determine where and to what extent underwater nondestructive testing should be carried out as a part of their underwater bridge inspection program.

Detection of Delaminations in Concrete Using a Wideband Radar

O. BÜYÜKÖZTÜRK AND H. C. RHIM

ABSTRACT

Radar method has a potential of being a powerful and effective tool of nondestructive testing (NDT) of concrete structures. Yet, not all of the available features of the method have been fully explored and current applications to civil engineering problems are limited to simplified systems. The advancement of the method can be achieved through the understanding of electromagnetic properties of concrete and the identification of optimum radar measurement parameters. For identifying optimal solution parameters for a given problem, radar measurements in parallel with accurate theoretical modeling of wave propagation and scattering are needed as well as implementation of signal processing schemes for imaging.

This paper presents a part of current research conducted at the Massachusetts Institute of Technology (MIT) as an interdisciplinary effort. The research work aims at the development of an effective methodology using radar as a tool for NDT of concrete structures and to develop specifications for radar equipment for this purpose.

The paper presents results of parametric studies to investigate the detectability of delaminations inside concrete through computer simulations and remote wideband radar measurements. Emphasis is given to the significance of transmitting waves over a wide frequency bandwidth. Two different frequency ranges have been used for the simulation; 3.4 to 5.8 GHz and 2.2 to 7.0 GHz, which have bandwidths of 2.4 GHz and 4.8 GHz, respectively. Detectability of 3.2 mm (1/8 in.) to 25.4 mm (1 in.) thick delaminations embedded at different depths inside concrete has been studied through simulation. For the comparison with simulation results, radar measurements are made at 3.4 to 5.8 GHz using a wideband inverse synthetic aperture radar on laboratory size concrete specimens with delaminations.

INTRODUCTION

The goal of any NDT techniques is to detect a delamination area or an object located at a certain distance below the surface in an optically opaque medium. In the radar method, the detectability of an object inside concrete and the penetration capability of a wave into concrete are affected by a variety of measurement parameters: center frequency, frequency bandwidth, polarization of the wave, measurement distance and angle, and geometric and

Oral Büyüköztürk, Professor, Department of Civil and Environmental Engineering, Massachusetts Institute of Technology, Cambridge, Massachusetts 02139

Hong C. Rhim, Postdoctoral Associate, Department of Civil and Environmental Engineering, Massachusetts Institute of Technology, Cambridge, Massachusetts 02139

material properties of a target (Rhim and Büyüköztürk, 1995). Bandwidth of a radar system affects the measurement resolution, which determines the detectability of delaminations with various thickness located at certain depths from the concrete surface. By increasing the system bandwidth, a high degree of range accuracy can be obtained (Mensa, 1981).

This necessitates a study to examine the influence of each parameter on radar measurement results and to establish optimum combination of the parameters. For the study, a numerical technique which can simulate the electromagnetic phenomena and provide information as to how the wave propagates through and scatters by a concrete target is needed.

SIMULATION USING FINITE DIFFERENCE-TIME DOMAIN METHOD

Study of wave scattering by a target which has known geometric and material properties is called a forward problem. A candidate approach for this purpose is the finite difference-time domain (FD-TD) solution of Maxwell's curl equations, which are the governing equations for the electromagnetic waves.

The FD-TD is a marching-in-time procedure which can simulate continuous wave propagation (Taflove, 1975). Results of such numerical simulations can be used as a basis for the parametric study to identify optimum radar measurement parameters. It is based on the discretization of the electric and magnetic fields over rectangular grids together with the finite difference approximation of the spatial and temporal derivatives appearing in the differential form of Maxwell's equations. The reasons for which the FD-TD methodology is selected for the present study include the relative ease of its implementation for complicated geometries with dielectrics, the need of only simple arithmetic operations in the solution process, and the flexibility for time- and frequency-domain analyses (Li et al., 1992).

In the modeling, a Gaussian pulse modulated sinusoidal wave is used as an excitation source. The wave can be centered at any frequency with the use of a carrier frequency. The Gaussian pulse modulated sinusoidal wave has a form of

$$V(t) = e^{-2(t-t_0)^2/T^2} \, cos(\omega t) \tag{1}$$

where t is time, t_0 is the center of the wave in time, T is the pulse width, and ω is the carrier frequency which shifts the center of the wave to a center frequency in frequency domain.

One- and two-dimensional imaging can be performed from the FD-TD modeling. In the following a one-dimensional signal processing scheme is summarized.

A modulated Gaussian pulse plane wave $V_{incident}(t)$ is directed to a concrete target to the cross-sectional area of a specimen. Using the FD-TD numerical modeling technique, the reflected wave from the concrete specimen is captured in the far field as $V_{reflected}(t)$. The reflected waves contain information about the concrete target in its interaction with the incident wave, as well as the characteristics of the incident wave. Thus, to extract information about the specimens out of the reflected waves, a numerical scheme is applied. The reflected wave in time domain, $V_{reflected}(t)$, is Fourier transformed to the frequency domain. The incident modulated Gaussian pulse $V_{incident}(t)$ is also Fourier transformed. Then, the reflected wave is normalized with respect to the incident wave in frequency. The normalized response is multiplied by a hamming window to examine the result with an emphasis of the initially given center frequency and bandwidth. Finally, the windowed normalized response in frequency domain is inverse Fourier transformed to time domain to obtain the true reflection from the concrete target. The procedure is summarized as,

$$F^{-1}\left[hamming\ window * \left\{ \frac{F\{V_{reflected}(t)\}}{F\{V_{incident}(t)\}} \right\} \right] \qquad (2)$$

The inverse Fourier transformed signal in V/m is plotted in dB scale for view.

$$magnitude\ in\ dB = 20 * log\{absolute(processed\ signal\ in\ V\ /\ m)\} \qquad (3)$$

SIMULATION FOR DETECTING DELAMINATIONS INSIDE CONCRETE

Detectability of thin delaminations is examined. Condition monitoring basically depends on the detectability of small abnormalities including cracks. The detectability of a crack is determined by center frequency, bandwidth, and the electromagnetic properties and size of the crack.

A concrete block which has cross-sectional dimensions of 304.8 mm (12 in.) x 101.6 mm (4 in.) is modeled. Three parameters studied are the size of delamination, location of the delamination from the surface of concrete, and the frequency bandwidth of the incident wave as summarized in Table 1. The frequency ranges are selected such that the center frequency is the same for both ranges while the bandwidth is different by a factor of 2.

Table 1. Parameters used for the delamination detectability study.

Size of the delamination	Distance from the surface to the center of a crack	Frequency of the incident wave
• 3.2 x 152.4 mm (1/8 × 6 in.) • 6.4 x 152.4 mm (1/4 × 6 in.) • 12.7 x 152.4 mm (1/2 × 6 in.) • 25.4 x 152.4 mm (1 × 6 in.)	• 25.4 mm (1 in.) • 50.8 mm (2 in.) • 76.2 mm (3 in.)	• 3.4 ~ 5.8 GHz (Bandwidth = 2.4 GHz) • 2.2 ~ 7.0 GHz (Bandwidth = 4.8 GHz)

For the simulation, the concrete specimen is modeled as a dielectric medium which has a dielectric constant of 4.8 and a conductivity of 0.11 mhos/m representing air dried concrete for the frequency range studied (Büyüköztürk and Rhim, 1993).

The simulation results suggest that 25.4 mm thick delamination can be detected at all three locations with both 3.4 to 5.8 GHz and 2.2 to 7.0 GHz waveforms. The thinner delaminations with 3.2 mm, 6.4 mm, and 12.7 mm thickness are difficult to be identified with the narrower frequency bandwidth of 3.4 to 5.8 GHz. However, as the frequency range increases to 2.2 to 7.0 GHz with bandwidth of 4.8 GHz, the delaminations were more visible. Sample results of the simulation for a concrete specimen with 25.4 mm (1 in.) delamination at 50.8 mm (2 in.) depth from the surface is shown in Figure 1 with 2.4 GHz bandwidth. This result is compared to the simulation with the wider bandwidth of 4.8 GHz in Figure 2. As shown in the figures, the sharper imagery is obtained in the wider frequency bandwidth simulation due to improved range resolution (Mensa, 1981):

$$\rho_r = \frac{\left(\dfrac{c}{\sqrt{\varepsilon_r}}\right)}{2B} \ in \ concrete \qquad\qquad (4)$$

where c is the speed of light (3×10^8 m/sec), ε_r is the dielectric constant of concrete, and B is the frequency bandwidth of the incident wave.

COMPARISON OF RADAR MEASUREMENT RESULTS TO SIMULATION

In comparison with the simulation, a laboratory size concrete specimen with dimensions of 304.8 x 304.8 x 101.6 mm (12 in. x 12 in. x 4 in.) is used for the radar measurement. The concrete target was placed on the top of a Styrofoam stand located 14.4 m from a monostatic antenna operating at frequencies from 3.4 to 5.8 GHz with VV (Vertical) polarization. The Styrofoam stand was used because it did not interfere with the electromagnetic waves due to its electromagnetic properties being close to those of air. VV polarization means transmitting and receiving the wave with the electric field oriented vertically. The measurement was conducted inside an anechoic chamber (Büyüköztürk and Rhim, 1995). The wave form of the incident wave generated by the antenna was gated continuous wave. The frequency of the continuous wave was swept from 3.4 to 5.8 GHz (Bandwidth = 2.4 GHz) with an increment of 0.1 GHz at different time steps. The pulse width of each incident wave was 20 ns. The reflected signals from concrete specimen at each frequency were recorded in amplitude and phase.

The image was obtained by inverse Fourier transform of the received signals in frequency to time domain, which were then converted to distance. The range axis in Figure 3 is relative to the center of the Styrofoam stand, which coincides with the front surface of the specimen. The amplitude of the received signal is plotted in dBsm (a decibel relative to 1 square meter of radar cross section). The radar is located at -14.4 m in the range direction in Figure 3. In Figure 3, one-dimensional imaging of the specimen obtained from the radar measurement is shown. The front surface reflection is followed by the reflection from the 25.4 mm (1 in.) delamination.

CONCLUSION

The electromagnetic wave modeling technique using the FD-TD method is applied to the simulation of wideband radar measurements. The measurement parameters of frequency bandwidth of the incident wave, size and location of delaminations are studied to predict radar measurement results. The simulation results provide the feasibility of detecting a delamination with a given combination of measurement parameters. Sample simulation is compared with actual radar measurement. Thus, the developed capability for computer simulation can be used for the prediction of the radar measurement and for optimizing the measurement parameters with better detectability.

ACKNOWLEDGMENTS

This work was supported by U.S. Army Corps of Engineers, Waterways Experiment Station, Vicksburg, Mississippi through Contract No. DACW39-92-K-0029. The authors would like to thank Dr. Tony C. Liu, Mr. Mitch Alexander, and Mr. Kenneth Saucier for the support provided.

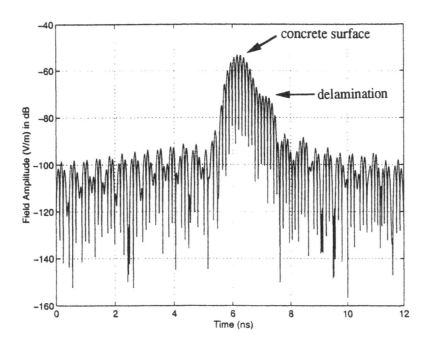

Figure 1. One-dimensional image obtained from modeling at 3.4 to 5.8 GHz for a 304.8 mm x 101.6 mm (12 in. x 4 in.) concrete block with a 25.4 mm (1 in.) x 152.4 mm (6 in.) delamination located at 50.8 mm (2 in.) from the surface.

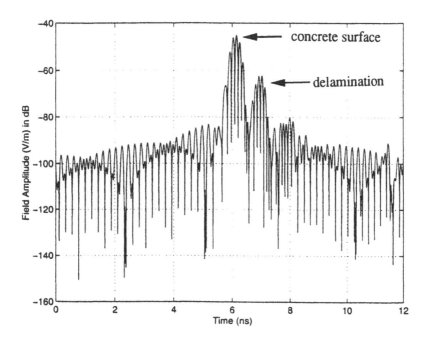

Figure 2. One-dimensional image obtained from modeling at 2.2 to 7.0 GHz for a 304.8 mm x 101.6 mm (12 in. x 4 in.) concrete block with a 25.4 mm (1 in.) x 152.4 mm (6 in.) delamination located at 50.8 mm (2 in.) from the surface.

59

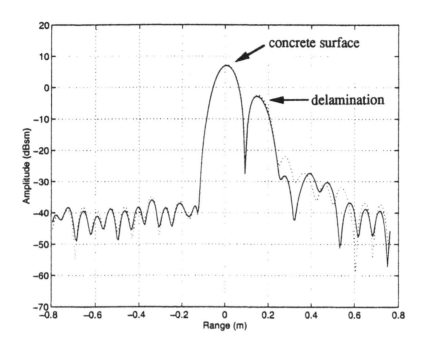

Figure 3. One-dimensional plot of the dry concrete block with a 25.4 mm (1 in.) thick delamination located at 50.8 mm (2 in.) depth from the front surface at 3.4 to 5.8 GHz.

REFERENCES

Büyüköztürk, O. and H.C. Rhim. 1993. "Electromagnetic Properties of Concrete for Nondestructive Testing," *Proceedings of the International Conference on Nondestructive Testing of Concrete in the Infrastructure*, Dearborn, Michigan, pp. 83-92, June 9 -11.

Büyüköztürk, O. and H.C. Rhim. 1995. "Radar Imaging of Reinforced Concrete Specimens for Nondestructive Testing," *Proceedings of the International Society for Optical Engineering (SPIE) Conference on NDE of Aging Dams and Structures*, Oakland, California, pp. 186-194, June 7.

Li, K., M.A. Tassoudji, R.T. Shin, and J.A. Kong. 1992. "Simulation of Electromagnetic Radiation and Scattering Using a Finite Difference-Time Domain Technique," *Computer Applications in Engineering Education*, Vol. 1 (1), pp. 45-63.

Mensa, D.L. 1981. High Resolution Radar Imaging, Artech House, Massachusetts.

Rhim, H.C. and Büyüköztürk, O. 1995. "Nondestructive Evaluation of Concrete Using Wideband Microwave Techniques," Research Report No. R95-01, Department of Civil and Environmental Engineering, Massachusetts Institute of Technology, Cambridge, Massachusetts.

Taflove, A. and M.E. Brodwin. 1975. "Numerical Solution of Steady-State Electromagnetic Scattering Problems Using the Time-Dependent Maxwell's Equations," *IEEE Transactions on Microwave Theory and Techniques*, Vol. MTT-23, No. 8, pp. 623-630, August.

Measurement of As-Built Conditions Using Ground Penetrating Radar

K. MASER

ABSTRACT

This paper provides an overview of the state-of-the-art of GPR technology as applied to the measurement of as-built conditions, with specific application examples and supporting research study results. Knowledge of the as-built conditions is an essential part of any non-destructive evaluation program. Existing records and drawings often do not represent what actually took place during construction, nor do they represent changes and repairs that have taken place after initial construction. Ground penetrating radar is a powerful NDE tool for the measurement of the internal dimensions of constructed materials. Existing capabilities include the determination of the thickness of asphalt, concrete, and granular material layers, and measurement of the depth and location of reinforcing steel. The equipment technology has been developed for data collection at up to normal traffic speeds (on pavements and bridge decks), and analytic techniques have been developed to automate the data processing.

The paper will describe the status of the GPR equipment technology that is available for this application, including equipment features, data acquisition methodologies, and data samples for different applications. The paper will also describe the basis principles for GPR data interpretation. Based on these principles, software has been developed to automate the processing of pavement layer thickness and depth and location of reinforcing. The software will be described, and example results from these applications will be provided. Results of research studies which have investigated accuracy of GPR based on comparisons to standard cores will also be presented.

DESCRIPTION OF THE GPR TECHNOLOGY

Ground penetrating radar systems generate short pulses of electromagnetic energy which penetrate into the pavement or bridge deck structure and reflect back from the material interfaces. The amplitude and arrival time of these return reflections are used to determine the thickness and properties of the material being evaluated, and the depth of reinforcing and other inclusions.

Kenneth Maser, President, INFRASENSE, Inc., 14 Kensington Rd., Arlington, MA 02174

The radar equipment described in this paper is geared specifically to mobile applications involving the coverage of large distances and areas (such as pavements and bridge decks). The equipment is based on a 1 GHz air-coupled horn antenna positioned from 0.3 to 0.5 meters (12-18 inches) above the surface. This is a non-contact arrangement which allows for road surveys to be carried out at any normal driving speed. For mobile applications, the horn antenna is superior to the more familiar ground-coupled antenna, since it permits driving speed surveys and since it provides a surface reflection for calibration of the surface material dielectric permittivity. Typical horn antenna systems generate 50 scans per second, with a pulse width of approximately 1 nanosecond. The radar analog signal is transmitted to a PC-based data acquisition system where it is digitized and stored to hard disk or tape. A distance measuring instrument (DMI) typically operated off of the survey vehicle transmission, provides position pulses which are encoded into the digitized radar for location referencing. The equipment used in most of the studies described in this paper is the RODAR™ system manufactured by Pulse Radar, Inc., of Houston, TX.

Since large quantities of data can be obtained quickly with this equipment, the key to an effective overall system is to have efficient software for mass processing of the data. The pavement and bridge deck surveys described in this paper have been based on two software systems: PAVLAYER© (PAVement LAyer Evaluation using Radar) and DECAR© (DEck Condition Assessment using Radar). Each of these programs moves sequentially through the digitized radar waveforms at a specified distance interval, computing the amplitudes and arrival times of the interface reflections which are related to pavement and bridge deck internal dimensions and properties. These amplitudes and arrival times are converted to layer thicknesses, rebar depths, and dielectric properties using ray transmission and reflection models (Maser, 1990; Maser and Scullion, 1992). Each program also includes software to assist in organizing the data analysis, and in presenting the output in a convenient and understandable format. The following sections describe specific applications of this survey system to the measurement of pavement thickness and depth of reinforcement.

PAVEMENT THICKNESS EVALUATION

Knowledge of pavement layer thickness is important in many areas of pavement management. Accurate thickness data is needed throughout the roadway and airfield network to predict pavement performance, to establish structural load carrying capacities, and to develop maintenance and rehabilitation priorities. For new construction, it is important to assure that the thickness of materials being placed by the contractor is close to specification. Until recently, the only acceptable method for pavement thickness measurement has been core samples and test pits. These are time consuming, destructive to the pavement system, dangerous to field personnel, and intrusive to traffic.

The highway radar system described above provides a low-cost, accurate alternative to coring. Accuracy studies using PAVLAYER have been conducted on over 150 pavement sections of varying materials and thicknesses located in fourteen US states and in three European countries. Over 800 cores and test pits have been used to verify the PAVLAYER predictions. The results show a typical accuracy of ± 7.5% for asphalt layers ranging from 51 to 508 mm (2-20 inches) thick, and ± 12% for granular base layers ranging from 152 mm to 330 mm (6-13 inches) thick.

Pavement layer thicknesses and properties may be calculated by measuring the amplitude and arrival times of the waveform peaks corresponding to reflections from the interfaces between the layers (see Maser and Scullion, 1992). The dielectric constant of a pavement layer relative to the previous layer may be calculated by measuring the amplitude of the waveform peaks corresponding to reflections from the interfaces between the layers. The travel time (t) of the transmit pulse within a layer in conjunction with its relative dielectric permittivity determines the layer thickness:

$$\text{Thickness (mm)} = (150\ t)/\sqrt{\varepsilon}, \tag{1}$$

where time is measured in nanoseconds and ε is the relative dielectric permittivity of the pavement layer. Computation of the surface layer dielectric permittivity can be made by measuring the ratio of the radar reflection from the pavement surface to the radar amplitude incident on the pavement. The incident amplitude on the pavement is determined by measuring the reflection from a metal plate on the pavement surface, since the metal plate reflects 100% of the incident energy. For example, using this data, one obtains the asphalt dielectric constant, ε_a, as follows:

$$\varepsilon_a = [(A_{pl} + A)/(A_{pl} - A)]^2 \tag{2}$$

where A = amplitude of reflection from asphalt, and A_{pl} = amplitude of reflection from metal plate (= negative of incident amplitude). A similar analysis can be used to compute the dielectric constant, ε_b, of the base material. The resulting relationship is :

$$\varepsilon_b = \varepsilon_a\ [(F - R2)/(F + R2)]^2 \tag{3}$$

where $F = (4\sqrt{\varepsilon_a})/(1 - \varepsilon_a)$ and R2 = the ratio of reflected amplitude from the top of the base layer to the reflected amplitude from the top of the asphalt. The above equations serve as the basis for analysis of the data collected during the various studies described below.

Systematic investigations of PAVLAYER have been carried out in conjunction with various federal, state, and local agencies in the US, the UK, Denmark, and Germany. In each of these studies, a comparison has been made between thicknesses computed from the radar data and the actual thicknesses which were subsequently obtained from cores. Table 1 summarizes the scope and the results of each of these studies. The Table lists the cooperating agency, the number and type of sections tested, the number of cores or test pits used for determining actual thicknesses, and the average deviation between radar and actual thickness at the core/test pit location.

Table 1 - Accuracy evaluation studies using a 1 GHz horn antenna[b] and PAVLAYER

Agency	Number of Sections			Number of Cores or Test Pits	Average Deviation (%)
	AC[a]	PCC[a]	AC/PCC[a]		
TexDOT[c]	12	1	--	90	5
Kansas DOT	11	--	3	73	7
Florida DOT	20	1	5	150	10
Wash. DOT	1	1	1	5	8
Wyoming DOT	9	--	--	36	10
Mn/ROAD	15	10	--	74	5
USA-SHRP	10	--	--	68	7
US Air Force	6	6	1	13	6
US FHWA	--	2	2	10	5
Pforzheim (Germ.)	26	--	--	35	8
Kent (UK)	5	--	--	76	5
TRL (UK)	3	1	--	115	6
Thüringen (Germ.)	9	--	--	28	10
TOTALS	127	22	12	817	7.5% (mean)

[a] AC = asphalt concrete; PCC = portland cement concrete; AC/PCC = AC over PCC
[b] Radar equipment for all of these studies was supplied by Pulse Radar, Inc., Houston
[c] DOT = Department of Transportation

Most of the above studies were carried out on a group of small sections, 100-500 feet in length. The radar data was analyzed for thickness prior to the availability of core data. The core data was then obtained and compared to the radar data at the core location. Exceptions to this protocol occurred in Florida, Kent (UK) and TRL (UK) projects, in which very long pavement sections were surveyed (10 - 30 km). In these latter situations, care was taken to identify distance control points in the radar survey, so that the core locations could be accurately located in the radar data.

A typical result for one section is shown in Figure 1 below. This result was taken from the Kansas DOT study. The radar data which generated this output revealed two distinct asphalt layers. The lower layer was the original pavement, and the upper layer was the most recent overlay. The figure shows that the radar-based results match very closely to the data obtained by coring.

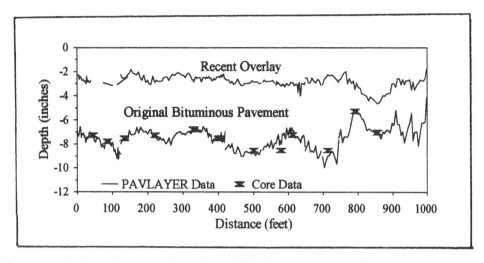

Figure 1 - Radar thickness data for a 1000 foot test site, with core thickness values
(Roddis, et. al., 1992)

The pavement thickness evaluation capability documented above is now being used on a routine basis for project and network level pavement evaluation. For project level work, the thicknesses are used to design pavement rehabilitation projects and to select the method of repair. For network level work, the thickness data provides an inventory for input into the agency Pavement Management System (PMS). This thickness data is used in modeling the future pavement performance, and thus provides a basis for long term planning and budgeting of pavement rehabilitation projects. Using GPR as described above, this inventory can be obtained at a greater speed and lower cost than it would through conventional coring. As an example of this application, the Florida DOT (FDOT) is now in the process of implementing this technology for an inventory survey of its 50,000 km. system of improved roads.

DEPTH OF REINFORCEMENT FOR BRIDGE DECKS

Repair and replacement of deteriorated bridge decks represents a major expense to many state highway agencies. The location and depth of the reinforcing steel plays a critical role in the performance of the deck, and in identifying the need for repair. Inadequate cover of reinforcing steel has been associated with premature deterioration due to the easy access of the top steel to corrosive attack by road salts.

A software system, called DECAR, has been developed for collecting and analyzing bridge deck data to determine deterioration and depth of reinforcement. Using DECAR, bridge decks are surveyed at normal driving speed with a sequence of parallel "passes" of the radar system. Each pass runs the length of the deck at a different transverse position. The transverse spacing of the passes depends on the level of detail of the survey. The DECAR software system organizes the bridge deck survey data collection, initiates and terminates data acquisition, records bridge identification numbers and pass locations with the collected data, and keeps a record of the survey. The software also organizes the data analysis so that: (i) the deck is distinguished from the neighboring pavement; (ii) passes on a given bridge are grouped and analyzed as a unit; (iii) analysis of multiple passes on multiple bridges can be carried out automatically; and (iv) the results are organized and presented in a convenient and understandable way.

The method for computing the depth of reinforcement is based on a calculation of the dielectric permittivity of the concrete, ε_c, and the detection of the arrival time of the reflections from the reinforcing steel (Maser, 1990). For an asphalt-overlaid deck, ε_c is computed using equation (3), with the pavement base replaced by the concrete.

This method was used to evaluated six Wyoming bridge decks, half with an asphalt overlay and half bare concrete (see Maser, 1994b), as part of a WTD research project. Ground truth was obtained by drilling cores into the deck, and determining the depth of asphalt (where there was an overlay) and the depth to the top reinforcement. The GPR based rebar depths were correlated with cores, and the results of this correlation are shown in Figure 2 below.

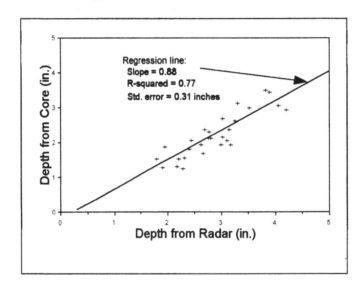

Figure 2 - Depth of Top Rebar: Correlation between GPR and Core Data

The scatter in the data is due to the fact that the horn antenna, operating at 18 inches above the pavement picks up reflections from 3 to 5 bars, depending on their spacing. The data, therefore, represents the average depth of these bars, and the cores represent the depths of individual bars. The 0.88 slope represents the fact that bars not directly under the antenna are at a greater distance than the bars directly below, thus producing an apparent radar depth that is greater than the actual bar depth. Once this relationship is know, it can be used for calibrating future data.

Figure 3 shows a contour plot of the depth of reinforcing steel for one of the concrete decks which were surveyed. The Figure highlights only those areas where the depth of steel was found to be below a two inch threshold. An inspection of the surface condition of the deck revealed that the areas shown to have low steel cover were also found to have spalling and developing delamination.

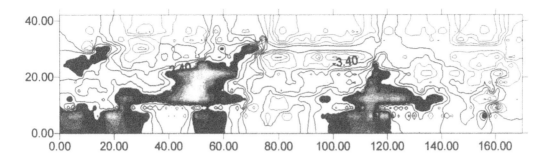

Figure 3 - Contour Plot of Depth of Reinforcement (Maser, 1994)
(areas where rebar is < 2" are highlighted in black)

CONCLUSION

This work has shown that highly accurate evaluations of as-built pavement and bridge deck can be carried out using the appropriate radar hardware and accompanying software. The air-coupled horn antennas used in this work were essential in providing a fast-moving non contact survey system, and in providing the short-pulse high resolution needed for these applications. The software proved very useful for organizing surveys and data analysis, and for extracting pavement physical properties from the raw radar data. The highway radar systems described in this paper are very recent developments, and it is likely that they will be further developed and extensively used for highway evaluation in the future.

REFERENCES

K. R. Maser (1990) "New Technology for Bridge Decks Assessment" Phases I and II Final Reports. Report No. FHWA-NETC-89-01 and 90-01. Work sponsored by the New England Transportation Consortium, Center for Construction Research and Education, Massachusetts Institute of Technology. Cambridge, MA. 1990.

K. R. Maser (1995) "Evaluation of bridge decks and pavements at highway speed using ground penetrating radar" *Proceedings of the SPIE Conference on Nondestructive Evaluation of Aging Infrastructure,* 6-8 June, 1995, Oakland, CA SPIE Publication 2456

W. M. K. Roddis, K. R. Maser and A. J. Gisi (1992) "Radar pavement thickness evaluations for varying base conditions" *Transportation Research Record No. 1355.* Transportation Research Board. National Research Council, Washington, D.C., 1992.

E. G. Fernando, K. R. Maser and B. Dietrich (1994) "Implementation of ground penetrating radar for network-level pavement evaluation in Florida" *Proceedings of the GPR '94, Fifth International Conference on Ground Penetrating Radar.* Kitchener, Ontario, CANADA. June 1994

K. R. Maser (1994) "Evaluation of ground penetrating radar applications of pavement and bridge management" Report FHWA/WY-94/01, Wyoming Transportation Department, January, 1994

Field Test Results of an Ultrasonic Applied Stress Measurement System for Fatigue Load Monitoring

P. A. FUCHS, U. HALABE, S. PETRO, P. KLINKHACHORN, H. GANGARAO,
A. V. CLARK, M. G. LOZEV AND S. B. CHASE

ABSTRACT

Fatigue loading in bridges due to traffic is an important factor in assessing the structures condition. Fatigue loading is measured by determining the applied stress seen by a structure. Conventional methods for applied stress measurement use strain gages, which are time consuming to install. A possible alternative for making the applied stress measurements is ultrasonic techniques using non-contact electromagnetic acoustic transducers (EMATs). This method requires no time consuming surface preparation and is easy to install. Some of the main difficulties of measuring applied stress fluctuations in bridge members are the relatively low (usually less than 14 MPa [2 ksi]) stress cycles and the frequency of the measurements, which are in the order of several Hertz. A prototype system that uses EMATs to measure applied stress cycles has been built for measuring fatigue loading in bridge structures. This paper discusses results of field tests conducted with this measurement system. This includes effects of an actual bridge environment, such as surface conditions, on the system.

BASIC SYSTEM OPERATION

Basic equipment to perform ultrasonic applied stress measurements consists of an ultrasonic pulser, an ultrasonic receiver, and some method to measure the ultrasonic wave time-of-flight (TOF), Fig. 1. The received ultrasonic signal can be digitized and processed in a computer to determine the time-of-flight of the ultrasonic wave. A strain conditioner is used to acquire data from strain gages. Measurements of strain from the strain gages and of ultrasonic time-of-flight data can be made at the same point in time for comparison purposes.

P. A. Fuchs and S. Petro, Constructed Facilities Center, West Virginia University, Morgantown, WV 26506

U. Halabe and H. GangaRao, Dept. of Civil and Environmental Engineering, Constructed Facilities Center, West Virginia University, Morgantown, WV 26506

P. Klinkhachorn, Dept. of Electrical and Computer Engineering, Constructed Facilities Center, West Virginia University, Morgantown, WV 26506

A. V. Clark, National Institute of Standards and Technology, Materials Reliability Division, Boulder, CO 80303

M. G. Lozev, Virginia Transportation Research Council, 530 Edgemont Rd., Charlottesville, VA 22903

S. B. Chase, Federal Highway Administration, 6300 Georgetown Pike, McLean, VA 22101

The transducer used to generate and receive ultrasonic waves is an electromagnetic acoustic transducer (EMAT). This is a non-contact transducer which can operate on surfaces without requiring surface preparation (Maxfield, Kuramoto, and Hulbert, 1987). The EMAT also does not require a couplant as is the case with a conventional piezoelectric transducer. In this application a Rayleigh wave (surface wave) is transmitted through the material with one transducer and received with another. On a bridge structure, the change in time-of-flight of this Rayleigh wave is measured to determine the applied stress (Clark, Fuchs, and Schaps, 1995). A sensor-head houses both the transmitting EMAT and the receiving EMAT in one contained unit. The sensor-head with 0.5 MHz EMAT coils was used for field testing. All of the equipment was packaged in a suitcase type enclosure, with dimensions 48 x 33 x 20 cm (19 x 13 x 8 in), for ease of use in the field. Power was provided by a portable generator.

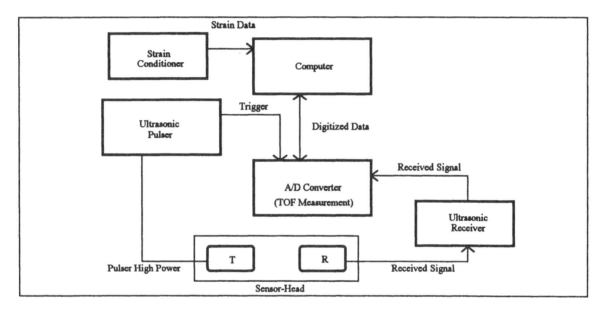

Figure 1. Ultrasonic Applied Stress System Block Diagram

FIELD TEST DESCRIPTION

Several bridges located close to the Constructed Facilities Center (CFC) in Morgantown, WV were examined with the ultrasonic applied stress measurement system. Another structure, located near Charlottesville, Virginia, was also used for testing. Only data from the bridge in Virginia is presented in this paper.

A slab-on-girder composite structure was chosen for a controlled field test in Virginia. The bridge was donated by the Virginia Transportation Research Council (VTRC), which also provided traffic control and support for the testing. The structure tested has four A36 steel girders, which are simply supported. The bridge is located on Rt. 691 west of Charlottesville, VA and passes over I-64. The testing was performed in the right-most lane in the west-bound direction. The span length was approximately 36 m (120') and the center of the span was 6 m (20') above the road surface. Live load measurements were performed at the center-span on the east-most two girders. A 11400 kg (25 ton) vehicle was used to apply stress to the structure. Strain gages were mounted on each of the two girders used for testing for comparison to the ultrasonic measurements. The girders were painted with lead based paint, which had a thickness of about 0.20 - 0.25 mm (8-10 mils) on the flange.

The testing on Rt. 691 was a controlled test. Therefore, traffic on the structure was limited to a test vehicle. The test vehicle was driven over the bridge at various speeds for evaluation of dynamic measurements with the ultrasonic applied stress measurement system. Representative data are shown in Fig. 2-4. The strain gage data and EMAT data are artificially offset in the figures so that the two plots may be easily distinguished.

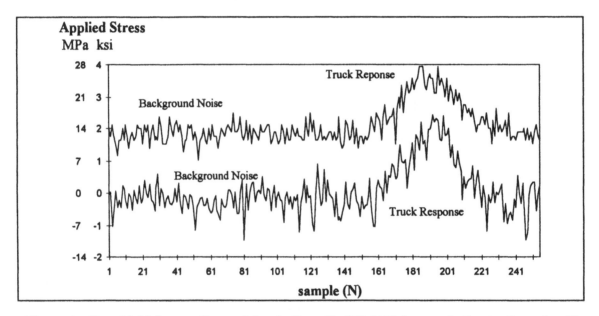

Figure 2. Test Vehicle traveling at 8 km/h (5 mph). [EMAT (bottom) Strain Gage (top)]

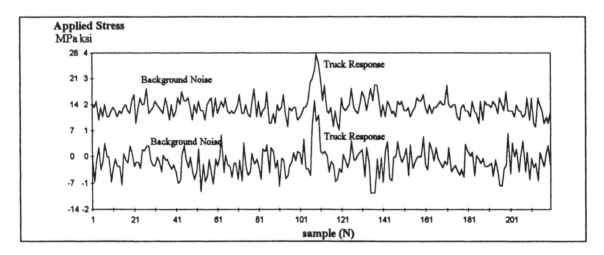

Figure 3. Test vehicle traveling at 40 km/h (25 mph).[EMAT (bottom) Strain Gage (top)]

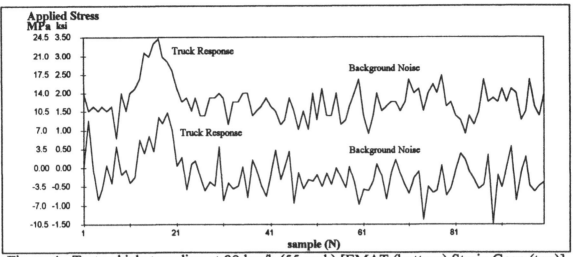

Figure 4.   Test vehicle traveling at 89 km/h (55 mph).[EMAT (bottom) Strain Gage (top)]

APPLIED STRESS CYCLE COUNTING

A comparison between the EMAT and strain gage data can be shown by examining the results of cycle counting the data. The raw data can be processed with a conventionally accepted technique of rainflow cycle counting (Downing and Socie, 1982). The results of the rainflow cycle counts can then be placed in a histogram for analysis. This histogram can be processed with an algorithm based on Miner's rule to determine the equivalent stress range represented by the particular histogram (Keating et al., 1990). As an example, a stress histogram resulting from processing the data in Fig. 2 is shown below (Fig. 5).

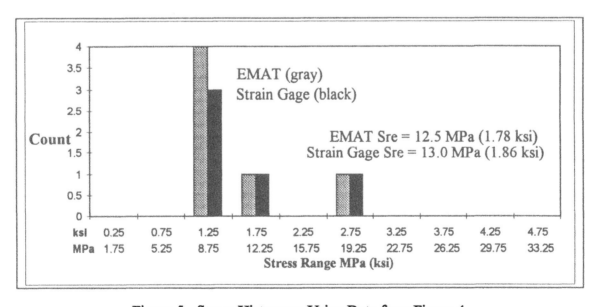

Figure 5. Stress Histogram Using Data from Figure 4.

Several sets of data collected from the I-64 bridge were processed with a rainflow cycle counting algorithm in order to produce stress histograms. The equivalent stress range represented by each histogram was then determined. Table I shows the value of S_{re} calculated for both the EMAT and strain gage data along with the corresponding percent error. Fifteen sets of data are given as a result of a test vehicle traveling at various speeds. A threshold of 7.0 MPa (1 ksi) was used to eliminate counts of stress cycles below the noise level. The percent error calculated from the data set **89 km/h (55 mph) A** was considered an outlier and as a result not included in the percent error average.

TABLE I. RESULTS OF CYCLE COUNTING I-64 BRIDGE DATA

	EMAT Sre MPa (ksi)	Strain Gage Sre MPa (ksi)	Error (%)
8 km/h (5 mph) A	13.2 (1.88)	13.1 (1.87)	0.5
8 km/h (5 mph) B	13.4 (1.92)	12.6 (1.80)	6.7
8 km/h (5 mph) C	13.2 (1.89)	14.0 (2.00)	5.5
8 km/h (5 mph) D	11.3 (1.61)	10.6 (1.51)	6.6
8 km/h (5 mph) E	13.1 (1.87)	15.1 (2.16)	12.0
40 km/h (25 mph) A	13.0 (1.85)	16.4 (2.34)	21.4
40 km/h (25 mph) B	13.2 (1.89)	11.1 (1.59)	18.9
40 km/h (25 mph) C	13.0 (1.85)	11.9 (1.70)	8.8
40 km/h (25 mph) D	13.6 (1.94)	12.1 (1.73)	12.1
40 km/h (25 mph) E	13.0 (1.85)	11.7 (1.67)	10.8
89 km/h (55 mph) A	22.8 (3.25)	13.2 (1.88)	72.9
89 km/h (55 mph) B	12.5 (1.78)	13.0 (1.86)	4.3
89 km/h (55 mph) C	11.3 (1.62)	13.0 (1.86)	12.9
89 km/h (55 mph) D	11.6 (1.65)	13.9 (1.99)	17.1
89 km/h (55 mph) E	13.1 (1.87)	13.2 (1.88)	0.5
Average Error			**9.9**

CONCLUSION

Typical noise for the EMAT system was about 7.0-10.5 MPa (1.0-1.5 ksi). The response time was about 8 Hz, which resulted from 50 signal averages per measurement. The strain gage noise was approximately 3.5 - 7.0 MPa (0.5-1.0 ksi), which was slightly better than the EMAT noise. The Rayleigh wave signals through the paint on actual bridge members were very good, and the noise measured in the field experiments was essentially the same level as in the laboratory when simulating field girder surface conditions. However, an improvement in the noise level would be beneficial. Also, the response time of the system will need to be improved.

The results of the applied stress measurements can be applied to various forms of analysis, such as determination of the fatigue cycles seen by a structure. For the particular data sets given, the stress histograms showed about a 10% difference between the EMAT and strain gage data. These histograms resulted from data only over a short period of time and represents only a small sample set. However, it was shown that the EMAT system was capable of performing applied stress measurements on actual bridge structures which could closely track results from a conventional strain gage.

ACKNOWLEDGMENTS

H. GangaRao, U. Halabe, S. Petro, P. Klinkhachorn, and P.A. Fuchs acknowledge the financial support of the Federal Highway Administration. M.G. Lozev acknowledges the financial support of the Virginia Transportation Research Council (VTRC). We would like to recognize the contributions of S.R. Schaps (NIST). Also, we would like to thank D. McColskey (NIST), T. Nguyen (NIST), C. Apusen (VTRC), J. French (VTRC), and P. Frum (West Virginia University) for their assistance with various phases of testing.

REFERENCES

Clark, A.V., Fuchs P.A., and Schaps S.R., "Fatigue Load Monitoring in Steel Bridges with Rayleigh Waves," accepted for publishing in Journal of Nondestructive Evaluation, 1995.

Downing, S.D., Socie, D.F., "Simple Rainflow Counting Algorithms," Int. J. Fatigue, Jan. 1982.

Keating, P.B., Kulicki, J.M., Mertz, D.R., Hess, C.R., "Economical and Fatigue Resistant Steel Bridge Detail," participant Notebook, National Highway Institute Course No.: 13049, June 1990.

Maxfield, B.W., Kuramoto, A., and Hulbert J.K., "Evaluating EMAT Designs for Selected Applications," Materials Evaluation, Oct. 1987.

Computerized Fracture Critical Bridge Inspection Program with NDE Applications

P. E. FISH

Abstract

Wisconsin Department of Transportation implemented a Fracture Critical Inspection Program in 1987. The program has a strong emphasis on Nondestructive Testing (NDT). The program is also completely computerized, using laptop computers to gather field data, digital cameras for pictures, and testing equipment with download features. Final inspection reports with detailed information can be delivered within days of the inspection.

The program requires an experienced inspection team and qualified personnel. Individuals performing testing must be licensed ASNT (American Society for Nondestructive Testing) Level III and must be licensed Certified Weld Inspectors (American Welding Society).

Several critical steps have been developed to assure that each inspection identifies all possible deficiencies that may be possible on a Fracture Critical Bridge. They include; review of all existing plans and maintenance history; identification of fracture critical members, identification of critical connection details, welds, & fatigue prone details, development of visual and NDE inspection plan; field inspection procedures; and a detailed formal report.

The program has found several bridges with critical fatigue conditions which have resulted in replacement or major rehabilitation. In addition, remote monitoring systems have been installed on structures with serious cracking to monitor for changing conditions.

Introduction

History In 1978, Wisconsin Department of Transportation discovered major cracking on a two-girder, fracture critical structure, just four years after it was constructed. In 1981, on the same structure, now seven years old, major cracking was discovered in the tie girder flange of the tied arch span.

This is one example of the type of failures that transportation departments discovered on welded structures in the 1970's and '80's. The failures from welded details and pinned connections, lead to much stricter standards for present day designs.

All areas were affected: Design with identification of fatigue prone details and classification of fatigue categories (AASHTO); Material requirements with emphasis on toughness and weldability, increased welding and fabrication standards with licensure of

Philip E. Fish, Specialized Bridge Inspection, Wisconsin Department of Transportation, 4802 Sheboygan Ave. Madison, WI 53707-7915

fabrication shops to minimum quality standards including personnel; and an increased effort on inspection of existing bridges, where critical details were overlooked or missed in the past.

FHWA inspection requirements for existing structures increased through this same time period, in reaction to the failures that had occurred. Obviously, many structures in Wisconsin were not built to the standards now required, thus the importance for quality inspection techniques. The new FHWA inspection requirements that are now being implemented throughout the nation, require an in-depth, hands on type inspection at a specified frequency, on all fracture critical structures.

Current Program Wisconsin Department of Transportation started an in-depth inspection program in 1985 and made it a full time program in 1987. This program includes extensive non-destructive testing. Ultrasonic inspection has played an major role in this type of inspection. All fracture critical structures, pin and hanger systems, and pinned connections are inspected on a five-year cycle.

The program requires an experienced inspection team and a practical inspection approach. Extensive preparation is required with review of all design, construction and maintenance documents. An inspection plan is developed from the review and downloaded to a laptop computer.

Inspection emphasis are on "hands on" visual and Nondestructive evaluation. Report documentation includes all historical data, design plans, pictorial documentation of structural deficiencies, nondestructive evaluation reports, conclusions and recommendations. All notes are collected on a field laptop computer. Pictures are taken on a digital camera which is downloaded to either the field laptop computer or the engineering work station. Nondestructive testing information is also downloaded in the same manner.

All final report documentation and historical documentation is executed with an engineering work station as a "single source" information file and reporting file for the inspection program.

Historical information such as design plans, construction history, maintenance history, inventory inspection, and routine inspections are scanned and downloaded to individual files for each bridge.

Fracture critical inspection report documentation is placed in the same file each time a bridge is inspected. It would also include all inspection data with pictures and nondestructive testing information.

Engineering work station files are accessible by persons requiring the information in central office and district offices. In the near future, it is planned to integrate this program with the PONTIS Bridge Management system.

Typical specifications for a Fracture Critical inspection are:

AASHTO Standard Specifications for Highway Bridges, 1989

AASHTO Manual for Condition Evaluation of Bridges, 1994.

AASHTO Guide Specifications for Strength Evaluation of Existing Steel and Concrete Bridges, 1989

AASHTO Guide Specifications for Fatigue Evaluation of Existing Steel Bridges, 1990.

AASHTO Guide Specifications for Strength Design of Truss Bridges (Load Factor Design), 1985

AASHTO Guide Specifications for Fatigue Design of Steel Bridges, 1989.

AASHTO Guide Specifications for Fracture Critical Non-Redundant Steel Bridge Members, 1986.

AASHTO Standard Specifications for Movable Highway Bridges, 1988.

FHWA Inspection of Fracture Critical Bridge Members, 1986.

FHWA Bridge Inspector's Training Manual 90, 1990.

FHWA Bridge Inspector's Manual for Movable Bridges, 1977.

FHWA Non-Destructive Testing Methods for Steel Bridges, 1986.

FHWA Recording and Coding Guide for the Structure Inventory and Appraisal of the Nation's Bridges, 1988.

FHWA Technical Advisory - Revisions to the National Bridge Inspection Standards, Sept. 1988.

FHWA Manual of Uniform Traffic Control Devices, 1988.

U.S. Government, National Bridge Inspection Standards, Code of Federal Regulations, Title 23, Part 650, Subpart C, Oct. 1988.

Wisconsin Department of Transportation, Maintenance Manual, 1991

Wisconsin Department of Transportation, Standard Specifications for Road and Bridge Construction, 1989

Nondestructive test methods typically used in the fracture critical inspection program are; visual, dye-penetrant, magnetic particle, ultrasound, acoustic emission and remote strain gauging.

Critical Steps In A Fracture Critical Inspection Process

Plan Review

It's important to search all possible data for a particular structure. Often drawings and data are stored in several locations. Typical locations for data in Wisconsin are; Bridge Design files, District Construction files, District Maintenance files, Central Office and District inspection files and local maintenance crew files.

After all design, shop, construction, and maintenance plans are assembled along with history & inspection data, a formalized review should take place. The review should include design engineers and the inspection team.

Identification of Fracture Critical Members

Fracture Critical members as defined by AASHTO and FHWA need to be identified by the design engineers and inspection team. This should include review of all design computations and shop drawings. Plan and elevation views showing location of Fracture Critical members should be developed from this review for use during the inspection process and implemented as part of the final report.

Identification of Critical Connection and Fatigue Prone Details

Selection of critical connection and fatigue prone details is based on current inspection specifications and practical experience of the inspection team. Details of plan, construction, maintenance, and inspection reviewed. Selection is made by design engineers and the inspection team. Identified details are noted on plan sheets along with along with criticality and type of inspection that will take place. These details are formatted for use during the inspection process and implemented as part of the final report.

Development of Inspection Plan
After all information has been gathered and all fracture critical, fatigue prone details, and critical connection details identified, the inspection team organizes the inspection plan. At this point, types of inspection are decided. An NDE inspection plan is developed along with a visual inspection plan. These plans are formatted and downloaded for field inspection to be implemented into the final report after the inspection.

Types of equipment for access are discussed and points of access are identified. The direction in which the structure will be inspected is determined and a traffic control plan is established. A safety plan is developed to protect the traveling public and inspection personnel.

A time schedule is established for public information bulletins to be distributed to the media.

Bridge Inspection
Bridge inspection is performed with a laptop PC. All formatted information is available in the PC file. All drawings are available in a 11" x 17" size for reference. All inspection notes are recorded on the field laptop. Extensive use of pictures is utilized to explain inspection findings. Pictures are stored on a digital disk camera for downloading. NDE inspection is performed per inspection plan and stored for downloading to the field laptop at the end of the inspection day.

Conditions found during the inspection that were not part of the inspection plan or were not anticipated, are implemented into the field inspection and made part of the inspection plan for the next inspection.

Formalized Report
A formalized report is developed from the format loaded for field inspection. A typical report contains the following information:
* Title Sheet with an overall picture of the bridge
* Table of Contents
* Introduction
        Purpose and Inspection Specifications
        Physical Data
        Construction and Maintenance History
* Design and Shop Plans
        Original Plans and any Update Plans
* Inspection Notes with Pictures And a Location Plan
* NDE Test Notes and Download Information
* Conclusion and Recommendations

The final report is reviewed by the inspection team and design engineers before distribution. All final comments of the team and design engineers are implemented into the report. The report is then distributed to design office, district design, maintenance, and inspection offices for review and action.

Examples of Deficiencies Located By Fracture Inspection Program

The program has been successful in locating some serious fatigue cracking and weld cracking. The following case studies are examples of types of deficiencies found in the past several years.

Case Study # 1

Location: Milwaukee, Wisconsin   Cemetery Access Road over IH 94.
Super Type: K-Frame, two girder with floor beams and stringers.
Built: 1957
In-Depth Inspection: September 1988, 1993
Unique Features: Fracture Critical
                        All connections field welded, including field splices.
Deficiencies: Cracked butt welds in girder flanges and webs.

Case Study # 2

Location: Milwaukee, Wisconsin   St. Paul Ave. and 26th St. over IH 94.
Super Type: Multi-girder with box girder supporting one span.
Built: 1960
In-Depth Inspection: June 1987, 1992
Unique Features: Fracture Critical Box Girder
                        All connections field welded, including field splices.
                        Pin and Hanger Joint
Deficiencies: Cracked butt welds in girder flanges and webs
                        Poor quality field welding on primary members
                        Bearing positioned wrong

Case Study # 3

Location: Milwaukee, Wisconsin   IH 94 over Menominee Valley
Super Type: Four Girder with Floor Beams and Stringers.
Built: 1960
In-Depth Inspection: 1986, 1987, 1988, 1990
Unique Features: Floor Beam to Girder Connections
                        Lower Lateral Bracing Connections
                        Deck removal saw cuts in the girder flanges.
Deficiencies: Cracked welds at floor beam connections to main girders
                        Saw cuts in girder flanges

Case Study # 4

Location: Milwaukee, Wisconsin   IH 94 over Oklahoma Ave.
Super Type: Multi-Girder with Box Girder Supporting Two Spans.
Built: 1967
In-Depth Inspection: July 1985, September 1988, April 1991
Unique Features: Fracture Critical Box Girder
                        Pin and Hanger Joints
Deficiencies: Frozen pin and hanger joints
                        Failed fixed bearings at abutments
                        Weld cracks (hydrogen) in cross girder

Case Study # 5

    Location: Prairie du Chien, Wisconsin   USH 18 over Mississippi River

    Super Type: Main Span; Tied Arch Approach Spans; Two Girder with Floor
        Beams and Stringers.

    Built: 1974

    In-Depth Inspection: 1978, 1981, 1985, 1989, 1994

    Unique Features: All Spans Fracture Critical
                     Pin and Hanger Joints
                     Electro-Slag Butt Welds

    Deficiencies: Girder web cracking at floor beam connections from
                        out-of-plane bending
                     Cracked electroslag butt welds
                     Banding in A441 flange plate steel (Arch Span)

Case Study # 6

    Location: Hudson, Wisconsin I 94 Over St. Croix River

    Super Type: Overhead Truss, Cantilever Truss, & Steel Continuous Deck Girder

    Built: 1951

    In-Depth Inspection: 1988, 1993, 1994, 1995

    Unique Features: All Spans Fracture Critical
                     Pin & Hanger supported - suspended truss span

    Deficiencies: Cracked floor beam connection angles
                     Failed stringer connection rivets
                     Cracked stringer connection angles

Case Study # 7

    Location: Sturgeon Bay, Wisconsin State Trunk 57 & 42 over Sturgeon Bay

    Super Type: Overhead Truss, Overhead Truss Bascule, Reinforced Steel Girder

    Built: 1930

    In-Depth Inspection: June, July August, September, October, November,
                           December 1994 and January, May, June, July 1995

    Unique Features: Fracture Critical Overhead Truss Spans
                     Fracture Critical Overhead Truss Bascule Span

    Deficiencies: Severe cracking on bascule girders from repair welding and
                        fatigue. Considerable wear on mechanical system.

NJDOT Structural Inspection Training Needs

R. J. SCANCELLA, G. TISSEVERASINGHE AND W. CARRAGINO

ABSTRACT

Private companies use the most qualified personnel available to perform intricate inspections and the New Jersey Department of Transportation is no exception. Because of downsizing of the Department over the last ten years, employees have retired without being replaced and personnel have been transferred to units where they are most needed. In order to give newly transferred or hired personnel the technological information required to perform the inspections, and to keep existing employees current with the latest advancements, in-depth training program is required. This paper presents the New Jersey Department of Transportation, Bureau of Materials Engineering and Testing, Inspection Unit training program.

INTRODUCTION

The Inspection Unit of the Bureau of Materials Engineering and Testing of the New Jersey Department of Transportation is required to inspect the fabrication of all structural materials prior to the product being shipped to a construction site. This unit also performs non-destructive testing on existing structures when requested by the Bureau of Maintenance or Bureau of Structures. This unit must also perform inspections on all materials used within a NJDOT project. The unit is divided into five (5) groups, which these are:

Administration - collects the data for each construction project regarding fabricators, manufacturers, or suppliers and forwards the appropriate paper work to the Project's Resident Engineer, Prime Contractor, and Fabricator/Manufacturer.

Robert J. Scancella,PE, New Jersey Department of Transportation, Project Engineer Materials, 1035 Parkway Ave., CN-600, Trenton, NJ 08625
Godfrey Tisseverasinghe, Senior Engineer, 1035 Parkway Ave., CN-600, Trenton, NJ 08625
Wayne Carragino, Senior Engineer, 1035 Parkway Ave., CN-600, Trenton, NJ 08625

Steel/NDT - inspects all structural steel both at the fabricators and on existing structures. This group also works closely with design, research, and maintenance units within the DOT in an effort to continually increase the life of steel structures.

Prestressed Concrete - inspects all pre-fabricated structural concrete whether is prestressed or precast. This group must also coordinate with the other inspection units for items such as reinforcement steel and steel plates.

Product Approval - this group is responsible for all non-structural manufactured material including items such as: reinforcement steel, traffic paint, structural paint, epoxy, joint material and casted metals.

Aggregate, Timber & Drainage - this group is responsible for maintaining the approved aggregate producer's list and inspection of timber and precast concrete drainage items.

At one time this unit consisted of over 40 employees. During the last ten years however, the unit has dwindled in size due to attrition, downsizing, and budget cuts. The unit now consists 27 employees. With a greater number of employees it was possible for specialization in various aspects of inspection. With the loss of employees the current workers are now responsible for multiple tasks. In addition, the amount of new construction projects has increased together with the maintenance requirements.

GOAL

State agencies, such as the NJDOT, can no longer train an employee with the idea in mind of specialized training. Because of the increase and diversity in projects, the Inspection Unit employees need to be cross-trained to perform intricate structural and non-structural inspections. Using a cross-training method, the employee becomes not only valuable to his particular section but can be used in other groups as needed.

The goal for this program is to economically train employees unfamiliar with structural materials and NDT, while at the same time increasing the knowledge of the experienced structural and NDT inspectors.

SOLUTION

A monthly Inspection Unit staff meeting decided that the best way to implement the new cross-training program was to train all employees to perform structural steel and prestressed concrete inspections. The Steel/NDT group requires the inspectors to be certified in the use of non-destructive testing methods and instruments, while the Prestressed Concrete group requires certification by ACI and PCI. These two groups perform most of the out of state inspections creating a need for more qualified inspectors in these two groups.

Presently the Prestressed Concrete group has sufficient staff to provide the minimum inspection required for the projects under construction. Training was therefore initiated with the Steel/NDT group. Employees above a Technician 4 level were eligible for the training. The Technician 3 level and above were chosen because of their experience, while the Technician 4 level and below did not have the necessary experience. The training was also made available to the present Steel/NDT inspectors as a refresher course. The inspectors were trained in visual, ultrasonic, dye penetrant, magnetic particle and radiographic testing/inspection. In order to maintain a high integrity of inspection personnel, technicians were trained in the classroom atmosphere and on-the-job to ensure certification for a period of three years after passing written and performance tests. The same requirements have been established for the present certified inspectors. The exception is that once inspectors are certified they are required to attend 40 hours of related training with a maximum of 16 hours in any given fiscal year.

IMPLEMENTATION

The first step in the implementation of our plan was to divide the available personnel into groups to maintain the normal work load. These groups were assigned two weeks of classroom training in each of the visual, dye penetrant and ultrasonic testing/inspection techniques since most inspection requires a ASNT level 2 certification.

Following the classroom training employees were assigned to a fabrication project with a certified inspector. Once new inspectors received enough hands-on training they were given the performance test. After passing the test, they were certified for three years. In the case of a failed performance test, more classroom and performance training were provided. The existing certified inspectors were given a written and performance test together with the new trainees. All of the existing inspectors passed both tests and were certified for three years. If any existing inspectors had not passed either tests they would have been required to take refresher training. All certified inspectors are required to have continuing education together with a performance test every three years to maintain their certification.

Once the training and performance testing requirements were met they became a part of the individual's "Performance Assessment Review" (PAR) and subject to annual appraisal. In the event of an unsatisfactory review, pay increments maybe withheld and possible disciplinary action may be taken. Unfortunately, this creates a punitive rather than incentive based program.

Finally, the newly certified employee becomes involved in the Steel/ NDT group. The Principal Engineer may use one of these inspectors when necessary. Presently three of these inspectors have been utilized and it is expected that the future work load will necessitate the use of more inspectors.

COST ANALYSIS

The cost for the classroom training of one individual for each NDT inspection technique is approximately $1500 to $2000. This cost includes travel and overnight expenses for the individual. Due to budgetary restraints only two inspectors could of been trained at that cost. Having only two inspectors available for cross-inspection would not significantly increase the efficiency of the unit. By bringing the training to New Jersey it was possible to train 8 individuals for the same cost of training two individuals. Classroom training for visual and dye penetrant was accomplished using in - house personnel, eliminating any cost to the unit or Department.

The total cost for the training of the eight inspectors was approximately $8000. The cost of hiring private inspection forces is approximately $70/day/project more than using State forces. Therefore, it would take 115 days of private inspection costs to equal the training expenses. At that point the trained inspectors are more cost effective. After eight months the new inspectors have used 90 days. The program expected to pay for itself within one year.

Presently the first group of newly trained inspectors are meeting our minimum needs of inspection requirements at the fabrication plants. However, because of attrition and possible future layoffs a second group of inspectors is being trained.

CONCLUSION

The new training caused some dissent among both new and existing inspectors. The existing inspectors were not comfortable the 40 hours of required classroom training and recertification procedures. Additionally, all of the inspectors were extremely distressed to have requirements added to their PAR. The inspectors were made aware that the cross-training was needed to fill vacancies. Newly-trained inspectors would only be used when the need couldn't be filled by the existing Steel/NDT inspectors. The reason for retesting and recertification is to keep the inspectors current with the latest in technology and techniques in the non-destructive testing field, and additionally, to give the Inspection Unit of the Bureau of Materials Engineering and Testing improved credentials.

It was a goal to make PARs incentive based instead of or punishment-based. Each Principal Engineer was asked to provide some incentive ideas to motivate the inspectors regarding the training and recertification program. Several initiatives were implemented but the first line managers (Project and Principal Engineers) were asked to continue to update the initiatives. The support of the first line managers was an important tool in influencing the inspectors regarding the program.

The first line managers decided to prepare a handbook for each employee of the Inspection Unit. The handbook describes all cross-training and certification requirements. It also gives examples of the required recertification training. It further provides information that helps the employees with the written and performance tests. Two Steel/NDT inspectors together with two of the first line managers prepared this handbook to ensure that it was user friendly. During the preparation of the handbook each employee will have the opportunity to review and comment on the material.

This cross - training and recertification program has helped the unit to better use its manpower resources for the fabrication of structural steel inspection. Since the inception of the additional training, there has been a marked improvement in the inspection services. The availability of more inspectors has prevented the necessity of excepting items on certification.   Experienced inspectors have found the recertification aspect to be most helpful and it has improved job morale and self-esteem.

It is unfortunate that cross-training can not be used for all personnel. Some resistance to the idea has been encountered. Some inspectors feel that specialization should not be relinquished, while others feel they are too old to be retrained. At this time the unit has only used volunteers for the cross-training program. In the future it may be necessary for all inspectors to obtain cross-training. In conclusion, it is the job of the managers to maintain tolerance and enthusiasm for the program.

REFERENCES

American Society of Non-destructive Testing - NDT certification requirements
American Concrete Institute - Concrete certification requirements
Prestressed Concrete Institute - Prestressed concrete certification requirements

ACKNOWLEDGMENTS

Henry Justus, Chief Bureau of Materials Engineering and Testing, NJDOT
Rich Smetanka, Principal Engineer Materials, NDT
Frederick Lovett, Principal Engineer Materials
Robert Skalla, Principal Engineer Materials
Warren Cummings, Principal Engineer Materials
Arthur Egan, Administrative Assistant

Time Domain Reflectometry Monitoring of Bridge Integrity and Performance

C. H. DOWDING AND C. E. PIERCE

ABSTRACT

Integrity of bridge supports, namely abutments, columns and piers, and foundations, can be diminished by scour, abutment and foundation movement, and seismic excitation, among other processes. Time domain reflectometry (TDR) analyses of signals pulsed through cables incorporated within the bridge structure can be employed to monitor suspected degradation of these supports. Two general approaches are described in this paper to locate and quantify damage of critical bridge support members using TDR. First, the overall stability of the structure can be assessed by monitoring movement of abutments and foundations with respect to the foundation materials. Second, columns and piers can be monitored for internal cracking. Both approaches utilize the same method of measuring deformation of metallic, coaxial cables caused by displacement or deformation of the surrounding medium. These sensing cables require an electronic monitoring system to collect both the global measurements of external displacement and local measurements of internal deformation. It is important that this system function during critical events, such as earthquakes and floods, when major damage is most likely to occur.

INTRODUCTION

Scour has been linked to nearly 95% of all severely damaged and failed highway bridges constructed over waterways in the United States (Lefter, 1993). The greatest loss of sediment to scour occurs at high water velocities, particularly during floods. Bridge pier movement can occur as a result of material loss beside and beneath the base of its footing, which produces undesired displacement of the bridge supports and may ultimately result in structural collapse. For these reasons, development of technologies to monitor scour and resulting structural movements is needed.

Recent earthquakes have demonstrated that many bridge columns and piers designed to resist earthquake shaking can be severely damaged yet remain standing. While physical damage can be obvious, often it is internal or obscured by exterior reinforcement. Thus it is important to quickly evaluate the integrity of many standing bridge piers immediately following an earthquake.

Time domain reflectometry (TDR) technology may offer a simple and rapid measurement

Charles H. Dowding, Professor, and Charles E. Pierce, Graduate Research Assistant, Department of Civil Engineering, Northwestern University, 2145 Sheridan Road, Evanston, IL 60208

technique for identifying external and internal changes of bridge piers and other supports with embedded coaxial cables and exterior electronics. This paper presents on-going investigations of installation techniques required for monitoring of bridge pier displacement and deformation. Cable placement in newly constructed columns and as part of a retrofit reinforcement program for existing columns is described. Early laboratory test results of detecting internal column damage during simulated earthquake loading are also discussed.

TIME DOMAIN REFLECTOMETRY BACKGROUND

TDR is an electromagnetic testing technique originally developed to detect faults along power transmission lines by pulsing voltage along the line and searching for voltage reflections caused by cable discontinuities. Reflection travel times are measured from the reflection site to the pulsing electronics attached at one end of the cable. Knowing the pulse propagation velocity of the transmission line, the accurate location of cable faults can be accomplished by converting measurements from the time domain to the lineal distance domain.

The amplitude of reflected voltage, called the reflection coefficient, at a cable shearing fault can be directly related to the magnitude of shear deformation (Dowding, Su, and O'Connor, 1989). TDR signals shown in figure 1 demonstrate that the reflection coefficient, measured as the percentage of reflected voltage and expressed in $m\rho$, increases as the cable is sheared incrementally. These initial findings have been extended by others to include extension and combined deformation modes. To accurately determine the extent of cable deformation, a correlation between reflection coefficient and deformation magnitude must be known for the selected cable(s).

A TDR measurement system provides several advantages for long-term monitoring of bridges.

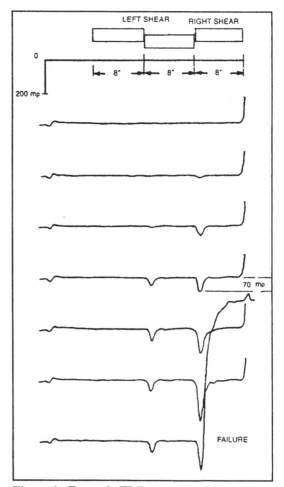

Figure 1. Example TDR signals.

Previous deployment of TDR-based instrumentation suggests that it is robust enough to withstand emplacement during heavy construction and should perform during potentially damaging conditions such as floods and earthquakes. A miniaturized, low power, intelligent TDR pulser is currently being developed for remote monitoring. This signal pulsing and recording instrument is designed to reduce the size, complexity, and cost of current cable testing electronics. Remote access to multiplexed cables has been demonstrated successfully in the field with current electronics and existing telecommunication lines (Dowding and Huang, 1994). These advantages show that a survey of all bridge supports can be performed remotely through one central unit, saving time and reducing costs.

EXTERNAL DISPLACEMENT: BRIDGE ABUTMENT AND PIER MOVEMENT

Measurements of scour at bridges founded on shallow footings indicate maximum scour occurs around the upstream side of footings during floods. As scour progresses and soil deposits are eroded, footings may be exposed and eventually undermined, leading to intolerable pier movement. Optimal placement of TDR cables would therefore be through the footing section on the upstream side. Most importantly, the cable monitoring system must operate at high water flows, when footing and pier displacement is most likely to occur as a result of scour.

Although bridge abutments may be less affected by scour, under certain circumstances they may need to be monitored for detrimental movements. Abutments and piers can be inspected using the same system designed to measure very small structural movements (on the order of 2 mm) relative to their foundation materials. Detection of such small displacements should provide an early warning of progressive movements which may reduce the stability of the bridge.

CABLE INSTALLATION

Several cable installation schemes have been investigated for monitoring pier movement (Dowding and Pierce, 1994b), but the most practical orientation of TDR cables is illustrated in figure 2. A single cable is shown extending from an accessible location on or below the bridge deck, down along the pier, and into a hole drilled through the footing and foundation materials. The cable must be encased in grout through the hole to ensure direct transfer of soil-structure displacements. In addition, the cable should be enclosed by a protective pipe along the entire length of the pier to screen the cable from debris and weather. A more durable transmission line can be connected to the monitoring cable at the top of the pier and extended to the electronics.

The number and arrangement of TDR cables installed on an individual bridge pier will depend on many factors, including the scour-critical classification, risk associated with structural collapse, and installation and monitoring costs, among others. It is expected that a TDR system can be installed on new bridges, since cables can be easily incorporated into new construction. However, the complexity and cost of cable installation on existing bridges are significantly greater, and installation techniques to minimize these problems are currently being researched.

DISPLACEMENT MEASUREMENTS

Movement of the pier footing can be detected from voltage reflections generated by local cable deformation. For instance, lateral translation or rotation of the footing would locally shear the cable-grout composite at the interface of the footing and foundation soils, as illustrated in figure

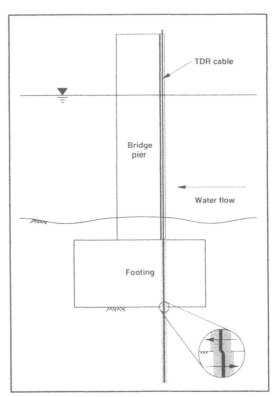

Figure 2. TDR cable measurement of footing displacement.

2. A minimum cable deformation of 2 mm produces a distinguishable voltage reflection in the TDR signal. Further footing displacement progressively shears the cable and produces increasingly larger voltage reflections at that point along the cable, creating similar TDR signals to those shown in figure 1.

To accurately measure these structural displacements, the cable and grout must be sufficiently weak to deform under the applied loads. A special cable-grout system is currently being developed to measure localized soil deformation in earth structures (Dowding and Pierce, 1994a). This project involves the design and production of an integrated cable and grout composite which exhibits low shear strength and compliance to deform uniformly with soft soils. In addition to embankments and excavations, this system can also be applied to monitor movements of footings or abutments founded on soft soils.

INTERNAL DEFORMATION: BRIDGE COLUMN CRACKING

Field observations of seismic failure of bridge columns in Northridge, California after the 1994 earthquake indicate shearing of the upper and lower connections occurred as a result of excessive lateral movements. In a recent field study, lateral loading of two-column bridge piers in East St. Louis, Illinois near the New Madrid fault revealed that tensile deformation occurs at the interface of the column and its base, and that vertical slippage or shearing occurs between overlapped, or spliced, steel reinforcement within the column (Lin, Gamble, and Hawkins, 1994). Splicing is a common construction practice which weakens the column against large moments that may develop during an earthquake. For this reason, remaining bridge columns are being retrofitted with external reinforcement along the spliced lengths to provide additional resistance.

To continue studies on the behavior of reinforced concrete bridge columns subjected to lateral loading, researchers at the University of Illinois constructed three model bridge piers, each with two columns per base beam (Lin, Gamble, and Hawkins, 1994). Each steel reinforced concrete column had a unique reinforcement geometry, but all exterior dimensions were the same. Columns were 12 ft high and 2 ft in diameter, with 6 ft center-to-center spacing on the beam. Steel reinforcing bars were spliced through the bottom 3 ft of each column. To investigate the response of the overlapping steel, columns were loaded under simulated earthquake conditions to failure.

CABLE INSTALLATION

To monitor internal deformation in the spliced region, four cables were attached to the inside of each reinforcement cage before columns were cast with concrete. After the columns were fully constructed, cables remained in the positions shown in figure 3. Two cables were aligned vertically to primarily monitor tensile deformations at the column-base interface. The other two cables were installed transversely through the column diameters and perpendicular to the vertical cables to monitor shearing between the spliced reinforcement. The dashed lines in figure 3 mark the cable segments encased in concrete; solid lines represent the exposed lengths of cable. Each vertical cable passes through the beam and exits on one side. Both ends of each transverse cable are exposed to allow pulsing from either end. Cable selection, preparation, and details of installation for these laboratory columns are described by Pierce and Dowding (1995).

Cables have also been installed on existing bridge columns during a seismic retrofitting project in East St. Louis, Illinois (Pierce, 1995). This work was performed solely to assess the

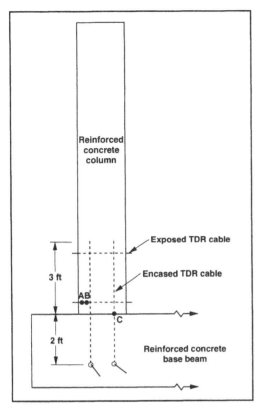

Figure 3. TDR cable orientation in column.

feasibility of cable installation under rigorous field construction conditions, and as such no measurements are expected at this site, although cables are accessible if monitoring becomes warranted. The field demonstration was effective, showing that cables can be easily affixed to the exterior steel reinforcement and remain intact after concrete placement. Further studies in this area are anticipated, including optimization of cable attachment and arrangement, and monitoring cable response to lateral loading.

DEFORMATION MEASUREMENTS

Laboratory columns were individually tested under reversed cyclic lateral loading to simulate earthquake conditions. Essentially, the free end (top) of the column was displaced laterally in controlled cycles, creating large bending moments at the column-base interface. Displacements were increased incrementally over time to failure, defined as a post-peak loading equal to 80% of the maximum measured load. A more thorough discussion of the testing procedure is given by Lin, Gamble, and Hawkins (1994).

To measure deformation, cables were pulsed when the column reached the maximum positive and negative displacements of each cycle. TDR signals were immediately collected and analyzed for signs of cable deformation, which only occurred during post-peak loading cycles. Not all of the cables deformed significantly to indicate internal damage, but several cables provided excellent results. Analyses showed that, in general, the vertical cables measured extension at the column-base interface, and the transverse cables sheared between the spliced steel as the bars slipped relative to each other, causing the interior concrete to shear and eventually crack.

In one column test, shear deformation was detected on the lower transverse cable at points A and B, shown in figure 3. Tensile deformation occurred along one of the vertical cables at the column-base interface in another test, designated as point C. Reflection coefficients measured from these cable deformations are compared to simultaneous strain gage measurements of the nearest reinforcing bars, as illustrated in figure 4. This plot shows that reflection coefficient, measured from both shearing and extension

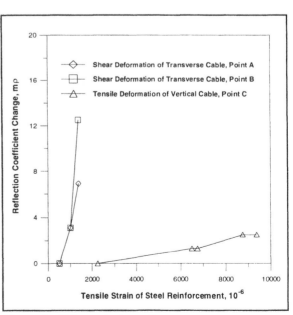

Figure 4. Column test results.

deformation modes, increases as the steel reinforcing bars were strained. The greater slope for cable shearing suggests that this cable was more sensitive to the strain environment than the vertical cable for these particular two tests. Tensile strain measured in the rebars does not necessarily correlate with cable shearing or extension, so a direct comparison cannot be made here. However, further studies can be conducted to quantify the reflection coefficient-cable deformation relationships for these cables.

SUMMARY

TDR technology may provide a means to inspect bridge piers and abutments by measuring internal deformation and external movement. Installation of metallic, coaxial cables in newly constructed laboratory columns and retrofitted field columns verified the cable robustness needed for monitoring long-term performance. Early results from laboratory tests showed that cables deform in high strain environments inside reinforced concrete columns. These observations point to the need to investigate placing cables in existing piers for monitoring scour-induced movement.

ACKNOWLEDGEMENTS

This project has been supported by the Infrastructure Technology Institute (ITI) at Northwestern University. The assistance and cooperation of Dr. William Gamble and Yongqian Lin of the University of Illinois, and their support by the Illinois Department of Transportation (IDOT), are gratefully acknowledged.

REFERENCES

Dowding, C.H. and F.C. Huang. 1994. "Early Detection of Rock Movement with Time Domain Reflectometry." *Journal of Geotechnical Engineering*, ASCE, v. 120, No. 8, pp. 1413-1427.

Dowding, C.H. and C.E. Pierce. 1994a. "Measurement of Localized Failure Planes in Soil with Time Domain Reflectometry." *Proc. of Symposium and Workshop on Time Domain Reflectometry in Environmental, Infrastructure, and Mining Applications*, Special Publication SP 19-94, U.S. Bureau of Mines, pp. 569-578.

Dowding, C.H. and C.E. Pierce. 1994b. "Monitoring of Bridge Scour and Abutment Movement with Time Domain Reflectometry." *Proc. of Symposium and Workshop on Time Domain Reflectometry in Environmental, Infrastructure, and Mining Applications*, Special Publication SP 19-94, U.S. Bureau of Mines, pp. 579-587.

Dowding, C.H., M.B. Su, and K. O'Connor. 1989. "Measurement of Rock Mass Deformation with Grouted Coaxial Antenna Cables." *Rock Mechanics and Rock Engineering*, v. 22, pp. 1-23.

Lefter, J. 1993. "Instrumentation for Measuring Scour at Bridge Piers and Abutments." *NCHRP Research Results Digest*, Transportation Research Board, No. 189, 8 pp.

Lin, Y., W.L. Gamble, and N.M. Hawkins. 1994. "Report to ILLDOT for Testing of Bridge Piers, Poplar Street Bridge Approaches." *Internal Report*, Department of Civil Engineering, University of Illinois at Urbana-Champaign, 64 pp.

Pierce, C.E. 1995. "TDR Cable Installation During Seismic Retrofitting of Bridge Piers in East St. Louis, Illinois." *Internal Report*, Department of Civil Engineering, Northwestern University, 6 pp.

Pierce, C.E. and C.H. Dowding. 1995. "Long-term Monitoring of Bridge Pier Integrity with Time Domain Reflectometry Cables." *Proc. of Conference and Exposition of Sensors and Systems*, Sensors Magazine, pp. 399-406.

Determination of Unknown Depth of Bridge Foundations Using Nondestructive Testing Methods

F. JALINOOS AND L. D. OLSON

ABSTRACT

This paper presents the test results of a research study on the feasibility of using nondestructive test (NDT) methods for the determination of unknown depths of bridge foundations. Of the over 580,000 highway bridges in the National Bridge Inventory, 104,000 of these bridges are over water and have both unknown foundations and consequently, unknown foundation scour risks. Foremost is the need to determine the foundation depth and then foundation type (footings or piles), geometry, and subsurface conditions. A comprehensive evaluation was made of potential NDT technologies that have relevance to this problem, of which only the surface-based Sonic Echo/Impulse Response and Flexural Wave methods for timber piles, and the borehole-based Parallel Seismic and Induction Field methods had been previously used to determine foundation depths.

This study documents the results for two acoustic NDT methods of Ultraseismic (a proposed new test method), and the Parallel Seismic test method which were found to have the broadest application to the investigated concrete, timber, and steel bridge substructures.

INTRODUCTION

The National Cooperative Highway Research Program (NCHRP) 21-5 project "Determination of Unknown Subsurface Bridge Foundations" (Olson et. al) was introduced to evaluate and develop existing and new technologies that can determine subsurface bridge foundation characteristics, where such information is unavailable. The project objectives were met in two stages. The first stage of the project consisted of the review and evaluation of existing and proposed technologies having promise for use in determining unknown subsurface bridge foundation characteristics such as depth, type, geometry and materials, followed by development of a research plan. The second part of the project included the performance of the research plan and submission of a final report documenting all research findings.

There are approximately 580,000 highway bridges in the National Bridge Inventory. For a large number of older, non-federal-aid bridges, there are no design plans, or else these plans have been lost over time; and, consequently, no information is available regarding the depth, type,

Farrokh Jalinoos, Senior Geophysicist, and Larry D. Olson, President, Olson Engineering, Inc., 14818 W. 6th Ave., #5A, Golden, CO 80401

geometry, or materials incorporated in the foundations. The best estimate of the population of bridges over water with unknown foundations, as of November, 1994, is about 104,000. These unknown bridge foundations pose a significant problem to the state DOTs because of safety concerns and consequently the Federal Highway Administration (FHWA) is requiring state transportation departments to screen and evaluate all bridges to determine their susceptibility to scour. The foundation depth information in particular is needed for performing an accurate scour evaluation at each bridge site, along with as much other information on foundation type, geometry, materials, and subsurface conditions as can be obtained. This NCHRP project was conceived to address these urgent concerns to find accurate, cost-effective nondestructive testing (NDT) methods to determine unknown foundation conditions.

This paper documents the results of two test methods which were found during the course of our study to have the broadest application to determining the unknown depths of bridge foundations. This include a new proposed Ultraseismic test method and Parallel Seismic test method as described below.

ULTRASEISMIC METHOD AND RESULTS

The Ultraseismic test method[1] was conceived during this study and is a broader application of the Sonic Echo/Impulse Response (Davis and Dunn, 1974) method and the Flexural Wave method (Douglas and Holt, 1993). However, the method can be used more broadly on bridge substructures as it is applicable to both columnar and more massive substructures. The Ultraseismic test involves impacting exposed substructure to generate and record the travel of compressional or flexural waves down and up substructure at multiple receiver locations on the substructure as illustrated in Fig. 1. The Ultraseismic data is then analyzed using seismic processing techniques adapted from the field of applied geophysics. Seismogram records are collected by using impulse hammers as the source and accelerometers as receivers that are mounted on the surface or side of the accessible bridge substructure at intervals of 1 ft or less. The bridge substructure element is used as the medium for the transmission of the seismic energy. All the usual wave modes travelling down or reflected back from foundation substructure boundaries (impedance changes, i.e. changes in the multiple of wave velocity x density x cross-sectional area, echoes from the bottom) are recorded by this method.

The Ultraseismic method is a broad application of geophysical processing to both the Sonic Echo/Impulse Response and Flexural Wave tests in that the initial arrivals of both compressional and flexural waves and their subsequent reflections are analyzed to predict unknown foundation depths. Two types of Ultraseismic test geometries have been specifically introduced for this problem:

---

[1]

The Ultraseismic method was researched and developed by Mr. Jalinoos and was named by Mr. Jalinoos at the suggestion of Dr. Phil Romig, Chairman of the Geophysics Department of the Colorado School of Mines in Golden.

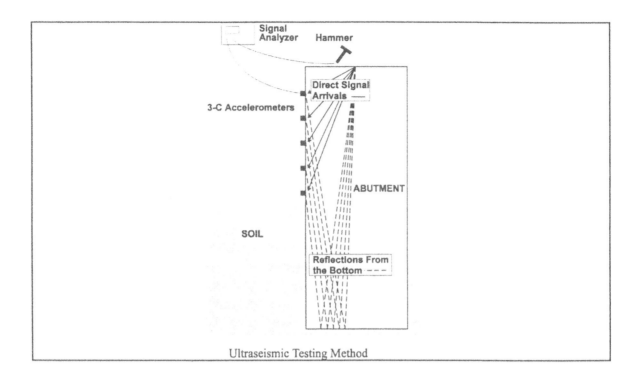

Figure 1-   Ultraseismic Test Method with Vertical Profiling Test Geometry.

1. For a one dimensional imaging of the foundation depth and tracking the upgoing and downgoing events, the term <u>Vertical Profiling (VP)</u> test method is used. In this method, the bridge column or abutment is hit from the top or bottom (both vertically and horizontally) and the resulting wave motion is recorded at regular intervals down the bridge substructure element. Typically, three-component recording of the wavefield is taken in order to analyze all types of ensuing wave motion. A VP line can be run in *both* a <u>columnar</u> (like a bridge pier or pile foundation) and a <u>tabular</u> (like a bridge abutment) structure.

2. For two-dimensional imaging of the foundation depth, the term <u>Horizontal Profiling (HP)</u> test geometry is used. In this method, the reflection echoes from the bottom are analyzed to compute the depth of the foundation. The source and receiver(s) are located horizontally along the top of accessible substructure, or any accessible face along the side of the substructure element, and a full survey is taken.

Discussed below are results from <u>Vertical Profiling Ultraseismic</u> tests. The source/receiver layout for Ultraseismic Vertical Profiling Tests on the South Column of Pier 4 of the Coors bridge is shown in Fig. 2. The source was located at the top of the pier which was struck both vertically on the pier top and horizontally (perpendicular to the axis of the column). Two horizontal hammer hits were used from opposite sides of the pier. Three-component accelerometers were mounted on the side of the column at 1 ft intervals from the top of the column down to near the grade surface. Thus, a full nine component dataset from 3 source

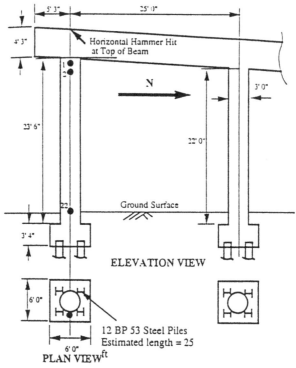

ELEVATION VIEW

PLAN VIEW

12 BP 53 Steel Piles
Estimated length = 25 ft

Figure 2-   Source/Receiver Layout for Ultraseismic Vertical Profiling Tests with Impacts at the Top of Pier 4, South Column, Coors Bridge.

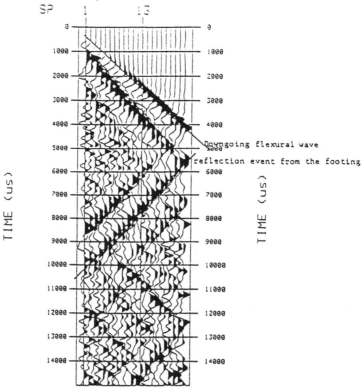

Figure 3-   Ultraseismic Vertical Profiling Records of Flexural Waves from Horizontal Impacts at the Top of Coors Bridge Pier 4, South Column with a 3-lb Hammer and 22 Horizontal Receiver Locations in the Same Vertical Plane on the South Column (see Fig. 2 for Test Geometry).

94

directions and 3-component recording were obtained. This allows for a near-full study of the dynamics of wave motion for the compressional and flexural wave modes of interest in this pier that consists of a beam on a column on a buried pilecap on 4 steel H-piles. Field data for a Vertical Profiling (VP) test done to measure flexural waves is shown in Fig. 3. The vertical axis is time in microseconds (1 μsec = 1 x 10^{-6} seconds) with time 0 being the time of the impact. The horizontal axis in Fig. 3 is the triaxial accelerometer receiver block location. The field records of Fig. 3 were made from horizontal impacts with a 3-lb hammer and the resulting horizontal receiver responses (perpendicular to the axis of the column in the same plane as the impact) which were obtained by moving the triaxial accelerometer to each position shown in Fig. 2. Each trace indicates the record from a different survey location on the column, with Trace 2 corresponding to 1 ft *below* the source location and trace 22 to 21 ft below the source location. Fig. 3 shows the first 15 milli-seconds (msec) of data at 4 μsec sampling interval. All data were de-biased to remove any DC shift, bandpass zero-phase filtered (0-0.5-3-4 kHz trapezoidal filter), and Automatic Gain Controlled (AGC). Several events are evident in Fig. 3: 1. Downgoing Flexural events are apparent (first arrivals) with a linear positive moveout with travel time increasing with distance (from top to bottom); and, 2. Upgoing Flexural (F) events (reflection events from the bottom of the footing) are apparent with a linear negative moveout with travel time decreasing with distance (from top to bottom). Analysis of the flexural wave reflector depth indicates a depth of 29.7 ft below the top of the pier beam. This compares to the depth of 31.1 ft from the top of the beam to the bottom of the pilecap. The compression (P) wave reflection event predicted a depth of 31.1 ft which agrees exactly with the depth of the pilecap. Both depth predictions are quite close to the actual depth. The advantages of being able to track wave travel up and down the bridge substructure are obvious, as it makes it clearer, and increases one's confidence in the analysis.

No evidence of the steel piles underlying the pilecap was apparent in the VP results. Vertical Profiling test lines can also be run from the side of a massive abutment and other tabular type structures. VP test lines are also used in tieing reflection events from the top and bottom of the structure to the events in a Horizontal Profile (HP) section.

PARALLEL SEISMIC TEST METHOD AND RESULTS

Typical Parallel Seismic (PS) test equipment includes an impulse hammer, hydrophone or geophone receiver, and dynamic signal analyzer or oscilloscope as shown. A portable PC - based digital oscilloscope was used to record the Parallel Seismic data in this study.

The Parallel Seismic (PS) method involves impacting the side or top of exposed bridge substructure with a 3 lb or 12 lb (preferred) hammer to generate wave energy which travels down the foundation and is refracted to the adjacent soil. The refracted wave arrival is tracked by a hydrophone receiver suspended in a water-filled cased borehole in the conventional approach to the test. A hydrophone receiver is sensitive to pressure changes in the water-filled tube, but it is also subject to contaminating tube wave energy. Research was performed during this project on the use of clamped three-component geophones in cement-bentonite, bentonite, and sand-backfilled, 4 inch ID, PVC cased borings to better examine the wave propagation behavior with reduced tube wave energy noise. The boring is drilled typically within 3 to 5 feet of the foundation edge and should extend at least 10 feet deeper than the anticipated and/or minimum required foundation depth for the depth to be determined. Example results from geophone use are presented below.

By using 3-component geophones in cased borings with good contact between the casing and soils (grouting preferred with a cement-bentonite mixture), improved quality of Parallel Seismic results was obtained at sites with variable soil velocity conditions. Hydrophones work well as receivers for soils of constant velocity (or saturated soils) surrounding the foundation, since the time arrivals are a function of the foundation length, and not the variation in the soil velocity. For soils with varying velocities, a break cannot be identified from the erratic arrival times unless the recorded traces are time corrected for the variation in the (measured) soil velocities. Typically, at the tested sites, the soil compressional velocities increased from about 1,000 ft/sec to 5,000 ft/sec at the water-table (100% saturated-soil which is the velocity of water) and at bedrock. Similarly, significant increases in shear velocity were observed at the bedrock. As a result, traditional Parallel Seismic (PS) analysis of the foundation data, by picking the first arrival travel times and plotting time vs depth, did not provide the clearest data at all of the sites with known foundation characteristics. Moreover, the bridge foundation shapes were mostly nonuniform (in cross-section), due to existence of a footing or pile caps and so forth, so that travel time plot vs. depth along the length of the foundation frequently deviated from a straight line.

It was observed, however, that the bottom of the foundation can act as a strong (secondary) source of energy, especially in more massive foundations. The foundation tip acts as a point diffractor in emitting both upward and downward traveling waves into the borehole. This diffraction event is best seen by using 3-component geophones in a cased, grouted borehole. The diffraction results in a steep V-shaped hyperbolic event in the recorded seismic section. The bottom of the foundation is then identified by noting the depth where the peak of the hyperbolic event occurs.

As a typical example for the Parallel Seismic test with geophones, AGC geophysical data is presented for the concrete pile pier of the Old Bastrop bridge in Fig. 4. The wave energy generated by the horizontal blow is both compressional and flexural, but the most energy is in the flexural wave in the caisson which becomes refracted as a shear wave in the borehole. The SP positions shown across the tops of the figures refer to depths of the records in feet from the top of borehole. Therefore, SP 3 refers to a receiver location of 3 ft below the top of the borehole. A sharp point diffractor event in the flexural wave energy from the foundation tip is indicated at a depth of 32 ft from the top of the borehole which predicts a depth of 32.7 ft below the pilecap for the pile bottom. This agrees fairly well with the plan depth of 34 ft.

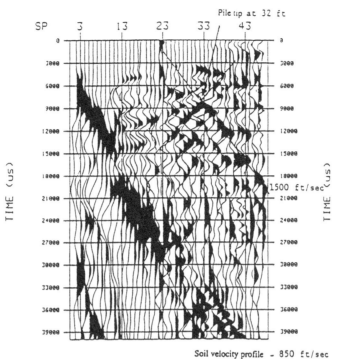

Figure 4-   Parallel Seismic Field Records (AGC) from the Concrete Pile Foundation from a 12 lb Horizontal Hammer Hit and Horizontal Component Geophone Recording at the Old Bastrop Bridge.

CONCLUSIONS

The results of this research indicate that of the surface tests (no boring required), the Ultraseismic test has the broadest application to the determination of the depths of unknown bridge foundations. The major limitation of the surface methods is that none of them are yet able to detect the presence of piles underlying a buried pilecap or more massive abutment or pier wall. Nonetheless, the new Ultraseismic method was able to determine the depths to the first significant substructure change (footing on soils or pilecaps, bottoms of abutments, bottoms of piers), especially for shallow footings.

Of the borehole methods, the Parallel Seismic test was found to have the broadest applications for varying substructure and geological conditions. The most accurate data was obtained by using clamped geophone receivers in grouted, cased borings as opposed to the previous use of only hydrophone receivers suspended in a water-filled borehole.

REFERENCES

Davis, A.G. and C.S. Dunn. 1974. "From Theory to Field Experience with the Nondestructive Vibration Testing of Piles," Proceedings of the Institution of Civil Engineers Part 2, 57: 571-593.
Douglas, R.A., and J.D. Holt. 1993. "Determining Length of Installed Timber Pilings by Dispersive Wave Propagation Methods," Report: Center for Transportation Engineering Studies, North Carolina State University.
Olson, L.D., F. Jalinoos, and M.F. Aouad. 1994. "Determination of Unknown Subsurface Bridge Foundations,"NCHRP Project21-5, National Academy of Sciences Final Report,Washington, D.C.

Impulse Response Evaluation of Drilled Shafts at the NGES Test Section at Northwestern University

S. L. GASSMAN AND **R. J. FINNO**

ABSTRACT

The impulse response test is a non-destructive evaluation technique commonly used for quality control of driven concrete piles and drilled shafts where the pile heads are accessible. When evaluating existing foundations, the presence of a pile cap or other structure often times makes the pile heads inaccessible and introduces uncertainties in the interpretation of the results. To evaluate the effects of intervening structures on impulse response results in a controlled environment, a test section was constructed at the National Geotechnical Experimentation Site at Northwestern University. The test section consisted of 5 drilled shafts constructed in 3 groups. The drilled shafts were constructed with diameters ranging from 2 to 3 ft and lengths varying from 40 to 90 ft. Pile caps vary in thickness from 2 to 5 ft. Two of the shafts were constructed with defects: a reduced cross-section and a thin, soil filled joint. Impulse response data were obtained in both the accessible and inaccessible head conditions. The ability to determine shaft integrity for inaccessible shafts depends on the thickness of the pile caps and the length-to-diameter ratio of the shafts. The toe of the shaft could be identified for inaccessible shafts with length-to-diameter ratios as high as 23 for cap thickness-to-pile diameter ratios up to 2.5.

INTRODUCTION

The impulse response method is a non-destructive evaluation technique commonly used for quality control of drilled shafts and driven concrete piles where the pile heads are accessible. When evaluating existing foundations, the presence of a pile cap or other structure often times makes the pile heads inaccessible and introduces uncertainties in the interpretation of the results. Understanding the effects of intervening structure on the impulse response signals will allow use of this simple procedure in the field to help identify unknown foundation types. A good example where this would be useful in practice is during inspection of older bridges when design drawings are not available. The purpose of this paper is to provide data to show the applicability of this technique to situations where the top of a drilled shaft is not accessible. This paper briefly describes the impulse response methodology, gives detail of the test section, and presents the impulse response results for the drilled shafts.

S. L. Gassman, Research Assistant, Department of Civil Engineering, Northwestern University, Evanston, IL 60208

R. J. Finno, James N. and Margie M. Krebs Professor, Department of Civil Engineering, 2145 Sheridan Road, Northwestern University, Evanston, IL 60208

IMPULSE RESPONSE METHOD

The impulse response method is a surface reflection technique where the top of a drilled shaft is impacted and both the impact force and compression wave reflections are measured on the impacted surface. This test was developed in the late 1970's in France as an extension of a vibration test which involved harmonically vibrating a known mass up to 2000 Hz at the head of a shaft and measuring the shaft response with a geophone (Higgs and Robertson, 1979).

The impact hammer is equipped with a load cell and is capable of generating transient vibrations with frequencies as high as 2000 Hz, depending on the material at the tip of the hammer. Vibrations at the shaft head are recorded by a vertical geophone which is activated upon hammer impact by means of a triggering device in a portable PC to which both the hammer and geophone are connected. The analysis of the data is undertaken in the frequency domain. A fast fourier transform is performed on both the force and the velocity signals to convert them from the time to the frequency domain. The velocity spectrum is divided by the force spectrum and the resulting quantities, called mobilities, are smoothed by applying a lowpass Blackman window filter which eliminated all frequencies with periods shorter than 2 data points.

An idealized graph of mobility versus frequency is shown in Figure 1. Quantitative information concerning the shaft length, low-strain stiffness and impedance can be obtained from this plot. Resonant peaks are clearly visible, and by measuring the frequency change between these peaks, Δf, the length from the geophone to the source of the reflection, L, can be found from:

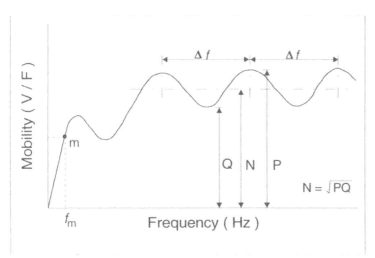

Figure 1. Idealized Impulse Response Result: Mobility vs. Frequency

$$L = \frac{v_c}{2\Delta f} \qquad (1)$$

where v_c is the longitudinal wave velocity in concrete.

From the low frequency portion of the mobility curve, the low strain stiffness, K', can be calculated. This parameter can be correlated to the initial stiffness in an axial load test (Davis and Robertson, 1975; Stain, 1982) and is found from:

$$K' = \frac{2\pi f_m}{V/F_m} \qquad (2)$$

where f_m and V/F_m are the frequency and mobility, respectively, at point m. The point m is selected on the initial linear portion of the mobility curve (Fig. 1).

To help evaluate the computed stiffness values, Davis and Dunn (1974) derived the theoretical minimum and maximum stiffnesses, K_{min} and K_{max}, respectively, for the shaft as:

$$K_{max} = \frac{AE}{L}\sqrt{\frac{P}{Q}}\tanh\sqrt{\frac{P}{Q}} \quad ; \quad K_{min} = \frac{AE}{L}\sqrt{\frac{Q}{P}}\tanh\sqrt{\frac{P}{Q}} \tag{3}$$

where A is the shaft cross-sectional area, E is the elastic modulus of the shaft, P is the maximum value of mobility and Q is the minimum value of mobility. K_{max} represents the shaft head stiffness when a shaft is supported on a rigid base while K_{min} corresponds to a shaft unsupported at its base.

The shaft mobility, N, is found by taking the geometric mean of the height of the resonant peaks (= √PQ; see Fig. 1) in the portion of the mobility curve where the shaft response is in resonance. It can also be defined theoretically as:

$$N = \frac{1}{\rho_c \, v_c \, A} \tag{4}$$

where ρ_c is the density of concrete. If the actual N is greater than the theoretical value, it is likely that there is a defect in the shaft due to a smaller than anticipated cross-section (i.e. a neck) or caused by poor concrete quality (low ρ_c and/or v_c).

NGES Test Section

To provide an opportunity to analyze the results of inaccessible head tests in a controlled environment, a test section was constructed at the National Geotechnical Experimentation Site (NGES) at Northwestern University. Impulse response, sonic echo, sonic logging and parallel seismic tests have been conducted at the site. As indicated on Figure 2, this test section consists of 5 drilled shafts with diameters ranging from 2 to 3 ft and lengths varying from 40 to 90 ft.

Reinforced concrete caps were cast to form 3 groups of shafts. A cross-section at the site is shown on Figure 3. Each shaft had a geophone embedded near its bottom to directly measure the longitudinal wave speed. Sixty cylinders were made from the concrete as it was placed to determine the strength and longitudinal wave speed variations over time. Access tubes for sonic logging were placed in two of the shafts, and three cased boreholes were installed so that parallel seismic tests could be conducted. Cross-hole seismic tests were also performed in these holes to evaluate the shear wave velocity of the soil at the site. Two of the shafts were constructed purposely with defects: a reduced cross-section and a thin, soil-filled joint. The former defect represents a typical construction deficiency while the latter represents a performance-related defect induced by excessive lateral loads. The main variables considered in the test section are the shaft length-to-diameter ratio (L/D), the pile cap thickness-to-shaft diameter ratio (B/D), and the bottom

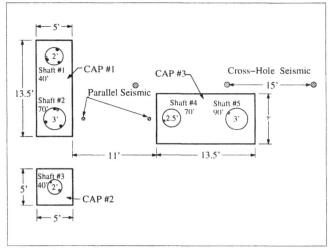

Figure 2. Plan of NGES Test Section

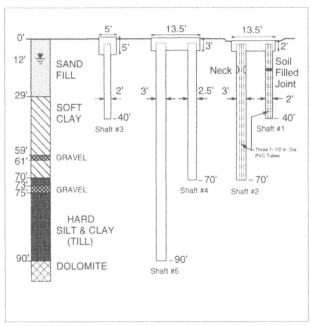

Figure 3. Cross-Section at NGES Site

condition of the shafts (soft clay, hard clay or dolomite).

After the five shafts were constructed, a period of six weeks was used to non-destructively evaluate the shafts using the impulse response, sonic echo and sonic logging methods. Thereafter, the reinforced concrete caps were poured and these tests were again conducted from the tops of the pile caps.

Care was taken to construct the shafts with uniform cross-sections. As such, construction began by drilling to the bottom of the surficial sand layer under a slurry head, placing an oversized casing for temporary support, removing the slurry and drilling the remainder of the hole in the dry. After a permanent liner was placed through the surficial sands, a reinforcing cage was set into the hole and the concrete was placed by free-fall. After the concrete had set at least one day, the annulus between the two casings was backfilled with sand and the temporary casing was pulled. The test section is more fully described by Finno and Gassman (1996).

ACCESSIBLE SHAFT HEAD RESULTS

Prior to cap construction, impulse response tests were performed on the shafts in the accessible head condition. As one might expect for shafts made under controlled conditions, the impulse response results were quite clear for most cases. The defects were clearly identifiable from the impulse response tests. Figure 4 shows the mobility plots for the two 70 ft shafts, shaft 2 and shaft 4, with diameters of 3 ft and 2.5 ft respectively. Reflections from the bottom of both shafts were clearly visible at frequencies below 600 Hz. A reflection from the neck in shaft 2 was also identifiable by the peaks at 540 and 1000 Hz. A secondary response from shaft 4 was identified around 500 Hz where the resonating peaks were forced into pairs. This secondary response indicates an unplanned defect, a bulge (verified by simulations), around 32 ft which corresponds to the bottom of the permanent liner. The mobility plots for the two 40 ft long, 2 ft diameter shafts are shown in Figure 5. The reflections from the toe of shaft 3 were nearly ideal

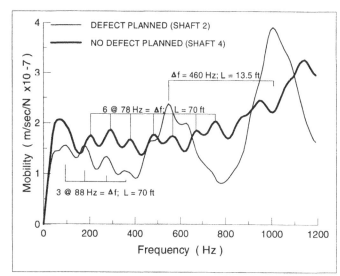

Figure 4. Mobility Plots for Shaft 2 and Shaft 4 with lengths = 70 ft.

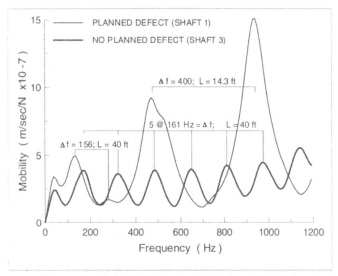

Figure 5. Mobility Plots for Shaft 1 and Shaft 3 with Lengths = 40 ft.

as a result of the careful construction procedures whereas the bottom of shaft 1 with the planned soil joint was barely discernible.   The impulse response results for the 90-ft-deep shaft contained indications of resonances from the bottom reflections, although not as clear as the other shafts. The magnitude of the mobility for shafts 3, 4 and 5 was less than the theoretical value for the accessible-head condition as shown in Table 1. This indicates that the as-built shaft impedance was higher than the design value, and thus was acceptable. In contrast, the mobility for shafts 1 and 2, both with planned defects, was greater than the theoretical value, and therefore the shaft has a lower impedance than designed. The adequacy of the shaft would have to be evaluated in light of the actual loading requirements.

The low-strain stiffness, K', was calculated from the low frequency portion of the each curve to provide an indication of the low-strain, soil-foundation interaction. The results are provided in Table 1. The two 40 ft shafts tipped in the soft clay have lower stiffnesses than the two 70 ft shafts tipped in the hard till. The larger stiffness values reflect a more rigid support at the bottom of the shaft.

TABLE 1. Summary of Accessible and Inaccessible Results

Shaft No.	L/D ratio	B/D ratio	Mobility, (m/s/N E-7)			Low Strain Stiffness, K', (MN/mm)	
			Accessible	Inaccessible	Theoretical[a]	Accessible	Inaccessible
1	20	1	4.6	1.0	3.57	0.61	0.9
2	23.3	0.67	1.9	.7	1.58	1.58	1.83
3	20	2.5	2.2	.5	3.57	0.9	1.32
4	28	1.2	1.6	.6	2.28	1.39	1.54
5	30	1	1.1	.5	1.58	1.76	1.93

[a] Based on v_c = 4000 m/s, ρ_c = 150 pcf

EFFECTS OF INTERVENING STRUCTURE

After the accessible-head tests were completed, the pile caps were constructed rendering the shafts inaccessible. Impulse response tests were conducted by impacting the pile cap and recording the vibrations with a geophone mounted to the pile cap surface. The accessible head and inaccessible head mobility curves for shaft 3 with its 5 ft thick concrete cap are compared in Figure 6. Reflections from the shaft toe were clearly apparent for both accessible and inaccessible conditions although the resonances for the inaccessible case were less pronounced than the accessible case, as a result of the signal attenuation from the pile cap. The average mobility decreased for the inaccessible head case as expected due to the increased mass from the pile cap. The inaccessible-head low-strain stiffness was greater than the accessible-head low-strain stiffness. The bottom of shafts 4 and 5 with L/D ratios of 28 and 30 and B/D ratios of 1.2 and 1, respectively were not identified whereas the bottom of shaft 3 with L/D of 20 and B/D = 2.5 was detected. Therefore, the limitation of the impulse response method applied to inaccessible shafts is a combination of the L/D ratio and B/D ratio.

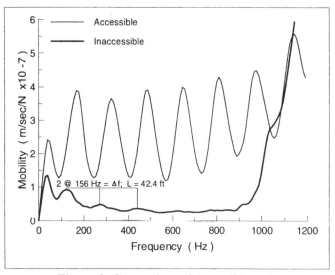

Figure 6. Comparison of Accessible and Inaccessible Mobility Curves for Shaft 3

SUMMARY

This paper summarized impulse response tests obtained at a test section at the National Geotechnical Experimentation Site at Northwestern University. Results were obtained both after the concrete drilled shafts were constructed and after pile caps were built. Five shafts were evaluated in both the accessible-head and inaccessible-head condition. The shafts ranged from 2 to 3 ft in diameter and 40 to 90 ft in length. The caps varied in thickness from 2 to 5 ft thick. The toe of the shaft could be identified for inaccessible shafts with L/D ratios as high as 23 with a B/D ratio of .67 and with L/D ratios as high as 20 with a B/D ratio of 2.5.

ACKNOWLEDGEMENTS

This work was funded by grants from the Infrastructure Technology Institute (ITI) at Northwestern University, the National Science Foundation and the Federal Highway Administration. The authors thank the Director of ITI, Mr. David Shultz, for his continuing support. The authors are grateful to Dr. Allen Davis and Mr. Bernie Hertlein of STS Consultants for sharing their knowledge of non-destructive testing techniques.

REFERENCES

Davis, A.G. and Dunn, C.S. 1974. "From Theory to Field Experience with Non Destructive Vibration Testing of Piles," *Proceedings of the Institution of Civil Engineers*, Part 2, No. 57, December, pp. 571-593.

Davis, A.G. and Robertson, S.A. 1975. "Economic Pile Testing," *Ground Engineering*, May, pp. 40-43.

Finno, R.J. and Gassman, S.L. 1996. "Analysis of Existing Drilled Shaft Foundations Using the Impulse Response Method" submitted to the *Journal of Geotechnical Engineering*, American Society of Civil Engineers.

Higgs, J.S. and Robertson, S.A. 1979. "Integrity Testing of Concrete Piles by Shock Method," *Concrete*, Oct. pp. 31-33.

Stain, R.T. 1982. "Integrity Testing," *Civil Engineering*, May, pp. 71-73.

Detection of Fatigue Cracks in Eyebars Using Time of Flight Diffraction

G. A. WASHER

ABSTRACT

This paper examines the development of a Time of Flight Diffraction (TOFD) ultrasonic method for defect detection and sizing in eyebars. Eyebars from the Pasko - Kennewick bridge are utilized as test specimens to determine eyebar fatigue performance, and to test the effectiveness of the TOFD method. The test setup for making crack depth measurements in eyebars is described. Measured results of the depth of machined slot test specimens and results of fatigue crack growth are presented.

INTRODUCTION

Eyebars are axial members utilized in the superstructure of some truss bridges built in the early part of this century. These eyebars are characterized by a throat section, a rolled bar of constant cross section, and a head forged at both ends of the throat. A large hole in the head section, called a bore, allows the eyebars to be pinned together in a chain. A typical eyebar - pin connection includes between two and eight eyebars mounted on a single pin.

The eyebar connection geometry presents a significant challenge to those charged with the maintenance and inspection of highway structures, due to the difficulty of inspecting the eyebar heads for fatigue cracking. Eyebar fatigue cracking has typically occurred radially from the bore, along a line perpendicular to the axial stress carried in the eyebar. This area is typically covered by other eyebars mounted on the same pin, making visual inspection impossible.

The most well known case of fatigue cracking in eyebars occurred in the Silver Bridge, between Point Pleasant, West Virginia and Kanauga, Ohio. This fracture critical structure suffered a catastrophic failure that resulted from fatigue cracks growing from the bore of one eyebar. The National Transportation Safety Board (NTIS, 1970) concluded that the failure was caused by two adjacent semicircular cracks, the first about 3.18 mm in radius, and the second about 1.59 mm.

Glenn Washer, P.E., Turner Fairbank Highway Research Center, FHWA, 6300 Georgetown Pike, McLean, VA 22101

Since this historic collapse, many inspection methods for detecting fatigue cracks in eyebars have been investigated. While these methods have been effective in detecting cracks to some degree, none is fully effective in detecting cracks of the size that caused the collapse of the Silver Bridge. In addition to the difficulties associated with accessing the eyebar surface, the poor production quality, irregular edges of the eyebar, and coarse grain structure make the inspection of eyebars with traditional nondestructive inspection techniques difficult.

ULTRASONIC METHODS

The application of traditional ultrasonic pulse echo techniques for eyebar inspection are among the most promising methods available. However, the irregular, non-circular eyebar edge surface makes beam path tracing difficult, and may lead to misinterpretation of results. Additionally, variations in the coupling of the transducer, eyebar surface finish, and crack orientation will affect the reflected wave amplitude.

The use Time of Flight Diffraction (TOFD) will eliminate these variabilities from the testing method. TOFD relies solely on time domain analysis of waves diffracted by the crack tip. The method accuracy not affected by variations in amplitude, provided the diffracted waves are of sufficient amplitude to be time resolvable. Additionally, the use of pitch - catch configuration and longitudinal waves eliminates the need to determine ray paths within the eyebar.

TIME OF FLIGHT DIFFRACTION

TOFD relies on the ultrasonic waves diffracted from the crack tip to locate and size defects. A typical TOFD setup include 2 transducers, one transmitter and one receiver. The spatial location of the crack tip may be resolved based on the time of flight of the signals diffracted by the crack tips, and those reflected by the specimen back wall.

The interaction of an ultrasonic beam with a crack generates several types of waves. A specular reflection of the ultrasonic wave will occur, following the basic laws of optics. A Rayleigh wave will travel along the crack edge, and be emitted at both ends of the crack. Diffracted or scattered waves will originate at the crack tips, and have a wide angular distribution. Mode conversion of shear to compression waves, and vise versa, will also occur (Ogilvy and Temple, 1983).

Additionally, the sound wave will be reflected by the back wall of the specimen and may be used as a time marker. There will also be a lateral wave, sometimes called

$$d = \sqrt{\left[\frac{t_c V_L}{2}\right]^2 - S^2} \qquad (1)$$

the near surface wave, that will travel along the shortest possible length between the two transducers, and may be utilized to determine the transducer separation.

The distance from the surface of the specimen to the crack tip can be calculated, assuming that the crack tip lies symmetrically between the transducers. This position is easily found be scanning of the transducer such that the time of flight of the diffracted wave is minimized (Charlesworth and Temple, 1989). The depth is found according to the equation 1, where d is the distance from specimen surface to the tip of the crack, and t_c is the time of flight of the wave diffracted from the tip of the crack (see figure 1). The horizontal distance between transducers is 2S, and the longitudinal wave velocity is V_l. The distance to the back wall may be found in a similar fashion. The crack length is then determined.

The use of longitudinal waves in the application of TOFD simplifies the data analysis process by eliminating spurious waveforms from the time domain being examined. Although shear waves may be used, the relatively slow speed of these waves results in the possibility that the waveform of interest will arrive simultaneously with mode converted longitudinal waves, or waves created by delay lines, inhomogeneous material, or other factors. The use of longitudinal waves reduces the possibility of misinterpretation because the wave diffracted at the crack tip travels by the shortest possible path between the transmitting and receiving transducers.

EXPERIMENT

The test setup for application of TOFD is shown in figure 1. The pitch catch transducer setup allows for a constant and known transducer separation critical to the crack measurement by the TOFD method. The transducers utilized in this study were 12.7-mm diameter standard transducers, with 2.25 and 5.0-MHz frequencies. Lucite wedges were manufactured with a slight incline, to guide the wave toward the critical

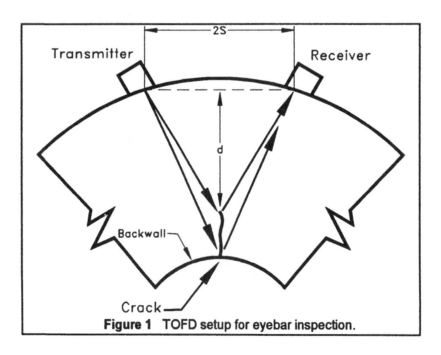

Figure 1 TOFD setup for eyebar inspection.

inspection area. By setting the transducers on either side of the critical inspection area, surface irregularity effects on the beam path are eliminated.

The TOFD method was first applied to machined slots 3.8 and 5.5 mm deep. The slots were cut across the eyebar thickness at the bore. The eyebar surface was finished with an 80-grit grinder to remove paint and level the surface across the eyebar thickness. The shape of the eyebar was not changed. Crack tip measurements by the TOFD method yielded the results shown in table 1. This proof of concept testing indicated that the principles of time of flight diffraction may be applied effectively to the complicated eyebar geometry. These results also indicated that the irregular eyebar surface conditions, which result in unknown ultrasonic beam incident angle on the crack tip, do not significantly effect results.

To make an accurate crack depth determination of the by the TOFD method, it is necessary to determine when the transducer is positioned symmetrically about the crack tip. This principle was tested by manually scanning in 12-mm steps, from (-x) to (+x), across the eyebar area containing a 5.5 mm slot. Figure 2 indicates the theoretical delay between the wave diffracted by the crack tip and the back wall echo. The crack tip is modeled as an infinitely small point 5.5 mm from the eyebar back wall. The transducer and receiver are modeled as a point source and receiver, respectively. Also shown in the figure are the manually scanned results. Differences in the calculated vs. actual results are caused by a number of factors, including irregular eyebar surface, finite transducer size and relatively large slot size. These results indicate that the characteristic elliptical TOFD data is not greatly affected by the irregular eyebar surface, and therefore may be used to locate the transducers symmetrically about the crack, and to discern crack signal from signal anomalies.

To determine the effectiveness of this method for the detection and sizing of fatigue cracks in actual eyebars, specimens from the Pasco - Kennewick were fatigue loaded at the Structures Laboratory of the Turner Fairbank Highway Research Center in McLean, VA. These eyebars were cycled under a constant amplitude fatigue load, providing a stress range of 84 and 61 MPa. The performance of the eyebar under fatigue loading correlated well with the S-N curves available in the literature (Albrect and Simon, 1981). Figure 3 indicates the results of the first three fatigue tests.

TABLE I Slot depth measured by the TOFD method.

Frequency (MHz)	Actual Slot Depth (mm)	Measured Slot Depth (mm)	Error (%)
2.25	3.8	3.96	+4.2
5.0	3.8	3.71	-2.4
2.25	5.5	5.66	+3.0
5.0	5.5	5.50	0

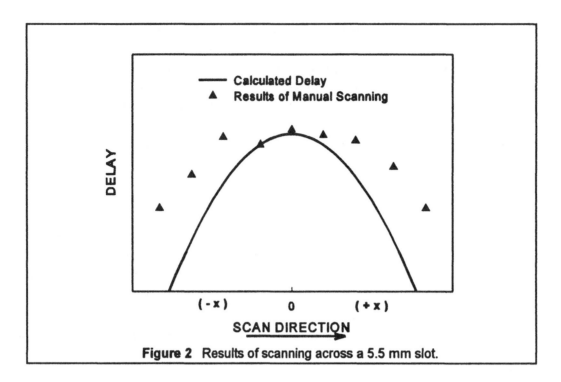

Figure 2 Results of scanning across a 5.5 mm slot.

The first eyebar subjected to fatigue loading (84 MPa) fractured before crack growth could be monitored. A second eyebar, subjected to a stress range of 61 MPa, experienced crack initiation at the surface of the eyebar. This eyebar was removed from the load frame and is currently being held for comparison with other inspection techniques, and to test the effectiveness of the TOFD method when further optimization of the technique is complete. The crack in this eyebar grew along the surface of the eyebar, to a surface length of 7 mm. The depth of the triangular shaped surface crack at the bore is approximately 3 mm. The third eyebar tested also had a crack initiate at the eyebar surface, and this crack is currently being monitored. Final results of this portion of the research are not available at this time. However, it is worth noting that the eyebar cracks in these eyebars are initiating at the surface of the eyebar, rather than at the inside of the bore, indicating bending about the weak axis of the eyebar is occurring. The varied location of the cracks with reference to the load frame would suggest that the load frame is not the cause of this bending. Fabrication errors in the eyebar, which result in the bore not being perpendicular to the face of the eyebar, are believed to be the cause of the surface cracking.

CONCLUSION

Ultrasonic TOFD has potential for the field inspection of eyebars. The accuracy of the method in determining the depth of cracks from the bore of the eyebar has been established. However, the amplitude of the signal diffracted from actual crack tips has not yet been demonstrated. Sufficient signal amplitude for the time resolution of diffracted waves is required to make the method effective for the inspection of eyebars. Research is currently underway to determine the amplitude of these signals, and to

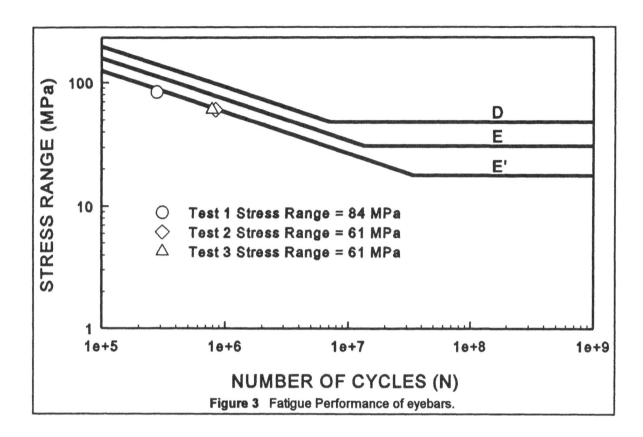

Figure 3 Fatigue Performance of eyebars.

establish the effectiveness of this method for the detection and measurement of eyebar fatigue cracks. Use of scanning and imaging technologies to aid in the analysis of signals is also being investigated.

REFERENCES

Albrect, P, Simon, S. "Fatigue Notch Factors For Structural Details", Journal of the Structural Division, AISC, July, 1981, pg 1279 - 1295.

Charlesworth, J. P., Temple, J. A. G., "Ultrasonic Time-of-Flight Diffraction", Research Studies Press, LTD. Taunton, England, 1989.

"Collapse of U.S. 35 Highway Bridge, Point Pleasant, West Virginia, December 15,1967", National Transportation Safety Board, Washington, D.C. 1970.

Oglivy, J. A., and Temple, J. A.G., "Diffraction of Elastic Waves By Cracks: Application To Time Of Flight Inspection", Ultrasonics, November 1983, pg 259 - 269.

Fatigue Evaluation of Highway Bridges Using Ultrasonic Technique

S. H. PETRO, H. V. S. GANGARAO AND U. B. HALABE

ABSTRACT

This paper presents an innovative approach for the evaluation of fatigue performance of highway bridges. The proposed method is a combination of analytical and experimental approaches. The steps include identification of critical components and ultrasonic stress measurements using a portable fatigue loading indicator which can be left unattended in the field under normal traffic. Miner's rule is used to compute an effective constant amplitude stress which is applied to estimate the cumulative damage for a given period of time in-terms of equivalent 10 ksi constant amplitude stress cycles. A prototype instrument has been built and tested. This paper describes the methods used by a fatigue loading indicator and discusses some of the preliminary testing results.

INTRODUCTION

Recent studies have projected the cost of returning our aging and seriously deteriorated highways and bridges to effective operating levels at over $3 trillion for the next ten years (Civil Engr. 1993). This estimate is much higher than the amount currently spent on these facilities. Short-term solutions cannot be expected to provide the rehabilitation level required; and even if unlimited resources were available, the old and traditional methods of inspection and rehabilitation cannot provide us with the desired performance and efficiency required for the future.

Truck traffic on highway bridges produces cyclic stresses that may lead to fatigue damage. Generally, this damage cannot be detected by visual bridge inspection normally carried out at two year intervals. Additionally, problems may develop between inspections which may not be perceptible by visual observation. A quantitative assessment of bridge condition is needed to adequately determine the safety of highway bridges. Stress measurement can be used to

Samer H. Petro, Engineering Scientist, Constructed Facilities Center, West Virginia University, Morgantown, WV 26506-6103

Hota V. S. GangaRao, Professor, and Udaya B. Halabe, Assistant Professor, Department of Civil and Environmental Engineering, Constructed Facilities Center, West Virginia University, Morgantown, WV 26506-6103

identify components most susceptible to fatigue cracking which is a major problem in steel highway bridges. This paper describes evaluation procedures to estimate the fatigue life of highway bridges using a prototype fatigue loading indicator which employs ultrasonic stress measurement to perform fatigue analysis on steel bridges.

FATIGUE LOADING INDICATOR

The fatigue life of the main longitudinal members in steel highway bridges is controlled by cyclic stresses caused by truck traffic. These stresses usually vary in amplitude and are normally below 5 ksi (35 MPA). Current stress measurement systems typically use bonded or welded foil-type resistance strain gages to measure applied stresses and fatigue crack growth rates. However, strain gaging can be time-consuming and labor intensive because surface preparation is required which may require paint removal and can consequently be a health hazard.

Another method for making applied stress measurements is the ultrasonic technique using non-contact electromagnetic acoustic transducers (EMATs). Ultrasonic stress measurements are based on the acoustoelastic effect which results in a change in ultrasonic velocity due to a change in stress (Hughes and Kelly 1953). The Constructed Facilities Center of West Virginia University developed a fatigue loading indicator which uses EMATs to facilitate field use. This instrument incorporates an electronic pulser, receiver, digitizer, and a computer to determine the time-of-flight (TOF) of the ultrasonic waves (Figure 1).

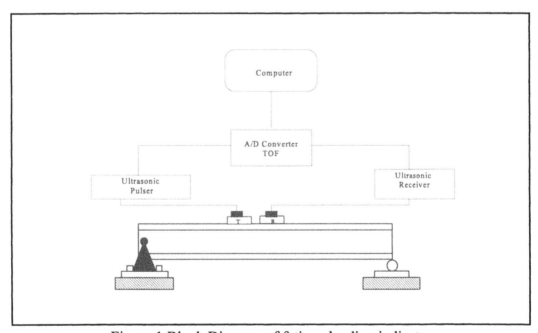

Figure 1 Block Diagram of fatigue loading indicator

The measured TOF can be used to calculate the velocity of the ultrasonic wave and hence the applied stress. This value of stress is used to perform fatigue stress analysis.

The assessment of fatigue damage of highway bridges usually under variable amplitude loading requires accumulation of fatigue damage and counting of stress cycles. The fatigue loading indicator uses Miner's rule for fatigue damage accumulation and a rainflow scheme to count stress cycles.

The number of stress cycles per truck passage is determined from the stress-time output of the proposed device using the rainflow counting scheme. The resulting number of stress cycles and stress ranges can be used to construct a stress histogram which represents the spectrum of variable amplitude stresses caused by normal traffic loads. For fatigue life analyses, the stress histogram data must be converted to an equivalent constant amplitude stress range that causes an equivalent damage sustained from the same number of variable amplitude stress cycles. The fatigue loading indicator will display a single number that represents the number of equivalent constant amplitude 10 ksi cycles experienced by the bridge. This number is determined in two steps. The first step involves calculating the equivalent constant amplitude stress range (S_{Re}) from the variable amplitude stress ranges (S_{Ri}) using Miner's equivalent stress equation (Miner 1945; Keating et al. 1990):

$$S_{Re} = (\Sigma \gamma_i S_{Ri}^3)^{1/3} \qquad\qquad (1)$$

Where $\gamma_i = ni / N$, $N = \Sigma ni$, $S_{Ri} = Smax - Smin$, and ni is the number of stress cycles at stress range S_{Ri}. The second step involves determining the cumulative fatigue damage in terms of the number of equivalent 10 ksi constant amplitude stress cycles, N_{10ksi} using:

$$N_{10\,ksi} = N \left(\frac{S_{Re}}{10\,ksi} \right)^3 \qquad\qquad (2)$$

For example, a single 5 ksi stress cycle would be equivalent to ($5^3/10^3$) or 0.125, 10 ksi equivalents and a single 15 ksi stress cycle would be equivalent to ($15^3/10^3$) or 3.375, 10 ksi equivalents. Reporting the cumulative fatigue damage in terms of 10 ksi equivalents enables a field engineer or technician to quantitatively and rapidly judge the level of deficiency (i.e., degradation) and identify appropriate remedial actions.

EXPERIMENTAL TESTING

Laboratory four-point bending tests were conducted using W8x24 A-36 steel I-beams that had a 3/8" (9.52 mm) thick and 6.5" (165 mm) wide flange. 1 MHZ Rayleigh Surface wave EMATs were used in these experiments because they are sensitive to stresses at the extreme fibers and their penetration depth extends about a wave length (e.g., about 3 mm for 1 MHZ sensor) in the specimen. The test specimens were loaded and unloaded with consistent increments and the corresponding change in time-of-flight was measured.

A typical concrete deck-steel girder highway bridge was tested to evaluate the accuracy of the fatigue loading indicator. Strain gages were also used for comparison with the ultrasonic measurements. Further details on field testing can be found in Fuchs et al. (1996).

RESULTS AND DISCUSSIONS

As mentioned earlier, applied stress levels due to truck traffic are below 5 ksi (35 MPa). Therefore, low levels of stress must be measured in steel with a resolution of approximately 0.3 ksi (2.1 MPa). For steel, the corresponding strain resolution required is about 10 microstrains. Since the stress-acoustic effect is small, this resolution requires measurements of changes in velocity on the order of parts per million (Clark et al. 1995). This resolution was achieved in a laboratory setting by averaging enough data to suppress the effects of noise (Fuchs et al. 1996). However, for the field testing, a stress resolution of 1 ksi (7 Mpa) was needed to eliminate the stress cycles below the noise level. Typical noise for the EMAT system in the field was about 1 to1.5 ksi (7 to 10.5 Mpa) (Fuchs et al. 1996). A better stress resolution is needed in the field for a better estimate of the remaining fatigue life.

Dynamic loads due to vehicular traffic usually generate frequencies around 10 to 15 Hz for concrete deck-steel stringer bridges. Because the frequency response is in the low hertz range, existing strain gages typically sample at 30-200 Hz (Christiano 1972). However, for the fatigue loading indicator, the repetition rate at which the ultrasonic signal is propagated and the number of averages dictate the overall sampling rate. Currently, the proposed instrument uses a repetition rate of 400 Hz and 50 signal averages which results in a sampling rate of 8 Hz. Higher repetition rates result in higher power consumption which is a concern.

SAFE LIFE ANALYSIS

Cumulative fatigue life analyses of highway bridges should consider past, present, and future service loads. Therefore, it is necessary to back calculate past traffic history and forecast future traffic based on the present age of the bridge. The accumulation of stress cycles depends on the frequency of occurrence (i.e., volume of traffic), traffic growth, traffic composition, lane distribution of trucks and truck weights. The number of traffic cycles experienced in the past and future are estimated by several methods. Those methods are also used to estimate the fatigue life exhausted at the end of each time interval. For example, the number of stress cycles per year are assumed to be proportional to the number of stress cycles recorded during the test period. With a certain degree of reliability, the stress measurements may be normalized on a yearly basis. With this method, the annual traffic volume remains constant for the life of the bridge (Hopwood 1994). Another method adjusts the fatigue cycles by a percentage (e.g., 5 percent) for past and future projections. Another method uses a linear regression of fatigue applications in previous years to determine an adjustment factor for future increase in traffic volume. These methods provide a range of traffic cycles for a given equivalent stress range that may be used to calculate the remaining life (Hopwood et al. 1994).

The proposed safe life analysis uses a compound growth equation to estimate future fatigue cycles and the remaining service life in 10 ksi cycles. In order to relate the cyclic life (10 ksi cycles) into the life of a structure in years, an estimate of the number of cycles per year is required. The number of cycles in a year can be estimated by annualizing the cycles gathered in the field collection period using the fatigue loading indicator. Table 1 presents a comparison between the remaining life of a typical highway bridge predicted using the proposed compound growth equation and AASHTO's remaining fatigue life. Article 3.2 (AASHTO 1990) uses a chart to estimate the lifetime average daily truck volume with a 30 year bridge life limit or an engineering judgement to account for growth rates to calculate the total fatigue life. Significant differences in the remaining fatigue life are shown as a result of different methods to account for growth rate.

Table 1. Comparison of Total Fatigue Life for Varying Percent Growth

Strain Gage	ADTT	S_R (ksi)	AASHTO Total Fatigue Life (years)			Proposed Method Total Fatigue Life (years)		
			R=0%	R= 2%	R= 4%	R= 0%	R= 2%	R= 4%
1	340	3.77	159	147	126	155	72	51
	170	"	317	294	252	308	100	67
	34	"	1586	1469	1259	1536	175	106
2	7887	2.10	39	37	32	38	30	25
	3944	"	79	74	63	76	48	37
	788	"	397	368	315	384	110	72

CONCLUSION

The prototype fatigue loading indicator which uses noncontact ultrasonic transducers is an effective means to obtain fast and reliable remaining life estimates of highway bridges. Several field tests of the prototype have been performed and the results have shown that the fatigue loading indicator yields reliable and consistent stresses. In addition, it was found that traffic growth rates have significant effect on the remaining life of a bridge and, therefore, should be included in the fatigue analysis. However, further enhancements to improve the stress resolution and sampling rate is required before this device can be used in the field.

ACKNOWLEDGMENTS

The authors gratefully acknowledge the financial support from the Federal Highway Administration (FHWA Project # DTFH61-92-C-00155).

REFERENCES

The American Association of State Highway and Transportation Officials , "Guide Specifications for Fatigue Evaluation of Existing Steel Bridges," Washington, D.C., 1990.

Christiano P.P., Goodman L.E., and Sun C.N., "Bridge Stress-Range History," Highway Research Record No. 382, Highway Research Board, 1972.

Civil Engg., "DOT Says Infrastructure Conditions are Improving," Civil Engineering, Vol. 63, No. 4, April 1993.

Clark, A.V., Fuchs P.A., and Schaps S.R., "On the Use of Rayleigh Waves for Fatigue Load Monitoring in Steel

Bridges," accepted for publishing in Journal of Nondestructive Evaluation,1995.

Dunker, K. F., and Rabbat, B. G. (1990). "Highway bridge type and performance patterns." Journal of Performance of Constructed Facilities, ASCE, 4(3).

Fuchs P.A., Halabe U., Petro S., Klinkhachorn P., GangaRao H., Clark A. V., Lozev M. G., and Chase S.B., "Field Test Results of an Ultrasonic Applied Stress Measurement System for Fatigue Load Monitoring," Structural Materials Technology, an NDT Conference , February 20 -23, 1996.

Hopwood T., "Strain Gaging to Supplement NDT Inspections ," Structural Materials Technology, an NDT Conference , Atlantic City, New Jersey, February 23 -25, 1994.

Hopwood, Theodore II., Hogan, Keith J., and Oberst, Christopher ,"Summary of stress evaluations of welded steel bridges on coal-haul routes," Kentucky Transportation Center, Frankfort, KY, 1994.

Hughes D.S., and Kelly J.S., Phys. Rev., Vol. 92(5), p. 1145, (1953).

Keating P.B., Kulicki J. M., mertz D. R., and Hess C. R."Economical and Fatigue Resistant Steel Bridge Details," Participant Notebook, National Highway Institute course No:13049, June 1990.

Miner M. A., "Cumulative Damage in Fatigue," Journal of Applied Mechanics, Transactions of the American Society of Mechanical Engineers, Vol 67, Sept. 1945, pp A-159-A-164.

Nondestructive Tests for Bridge Fatigue Damage Detection

S. ALAMPALLI, G. FU AND E. W. DILLON

ABSTRACT

Current bridge inspection methods rely largely on visual examination to evaluate bridge condition. Remote bridge monitoring systems (RBMS) using measured structural vibration have been perceived to assist in bridge inspection. However, sensitivity of measured modal properties for RBMS is critical for practical application. A study was conducted at New York State Department of Transportation (NYSDOT) to examine sensitivity of modal parameters in detecting fatigue cracks, including frequencies, mode shapes, and their derivatives. Modal tests were conducted on a 1/6 scale model of a multiple-steel-girder simple-span bridge and on a simply supported fracture-critical steel bridge, including both intact and damaged states. Statistical methods were used in analyzing the results for detecting practically visible damages. Results indicate that modal frequencies in conjunction with mode shapes may be used to identify the existence of bridge damage as small as 6 cm long. However, it is difficult to identify the damage location using these modal parameters.

INTRODUCTION

A significant number of highway bridges in US are structurally deficient or functionally obsolete. Periodic inspection has become a major component of maintaining safety of these bridges, to identify potential hazardous conditions for preventing possibly serious consequences. Currently, the Federal Highway Administration (FHWA) requires bridge inspection at least every two years (National, 1987), but some deficiencies may develop and cause serious failure much more rapidly. Near failure of a viaduct in Rhode Island and other recent bridge failures in several states, similar to the Schoharie Creek Bridge in New York, prompted researchers to look for new inspection tools. Developing new methods for inspection and evaluation of bridges has recently received considerable attention. Bridge condition monitoring using modal properties has been suggested to supplement or even replace current inspection. On the other hand, it is not clear what types and sizes of damage are suitable for potential application of this

Sreenivas Alampalli, Gongkang Fu and Everett W. Dillon, Transportation Research & Development Bureau, New York State Department of Transportation, Bldg. 7A, Room 600, 1220 Washington Avenue, Albany, NY 12232-0869

technique. This includes the issue of detectability for small, possibly visible damage (such as fatigue cracks) using measured modal properties with inevitable noise.

A recently completed study undertaken at NYSDOT was aimed at investigating the feasibility of measured vibration for bridge inspection and evaluation. Sensitivity of measured modal parameters to damage was a major focus. The approach was applying modal-testing techniques using commercially available modal-testing instrumentation and latest analytical techniques on a scaled-model steel bridge and a full-scale (abandoned) steel bridge, for diagnosis of simulated fatigue damage. The impact test method was used because its reliability is widely accepted (Ewins, 1984).

The project included the following steps: 1) measurable and stable modal parameters were identified, considering random variation observed in repeated tests, 2) baseline modal parameters were identified, including their random variation, 3) several commonly observed fatigue damages were simulated in the test structures by introducing sawcut and at each stage modal testing was repeated to obtain the modal parameters, and 4) statistical analyses were performed using the obtained modal parameters for damage diagnosis, testing the hypothesis of whether the parameters have been changed. This simulated practical application of modal testing to bridge inspection.

TEST STRUCTURES

Model Bridge: The test structure was a 1/6-scale model of the prototype bridge. It had five girders over a span of 3.72 m, composited by shear studs with a 3.65 cm concrete deck reinforced by No. 12 gage wire in both longitudinal and transverse directions. The girders were made of standard M6x4.4 beams spaced at 46.51 cm with cover plates 6.3 mm, 3.33 cm wide, and 2.48 m long. The diaphragms were connected to stiffeners welded to the girders with two bolts at each end. All bolts connecting the girders with the diaphragms were tightened to a uniform torque of 27.12 N-m. The structure rested on supports made of heavy steel girders anchored to the floor. One end of each bridge girder had a fixed support and the other a roller. A total of 65 data points were chosen for modal testing measurements (Alampalli, Fu, and Dillon, 1995).

Field Bridge: A fracture-critical bridge with two steel girders was also included in the test program for field test, referred to hereafter as "the field bridge". Located over Mud Creek on Van Duesen Road in Claverack, NY, the bridge was built in 1930 and was closed to service in 1988. It had two W18x64 steel beams supporting floor beams and a reinforced concrete deck. Its floor beams were fastened to the main beams by bolts. The main beams appeared to be embed in the concrete abutments at both ends. A total of 54 data points were chosen for modal testing measurements (Alampalli, Fu, and Dillon, 1995).

MEASURED MODAL PARAMETERS

The following structural signatures were used in this study for damage detection, based on the pilot tests conducted on intact structures to evaluate their consistency: 1) modal frequencies, 2) mode shapes, 3) modal assurance criterion (MAC) factors, and 4) coordinate modal assurance criterion (COMAC) factors. The first two are inherent modal parameters of the structure. The other two are derived from the mode shapes. Modal analysis techniques were used to obtain modal parameter (modal frequencies and mode shapes), and are described in Alampalli, Fu, and Dillon (1995).

MAC (Allemang and Brown, 1982) was used to indicate correlation between two measured mode shapes from two different tests, but from the same mode. It varies from 0 to 1, with 0 for no correlation and 1 for full correlation. MAC was used in this study to detect existence of damage, by identifying MAC values altered from their original values near 1 for individual modes. Similarly, COMAC (Lieven and Ewins, 1988) was used to identify locations where mode shapes from two tests do not agree, indicating damage locations.

TEST PROGRAM

Fatigue damage has been widely observed in steel bridges, and has been responsible for several cases of bridge collapse. To study sensitivity of the structural signatures to possible fatigue damage, fatigue cracks were simulated in the model and field bridges. For the model bridge, three different damage scenarios were considered. They progressively simulated the following typical fatigue-related damage scenarios observed in steel bridge structures: 1) Gusset plate weld crack (Cases 1-5), 2) Longitudinal stiffener bottom weld crack (Cases 6-10), 3) Crack at a cover plate toe (Cases 11-15). For the field bridge, three simulated damage scenarios were introduced by sawcut: 1) Flange cracking at midspan, 2) Flange cracking at one-third-span, and 3) Web cracking at midspan.

RESULTS

The modal parameters for the first 12 modes were obtained by testing. Mean modal frequencies and corresponding mean MAC and COMAC values were estimated and found to be inadequate for damage diagnosis, because they indicated no definite trend. Thus statistical techniques were used to analyze the data, to account for such variation (Alampalli, Fu, and Dillon, 1995).

The two-sample t-test (Box, Hunter and Hunter, 1978) was used to quantify changes in measured structural signatures for damage diagnosis. It tests whether the means of two populations are identical, using the means and standard deviations of two samples from respective populations. This hypothesis test was used here to evaluate probability of damage, also interpreted as damage detectability, using two samples of structural signatures measured before and after a damage. It was intended to determine whether the damage introduced was significant enough to be detected, or if random variation was even more dominant so that the damage could not be diagnosed.

This test was applied to modal frequencies for the various damage cases, resulting in the probability that the two samples from different structural conditions indeed had different mean values of population, as shown in Table 1. Every scenario was treated as a separate damage, and modal frequencies obtained at the beginning of each scenario (intact, Case 5, and Case 10) were taken as baselines for that scenario. Table 1 essentially gives the probability of damage diagnosis or detectability for each damage case, with reference to the scenario's intact state. For example, the first five columns give the probabilities that Cases 1 to 5 were indeed damage states from the intact state, for all modal frequencies. Probability values are expected to increase from left to right within each scenario for any given row (modal frequency), as damage severity increases in that direction. Table 1 shows that this consistency existed for most modes.

Note that damage detectabilities in Table 1 are generally higher for cases in Scenario 1 than for Scenario 2, which are higher than for Scenario 3. This shows that the first damage scenario caused more changes of the modal frequencies. This was attributed to Scenario 1 damage being in a bottom flange at midspan, and thus more critical and significant for bridge stiffness. Similarly, Scenarios 2 and 3 were not equally detectable, because they did not equally alter structural condition. Hence, it was concluded that although major damage (such as the last case of each scenario) can be detected with high probability, minor damages may not be detectable with high confidence.

Similar conclusions were reached based on damage detectability for MAC values ((Alampalli, Fu, and Dillon, 1995). These detectabilities are comparable to those of frequencies, indicating similar sensitivity of MAC for damage diagnosis. Thus detection based on MAC values can be used to supplement diagnosis decisions made using frequencies.

Since modal frequencies and MAC refer to individual modes of vibration, diagnosis using these structural signatures can only indicate whether damage or deterioration exists. In practice, when this question is answered positively, the next step is to identify where the damage is or at least in what area the damage has occurred. Without that information, the diagnosis may be considered incomplete. Mode shapes and their derivatives (e.g., COMAC) as structural signatures are natural choices for this purpose, because they provide information on vibration patterns for individual points in structures.

Damage detectability using COMAC for the model bridge was studied by the two-sample t-test. However, the results couldn't lead to reliable identification of damage locations. Results also indicated a high probability of false identification of damage location (i.e., the actual damage location missed and wrong locations identified). Same conclusions were drawn based on the results obtained using the mode shapes directly (Alampalli, Fu, and Dillon, 1995). Aside from noise in the data leading to possible false identification of damage location, these behaviors were mainly attributable to compatible mode shape changes at various data points caused by local damage. This is shown in Figure 1. For an example of Damage Case 5, the graph shows variation of Mode 1 of Girder 3 (Points 27 to 39) of the model bridge before and after damage, described by the median, 25 th percentile and 75 th percentile normalized by the median of the intact mode shape. These curves show that the mode shape changed comparably at all locations on the girder, indicating that the damage affected mode shapes not only in the damage vicinity but also other locations. This was also observed in other modes, damage cases, and the field bridge.

Using mode shapes or their derivatives for damage-location identification is essentially based on an assumption that local damage would change mode shapes more significantly at the damage location or vicinity than in other areas. Figure 1 shows that this assumption is not verified, and instead demonstrates proportional changes of mode shape at various locations. This is because, upon local damage, adjacent structural elements autogenously distribute load effects through an altered load-path system, and local damage effects are thus distributed to other elements of the structure, and is believed to have been provided by intentional and unintentional redundancy. Results obtained from the field bridge confirmed those obtained in the laboratory studies.

CONCLUSIONS

Modal frequencies may be used to detect the existence of damage or deterioration of highway bridges. Cross-diagnosis, using multiple signatures such as mode shapes, MAC, and COMAC, is warranted for such detection, because a single signature may not be conclusive due to the inevitable variation of measured data. Criteria for warning triggering using these signatures need to be determined by taking into account how their random variation affects sensitivity of detection. Based on a 95-percent confidence, presence of a crack of 6 cm (2.4 in.) or longer can be detected using modal frequencies supplemented by MAC, both obtainable using commercially available instrumentation, in a 21.95 m (72 ft) long bridge. However, the crack location may be hard to find using mode shapes and their derivatives. This is primarily because mode shapes at undamaged locations and the damage location are comparably affected by the damage, making it difficult to differentiate damaged from undamaged areas.

ACKNOWLEDGMENTS

Many NYSDOT personnel contributed to make this study possible. Thanks are extended to Idris Aziz and Wilfred J. Deschamps for assistance in data collection and analysis, Julie Tarter, Sherif Boulos, and Wilfred J. Deschamps for model bridge construction, and John Minor, Raymond Steive, Abbas Sanobari, Sherif Boulos, Deniz Sandhu, and Amita Agarwal for their suggestions.

REFERENCES

Alampalli, S., Fu, G., and Dillon, E.W. 1995 "Measured Bridge Vibration for Damage Detection," (in print) Final Report to FHWA, New York State Department of Transportation, Albany, NY.

Allemang, R.J., and Brown, D.L. 1982 . "A Correlation Coefficient for Modal Vector Analysis," Proc. First Intl. Modal Analysis Conf., Bethel, CT, 110-116.

Box, G.E.P., Hunter, W.G., and Hunter, J.S. 1978. "Statistics for Experiments," John Wiley, New York, NY.

Ewins, D.J. 1984. "Modal Testing: Theory and Practice," John Wiley, New York, NY.

Lieven, N.A.J., and Ewins, D.J. 1988. "Spatial Correlation of Mode Shapes, The Coordinate Modal Assurance Criterion (COMAC)," Proc. Sixth Intl. Modal Analysis Conf., Kissimmee, FL, 690-695.

"National Bridge Inspection Standards; Frequency of Inspection and Inventory," 1987. Federal Register, Department of Transportation, Federal Highway Administration, 52(66).

TABLE I. Damage detectability in percent using modal frequencies for model bridge.

M o d e	Damage Case with respect to Intact					Damage Case with respect to Case 5					Damage Case with respect to Case 10				
	1	2	3	4	5	6	7	8	9	10	11	12	13	14	15
1	99.9	100	100	100	100	100	100	97.8	100	100	79.7	88.2	11.9	93.5	99.0
2	99.6	100	100	100	100	100	100	97.2	100	100	70.3	79.5	28.3	100	99.4
3	36.2	98.1	100	100	100	100	99.3	92.3	99.0	100	96.8	88.5	11.9	86.8	100
4	90.7	99.0	95.0	98.5	77.6	94.5	63.3	100	99.9	100	18.5	94.9	85.7	99.5	72.3
5	83.1	58.4	53.4	100	99.9	34.5	60.5	87.4	89.0	90.0	78.4	73.7	37.9	16.5	21.1
6	47.9	99.9	100	100	100	94.2	99.2	65.6	99.1	87.6	95.4	75.8	82.9	25.6	99.0
7	99.8	99.8	92.9	100	100	97.8	0.0	97.6	79.2	98.7	90.3	73.6	81.2	36.6	98.4
8	60.2	92.9	60.3	99.2	85.5	100	100	98.5	76.8	54.0	77.6	93.2	99.2	99.5	96.8
9	85.7	100	100	100	100	60.2	65.4	58.6	59.5	99.9	87.1	99.5	11.2	56.8	8.4
10	31.2	83.2	81.8	100	100	99.9	99.9	87.6	99.0	10.3	82.8	91.4	84.3	97.0	99.9
11	44.4	100	100	100	99.9	81.9	90.2	6.1	87.3	6.9	96.6	72.4	70.8	21.0	98.7
12	7.3	99.1	98.9	100	99.7	96.9	96.9	29.4	96.2	79.3	90.2	58.6	94.1	30.2	98.0

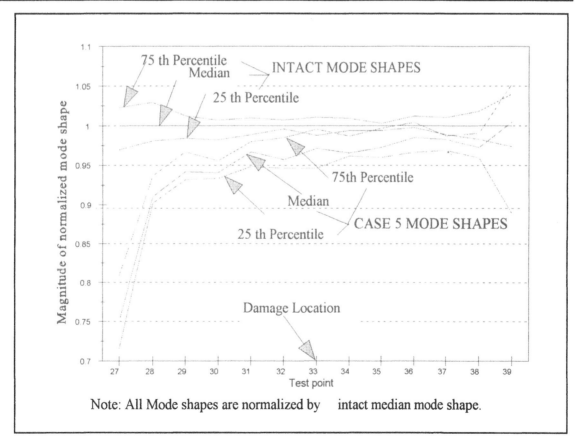

Figure 1. Variation in normalized mode shape due to damage.

A Nondestructive Technique for Monitoring and Evaluating the Stability of Concrete Highway Bridges

M. C. ISOLA

Abstract

The Michigan Department of Transportation's (MDOT) Structural Research Unit is often called upon to conduct nondestructive testing and evaluation of the state's highway bridges. Our testing and research over the last several years has included monitoring bridge performance for load restriction requirements. The nondestructive testing and evaluation of MDOT's highway bridges is of critical importance when considering today's reduced budget for the maintenance of these structures.

Introduction

In October 1994, the Bridge Management Unit of the MDOT requested a load test of the 68-year-old bridge carrying M-37 over the CSX Railroad, just south of Bailey, Michigan. At that time, a contractor was replacing part of the deck on the reinforced concrete T-beam bridge. The bridge had a load restriction of 45 tons for a Michigan type, two-unit, 77-ton vehicle, but due to extensive deterioration, the Bridge Management Unit of the MDOT was considering reducing the load carrying capacity to 34 tons. This action, however, would have presented a hardship to local manufacturers who use this heavily travelled route.

The Bridge Management Unit requested a load test of the bridge to examine the assumptions they used during their analysis of the structure. The Structural Research Unit, after reviewing the feasibility of various types of load tests, decided to "proof load test" the bridge to determine its response and capacity. This type of test "proves" the ability of the structure to carry its full dead load plus a magnified live load that is larger than the bridge is expected to carry when in service. During this test, the Structural Research Unit monitored the structure for any signs of distress or overloading.

At the time of the request for the load test, the contractor had the bridge closed for rehabilitative construction, during which the existing bituminous overlay, the concrete wearing course, and $2\frac{1}{2}$ inches of the existing deck were being removed. The contractor then replaced the $2\frac{1}{2}$ inches of existing deck and added a new $5\frac{1}{2}$ inch reinforced concrete wearing course. The designers of the rehabilitation extended the steel reinforcement through the expansion joints making the bridge spans continuous rather than simply supported.

Michael C. Isola, Michigan Department of Transportation

Test Procedure

The test procedure employed by the Structural Research Unit followed the guidelines in the NCHRP Project 12-28(13) A, "Bridge Rating Through Nondestructive Load Testing". Basically, this test involved placing increments of load on a structure and measuring its response. The structure was continually loaded until distress was indicated or until the estimated "proof load" was achieved. Continuously updated load versus deflection diagrams were monitored for any sign of non-linear displacement that would show that the elastic limit of the beams had been surpassed. Using these diagrams, the actual stiffness of the beams was compared to their predicted stiffness. Stiffness, a function of the moment of inertia of the beam, the modulus of elasticity of the materials, composite action between the beam and slab, and continuity between the spans, can be directly related to strength.

The Structural Research Unit established two target proof loads for this structure. The first target, a proof load of 62 tons, was based on the 34 ton, two-unit vehicle capacity (i.e. the reduced load rating the Bridge Management Unit considered). The second proof load was 82 tons, based on the 45 ton, two-unit vehicle capacity, which corresponded to the existing load rating on the structure before the rehabilitative construction. The idea behind this testing was to, first, target the proof load for the 34-ton rating and if the bridge exhibited signs of greater strength with no distress, target the proof load for the 45-ton load rating.

Pre-Test Analysis - Inspection

To ensure safety, a pre-test analysis of the structure, which included a shear and moment analysis of the beams and a check of the load capacity of the piers was completed. This analysis was compared to the Bridge Management Unit's estimate for the interior beams of the structure. Due to the poor condition of the structure, particularly the fascia beams, the analysis of this structure was extremely conservative. Based on this analysis, it was determined that the bending moment at the middle of the center span controlled the load carrying capacity of the structure and was chosen for the load test.

Before load testing, an inspection of the bridge was conducted which included a general inspection of the structure and a detailed inspection of the beam stems of the center span. The width of several vertical cracks near the midspan fascia beams were measured. In addition, rust was removed from the exposed reinforcing steel of these beams and the width of the remaining steel area was recorded. The bottom row of reinforcement in the south fascia beam was fully exposed, a clear indication that this beam was the weakest load carrying component of the structure.

Test Load and Weighing Procedure

It is important that the load configuration of the test vehicle is similar to the vehicle used in posting the bridge for load rating. Load distribution to the structure varies with spacing of the axles, tire pressure, pressure in the air-suspension system, width of the vehicle, and position of the load on the vehicle's trailer. For this load test, a two-unit, 11 axle, 77 ton (gross) vehicle with a flat-bed trailer was used. Additional loads were applied to the trailer with temporary concrete barriers. Each barrier to be used in the test was weighed using portable scales provided by the Michigan State Police.

It was important to know both the amount and distribution of the load to be placed on the structure. Two different weighing methods were employed to provide an independent check of the weighing procedure. For the first method, each axle of the empty vehicle was weighed using the portable scales, while for the second method, each axle of the empty vehicle was weighed at a certified truck weigh scale. A comparison of the axle weights and the gross vehicle weight revealed a discrepancy between the two weighing methods of about 6,500 pounds. It was concluded that one of the portable scales was "under-weighing" by about six percent. In addition, it was found that changes or variability of air pressure in a truck's air suspension system will alter the load distribution to the axles. These errors were accounted for by adjusting the vehicle's gross weight and axle loads.

During the load test, the temporary concrete barrier was applied to the structure in ten increments (using two 5,000 pound temporary concrete barriers per case), designated as load cases. The vehicle gross weight for each load case was adjusted by adding the adjusted temporary barrier weights to the empty vehicle weight as recorded at the certified weigh scale.

Deflection Measurements

Deflections of the concrete T-beams were measured using six linear variable differential transformers (LVDT). The core of each LVDT was suspended from the bridge beams with 1/16 inch diameter aircraft cable, a ten pound weight (to keep the line taught), and a swivel connection between the LVDT core and the weight (to allow for minor imperfections in alignment). The barrel of the LVDT was attached to a 50 pound weight, which rested on the railroad bed under the structure. It was determined that the LVDT's in this configuration had an accuracy of 0.001 inches.

Deflection measurements were also collected with an electronic level to check the data received from the LVDT's. The level that was used has a "bar code" type rod and has an accuracy of approximately 0.012 inches. Level rods were attached to each of the nine beams of the structure and readings taken for each load case. After review of the data, it was determined that the accuracy of the level was not adequate for this application because the change in beam deflection for each load case was in the range of 0.002 to 0.004 inches. Therefore, the level rod data were not used for interpretation of the results.

Data Collection

During the load test, deflection data was entered directly into a lap top computer. Load versus deflection diagrams for all instrumented beams as each load case was placed on the structure were plotted. These diagrams were monitored for possible elastic, non-linear behavior and relative stiffness of the beams. This computational ability to monitor the structure's response and to compare it to theoretical response, was critical to the successful and safe completion of the load test.

A high powered telescope was used to observe the fascia beams during the loading sequence for changes in crack dimensions. The new bridge deck was also monitored for changes of crack widths and initiation of new cracks.

Loading Procedure

The 82 ton proof load was applied to the bridge in ten increments of approximately 10,000 pounds each by placing two temporary concrete barriers on the test vehicle for each load case. In order to achieve maximum moment, the bridge was loaded with eight axles of the trailer centered on the 41' middle span of the structure. In this alignment, however, the semi-tractor was actually off the span. Therefore, its weight was not included in the pre-analysis (simple span) moment calculations. The load test took place over two days with the following procedure:

1. Initial deflection readings with no live load on the structure were taken.

2. The test vehicle was parked on the bridge with the trailer centered on middle span with the trailer wheels flush to the curb. This position was chosen in an attempt to distribute the maximum load to the south fascia beam, which was clearly the worst beam on the structure. In addition, this load position would also give the largest percentage of load distribution to any one beam. The contractor then placed two temporary concrete barriers on the trailer for each load case. For example, load case 1 consisted of the empty test vehicle parked on the bridge, whereas load case 2 had two temporary concrete barriers loaded on the vehicle's trailer, etc. Finally, load case 10 had 18 barriers loaded on the trailer for a gross vehicle weight of 139,000 pounds.

3. After the contractor placed load case 10, four temporary barriers were removed from the test vehicle to return to load case 8. The test vehicle was then driven off the bridge so it could be weighed.

4. The contractor drove the test vehicle back onto the bridge and parked it in the center of the driving lane on the middle span. Deflection readings were then recorded again.

5. The contractor removed the test vehicle from the bridge. The bridge was allowed 15 minutes to rebound before the final (no-load) deflection readings were recorded.

Test Results

The deflections recorded of the beams were far less than estimated, indicating greater stiffness and thus, greater load capacity. The bridge was loaded through load case 10 with no signs of distress. In fact, while monitoring the deflection diagrams for large distresses, it was discovered that the diagrams remained linear for each load case thus proving that the structure remained within its elastic load range. It was determined that the maximum distribution of load to any beam is approximately 26 percent, whereas the AASHTO distribution factor used for analysis is 31.7 percent. It was also determined that the maximum deflection occurred in the south fascia beam (0.04 inches). After the final load was removed from the structure, and after a 15 minute relaxation period, deflection measurements showed that the south fascia beam had a deflection of 0.011 inches. This was an indication that permanent deformation may have resulted in this beam or, perhaps the beam needed more time to rebound.

The load test was repeated on the north half of the structure with the knowledge that the axle weights from the first day of testing were unequally distributed from the front to the rear of the trailer. Because it was desired to equally distribute the load to all axles of the trailer, the barrier loads were set closer to the center of the trailer. Again the bridge was loaded through load case 10 with little or no sign of distress. After the final load was removed from the structure, and after a 15 minute relaxation period, deflection measurements showed that the north fascia beam had a deflection of 0.006 inches. This was an indication that permanent deformation may have resulted in this beam or, perhaps the beam need more time to rebound.

No measurable change was found in the cracks that were marked on the beams during the pre-test inspection. However, during the test, new cracks developed in the slab over the piers. Because the slab had not been loaded previously, it was found that negative moment at the piers caused tension and cracking in the now continuous slab. These cracks were monitored for changes in width as the test progressed, but of the six cracks that were monitored, only one crack increased in width (from 0.012 to 0.016 inches).

Rating of the Structure

Both fascia beams on this structure were rated in serious condition before the load test, with the south fascia beam in the worst condition. The bottom row of reinforcement in this beam was completely exposed and consequently cannot be relied upon for strength because of its lack of bond to the remainder of the beam. As a result, this can cause a significant loss in the load carrying capacity of the beam. In fact, this was verified during the load test. It was found that the load carrying capacity of the south fascia beam has been significantly reduced due to the deterioration of the reinforcement. Therefore, the live load normally carried by this beam is being redistributed to the rest of the structure which is in poor condition.

As weak as both fascia beams appear to be in this structure, the load test demonstrated that the beams still have far greater strength than the Bridge Management Unit estimated. This can be attributed to the following:

1. During the load test, the new reinforced concrete deck acted fully composite with the existing section.

2. Load distribution to the beams was better than estimated.

3. Loss of reinforcing steel due to corrosion was less than estimated.

4. Reinforcing steel in the new deck, passing through the expansion joints at the piers allowed the structure to act continuous rather than simply supported.

Post Test Analysis

The estimated distribution of load to the trailer axles and the actual axle spacings of the test vehicle were used to compute the maximum moment of the center span based on a simply supported span. This moment was divided by the moment produced by the two-unit, 11 axle, 77 gross ton capacity vehicle and then multiplied by the gross weight of the vehicle to compute the "equivalent proof load".

In an attempt to explain why such small deflections resulted when compared to pre-test estimates, the information gathered during the detailed inspection of the structure and the load test was used to estimate a more accurate beam resisting section. For this estimate, the reinforced concrete wearing course was made composite with the existing deck and 100 percent of the steel reinforcement was included, regardless of obvious corrosion section loss. Given the maximum deflections and estimated distribution factors, the corresponding stiffness (or strength) and moment of inertia were back-calculated. This resulted in stiffness and moment of inertia estimates that closely agreed with the post-test estimated section.

Conclusion

Because the bridge successfully carried the 82-ton proof load with no detectable distress during the load test, it was recommended that the Bridge Management Unit keep the existing posted loading of 45 tons for a two-unit, 11 axle vehicle. However, it was determined that the posting should not exceed 45 tons, due to the condition of the fascia beams. The Bridge Management Unit concurred with this recommendation and the bridge was reopened to traffic in November 1994.

In May 1995 the Structural Research Unit inspected the bridge to find the concrete overlay deteriorating rapidly and separating from the original deck at the piers. This prompted yet another analysis of this structure using new assumptions. As a result, it was found that the 45 ton, two-unit vehicle posting is still adequate. However, if the structure is to remain in service for more than five years, it was recommended that the beams, particularly the fascia beams, have extensive rehabilitation as well as retesting the structure to detect further deterioration and consequent loss of load capacity.

It was further recommended that an effort be made by the MDOT to replace this bridge due to the poor condition of the beams, piers and already deteriorating overlay. As a result, maintenance crews will be inspecting this structure every six months and reporting, in detail, the condition of the entire structure.

References

1. Isola, M.C. and Juntunen, D.A., 1995. "Proof Load Test of R01 of 61131 - M-37 Over CSX Railroad, South of Bailey, Michigan." Michigan Department of Transportation Research Project 94 TI-1731 (Report No. R-1336).

An Efficient Method for Evaluating the Load Response Behaviour of Steel and Prestress Concrete Bridges

J. STEPHENS, L. JOHNSON, S. PATTERSON, J. SCHULZ AND B. COMMANDER

ABSTRACT

Four bridges on the interstate system in Montana were field tested under vehicle loads as part of a study on the effects of heavy Canadian style trucks on the highway infrastructure. Three prestress concrete and one steel bridge were tested in four days under various moving loads. The data were used to study bridge behavior and calculate bridge load ratings.

INTRODUCTION

A study conducted by the Transportation Research Board (TRB) found that total truck transportation costs should decrease if truck weight limits are increased (TRB, 1990). Of the various truck size/weight scenarios considered by TRB, the greatest net savings in costs was realized by the adoption of Canadian Interprovincial Limits. As might be expected, the study found that these increased truck weight limits will result in increased demands on the highway infrastructure. The Canadian configurations are both shorter and heavier than U.S. vehicles, and thus the increase in demands on the bridge system are of special concern. The heaviest Canadian vehicle, the B-train, weighs up to 138,000 pounds at a length of 64 feet.

The objective of this effort was to assess the impact that Canadian vehicle configurations will have on the bridge system in Montana. While this assessment could have been done strictly by analysis, the decision was made to test a sample of bridges to obtain a direct indication of their response. Attention initially focused on the effect of these vehicles on simple spans up to 75 feet long; most of the bridges on the state highway system are composed of these spans. Live load moments in these spans will increase by up to 20 percent under Canadian vehicles. This increase should not result in immediate failure of these bridges, but it could influence the durability of concrete structures and/or shorten the fatigue life of steel structures. Therefore, three prestress concrete and one steel bridge were studied both by field testing and analysis. Field testing, accomplished in only 4 days (1 day per bridge), consisted of measuring strains at 32 locations on each bridge. The data were used in estimating any durability problems that the bridges may experience and in calculating load ratings. Analytical models of the bridges were developed from the data as part of this process.

Jerry Stephens, Larry Johnson, Steve Patterson, CE Dept., Montana State University, Bozeman, MT 59717
Jeff Schulz, Brett Commander, Bridge Diagnostics Inc., 5398 Manhattan Circle, S.100, Boulder, CO 80303

TEST SETUP

Salient features of the three prestress concrete and one steel stringer bridge evaluated in this study are summarized in Table 1. These structures were selected from a collection of 8 bridges on I-15 in northern Montana upon which trucks are presently allowed to operate at full Canadian weights. The ends of the stringers on the three prestressed concrete bridges rest directly on abutments or piers. One end of the stringers in the steel bridge rest on a pier, while the other end is supported by a transverse, simply supported girder. All stringer end supports would be modeled as pins in conventional analyses. The bridges all had exposed concrete decks designed to act either composite or non-composite with the stringers, as listed in Table 1.

Strain data was collected for each bridge under several load events. Instrumentation and data collection was directed by Bridge Diagnostics Inc. (BDI) using equipment provided by BDI. A total of 32 strain transducers were used on each bridge. The transducers were primarily positioned to characterize the flexural response of the stringer systems. Analysis indicated that this aspect of bridge behavior controlled the allowable capacity of all the bridges. Transducers were mounted on every stringer at various points along the length of the span; these transducers were oriented along the longitudinal axis of the member. At many of these locations, gages were installed near both the top and bottom edges of the beams to determine neutral axis position and evaluate the degree of composite action of the stringer/deck system. A few transducers in each test were used to monitor the response of other elements of the bridge, such as the deck and piers.

Each transducer consisted of strain gaged steel coupon approximately 3 inches in length, 2 inches in width, and 1/4 inch thick. A full bridge strain gage configuration was used for the transducers, which offers 3 times the sensitivity of a 1/4 bridge installation. Each end of the transducer was bolted to a stud, which was in turn either glued or clamped to the bridge. In a typical installation sequence, the mounting studs were bolted on the ends of the transducer, the base of the studs were coated with a rapid setting epoxy, and the assembly was pressed into position against the surface of the bridge. The surface of the bridge was lightly sanded at the gage locations with a small grinder prior to gage installation. The transducers were removed by unbolting the strain gaged coupons from the studs, and then removing the studs with a gentle tap.

Analog signals from the transducers were converted to digital form (32 Hz sampling rate) and stored on a personal computer. The acquisition system offered real time display of the data from each channel in bar graph form. This display was used to verify that all transducers were actively transmitting data. Review of the data after the tests was facilitated by using a program developed by BDI that allowed for multiple strain histories to be displayed simultaneously on the screen. The program also could search out and save the maximum measured response during any test for all transducers. Preprocessing of the raw data consisted of running a simple digital filter to remove high frequency noise and removing any baseline shift.

TABLE I--DESCRIPTION OF BRIDGES

Bridge	Stringer type	Span length (ft)	Geometry	Deck	Year built
379S	Prestressed	36	Straight	Non-composite	1964
389S	Prestressed	65	Straight	Composite	1961
390N	Prestressed	65	Skew	Composite	1977
390S	Steel	53	Straight	Composite	1961

The bridges were tested using a loaded 3 axle dump truck and an 8 axle Canadian B-train. The test vehicles crossed the bridge at either slow speeds (approximately 4 miles per hour) to create quasi-static conditions, or at highway speeds to more realistically model in-service conditions. Tests were run with the test vehicle in each lane of the bridge, at both low and high speeds. In slow speed tests, the position of the vehicle was monitored with respect to fixed reference points on the bridge, allowing the strains to be recorded as a common function of vehicle position for all channels and tests. Use of moving loads rather than stepped static testing reduced the time required to run the tests and provided a continuous view of element behavior.

Each bridge was tested in a single day. Instrumentation of each structure was completed in 3 to 4 hours, depending on the accessibility of the stringers. Low and high speed tests were then run with the control vehicles in various lane positions. These tests were completed in 2 hours. Response under casual truck traffic was then recorded for 3 hours. It took approximately 1 hour at the end of the day to remove the instrumentation. The instrumentation crew consisted of 5 people, 2 from BDI and 3 from Montana State University. The Montana Department of Transportation provided traffic control and general support. At least one lane of the highway was open at all times.

ANALYSIS OF RESULTS

Typical strain histories collected from a pair of transducers mounted longitudinally on the bottom and near the top of a prestressed beam are shown in Figure 1A. These strains were measured in the center of the span as a loaded B-train traversed the bridge. The three peaks in the response correspond to the points at which the heavy axle groups of the B-train were in a position to generate the maximum response at the gage location. These strain values were used in conjunction with the distance between the gages to estimate the position of the neutral axis. In this case, the neutral axis was close to the top of the beam, clearly indicating composite action between the deck and the beams. The position of the neutral axis was checked at different load levels to determine if element behavior was dependent on level of response. Notably, deck

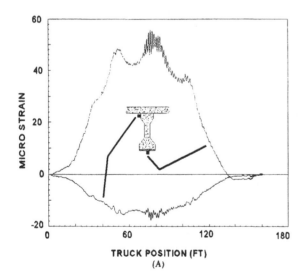
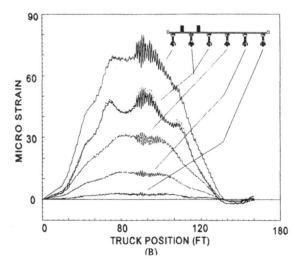

Figure 1. Typical strain histories, quasi-static test

systems that act compositely at low load levels have been observed to act non-compositely at higher load levels. In the short span prestress bridge that was tested (MP379S), for example, the neutral axis location indicated composite action between the deck and stringer system, even though the structure was designed non-compositely. Major variations in the neutral axis position between stringers can be indicative of oddities in behavior that merit further investigation.

Typical strains measured at the top of every prestressed beam in one of the bridges, as a loaded B-train traversed the bridge, are presented in Figure 1B. Families of strain histories of this type, indicating a progression of strain values (either for all stringers at a common section transverse to the span, or for a single stringer at all sections along its length) were qualitatively reviewed for irregularities in behavior. Due to the basic symmetry and simplicity of the stringer system, any irregularities in the over-all patterns of these strains would indicate odd member performance or unusual load flow that would merit further investigation. The histories in Figure 1B are very regular in nature, with maximum load carried by the stringers under the wheel paths.

Strains measured during the tests were used to directly estimate load distribution and impact factors for the stringers in the straight bridges. In bridges with identical stringers and straight spans (no skew), the fraction of the load carried by each stringer is proportional to the strain in the stringer. Peak strain values measured at the centerline of each stringer for passes of the test vehicle in the driving and passing lanes were superimposed to estimate the load distribution situation for two vehicles crossing the bridge, side-by-side. These distribution factors, expressed in terms of wheel line loads, are reported in Table 2 with the factors determined following standard AASHTO (1994) procedures. The distribution factors derived from these tests were all within 20 percent of the AASHTO values. In all cases, the distribution factors derived from the test data were less than the AASHTO calculated values.

Impact factors were estimated for each structure from the test data and are presented with the corresponding AASHTO calculated impact factors in Table 2. Estimates of impact factors were obtained by comparing response records for identical tests run at low and high speeds. While several ways exist to perform this seemingly simple comparison, the values in Table 2 were obtained by simply dividing the peak response in the dynamic test by that in the static test. These impact factors range from 12 to 77 percent of the corresponding AASHTO values.

Maximum expected live load strain and stress levels in the bridge stringers under future service loads were extrapolated from the test data using the distribution factors described above and simple flexural analyses of the stringer systems. The results of these analyses are presented in Table 3. The measured strains are for the response of a single test vehicle crossing the bridge. For actual service conditions, these values had to be modified to consider the effects of different vehicle types (other than the test vehicle) operating at different loads, and for two vehicles crossing the bridge, side-by-side. Expected strains for any type of vehicle crossing the bridge

TABLE II--DISTRIBUTION AND IMPACT FACTORS

Bridge	Distribution Factors			Impact Factors		
	AASHTO	Test	Ratio, Test to AASHTO	AASHTO	Test	Ratio, Test to AASHTO
379S	1.49	1.30	0.87	0.30	0.05	0.16
389S	0.91	0.82	0.90	0.26	0.03	0.12
390N	1.36	1.25	0.92	0.26	0.20	0.77
390S	1.51	1.26	0.83	0.28	0.11	0.39

TABLE III--MAXIMUM LIVE LOAD STRAINS UNDER VARIOUS CONDITIONS

Bridge	Measured strain, B-train test vehicle on bridge (x 10^{-6})	Strain extrapolated from test data, two B-trains on bridge (x 10^{-6})	Calculated live load strain for cracks to open (x 10^{-6})	Ratio, two B-Train strain to live load strain required to open cracks
379S	34	50	213	0.23
389S	67	107	324	0.33
390N	60	100	324	0.31
390, Stringer	131	169	Not applicable	Not applicable
390S, Girder	98	235	Not applicable	Not applicable

by itself were calculated by multiplying the measured strains by the ratio of the bending moment caused by the vehicle under study divided by the bending moment caused by the test vehicle. Expected strains for two vehicles on the bridge (side-by-side) were calculated by further adjusting these values using superposition of the measured strains for alternate lane events.

ANALYTICAL MODELING

The usefulness of the test data can be considerably enhanced by using the data to develop an analytical model of the structure. Use of such a model is essential to study the response of bridges with any sort of irregularity, such as skew orientation or non-uniform stringer sizes or strengths. The model, once developed, can be easily used to evaluate the response at various points in the structure (not just the instrumented locations) and under any loading. A structural analysis package has been written by BDI specifically for developing analytical models of bridges from the strain data collected during the tests. The program has a modeling module, in which a finite element representation of the bridge is created, the instrumentation locations are input, and the test/load vehicles are defined. Information from this module is passed to the analysis module, which can be used to optimize the model based on the test results, analyze an existing model under user specified conditions, or perform a bridge load rating for any vehicle. A model of this type was used to help analyze the behavior of the skew bridge tested herein.

ASSESSMENT OF LONG TERM DURABILITY AND STRENGTH

Adequate serviceability was expected from the prestress concrete structures if the concrete at the bottom of the stringers remained in compression at service load levels. This condition was checked by comparing the estimated live load strains under the maximum demand with the theoretical live load strains at which the bottom fibers of the stringers would go into tension. The live load strain limit was calculated using conventional analysis procedures. The maximum expected actual live load strain and the live load strain at the transition point from tension to compression at the bottom of the beam are summarized in Table 3 for each bridge. In all cases, the actual live load strains are a maximum of 33 percent of the calculated live load strain at which the bottom fibers of the stringers will go into tension.

Possible fatigue problems in the steel bridge were simply evaluated by comparing the estimated stress range under the maximum live load demand to the allowable stress range in the

AASHTO code. The maximum live load stress range in the stringers at the critical detail was estimated to be 4.90 ksi from the strain data (Table 3) by multiplying the strains by Young's modulus. The allowable stress range determined for this member and structural detail from AASHTO was 23 ksi. A stress range of 6.8 ksi was estimated for the critical location in the transverse support girder under the maximum live load demand based on the test data. The allowable stress range for this member, based on AASHTO, was 16 ksi.

While these bridges were all believed to have adequate strength to support Canadian truck configurations, load ratings were done for each structure. Ratings were done using standard AASHTO procedures (allowable stress and load factor). The inventory ratings calculated using the allowable stress approach are presented in Table 4. In each case, the ratings were done using standard inputs for the distribution and impact factors, and then they were repeated using the values for these factors determined from the tests. The prestressed concrete bridges were found to be strong enough to support Canadian configurations at inventory load levels. The steel stringer bridge was found to be marginally understrength at inventory load levels using this rating procedure. (Note that higher inventory ratings were obtained for this bridge using other load rating approaches). The load ratings obtained using the factors determined by test were from 15 to 42 percent higher than those obtained using the standard inputs.

SUMMARY AND CONCLUSIONS

Using a combination of analysis and field testing, the performance of three prestressed concrete and one steel stringer bridge under Canadian Interprovincial weight limits was investigated. Bridge testing, accomplished in only 4 days, provided a confirmation of the load transfer mechanism in the bridges and a direct indication of the strain and stress levels they may experience under service loads. This investigation revealed that bridge durability and strength should not be compromised by the increased loads.

TABLE IV--INVENTORY RATINGS FOR B-TRAIN, AASHTO ALLOWABLE STRESS APPROACH

Bridge (Milepost)	Conventional AASHTO Rating	Rating Using Test Modified Distribution and Impact Factors	Ratio, Test Modified Rating to Conventional AASHTO Rating
379S	1.32	1.87	1.42
389S	1.73	2.09	1.21
390N	1.07	1.23	1.15
390S Stringer	0.72	0.99	1.38
390S Girder	0.74	0.93	1.26

REFERENCES

American Association of State Highway and Transportation Officials. 1994. *Manual for Condition Evaluation of Bridges*. Washington, D.C.

Transportation Research Board. 1990. *Truck Weight Limits: Issues and Options*. National Research Council Special Publication 225. Washington, D.C.

Evaluation of a New Strain Monitoring Device

M. A. TAYLOR, P. J. STOLARSKI, P. MAREK AND F. REED

ABSTRACT

A new strain monitoring device is described. It avoids several problems with existing systems. Large amounts of data may be analysed in situ by means of a Rainflow algorithm. The system was used to monitor strains in primary (plates) and bracing members on one span of a CALTRANS 3 lane multispan, steel trapezoidal box girder with composite concrete deck.

INTRODUCTION

One serious problem with the monitoring of major bridges is the large amount of data generated by multiple lanes and high traffic flows. This paper describes the use of a new device (the RE49), developed in the Czech Republic, which offers some advantages over previous systems. The device is small (approx 250mm max. dimension) and light (approx 4 kg), can be powered by an automobile battery, and has an onboard computer which (at regular intervals) condenses large amounts of data into a compact dataset of equivalent cycles using the Rainflow method.

DESCRIPTION OF BRYTE BEND BRIDGE

The bridge is a multi-span, 3 lane, trapezoidal steel box girder with composite concrete deck. Constructed in 1970, the bridge carries I-80 highway traffic over the Sacramento River. There are both continuous and simple spans. The span selected for study (19L) has a simple span of 40m, width of 11m and depth of 2.5m.

Michael Taylor, Professor Emeritus of Civil Engineering, University of California, Davis., Davis CA 95616.
Phil Stolarski, Project Manager, Caltrans. Sacramento, CA
Pavel Marek, Visiting Professor, San Jose State University, CA 95192-0083
Frank Reed,Project Manager, Caltrans, Sacramento, CA

Ten strain gages were monitored in this study. Four of these were mounted on primary load carrying members while six were located on both chords and diagonals of the framed bracing system. There are six internal bracing frames per span. Gage selections were chosen to coordinate with previous and ongoing Caltrans studies of the bracing system.

OUTLINE OF STUDY

Testing extended throughout 1993. The model of RE49 employed can read 4 channels so the ten gages were studied in groups of three or four (three when a "dummy" gage was used). The usual period of data collection was three weeks. For full details of the study see Taylor et al (1995).

The procedure involved installation followed by calibrations and then the device was simply left to record data and to condense it. Three weeks later an engineer would return and download the data from the device to an 8mm disk or to a laptop computer or both. Each channel was read at a scan rate (multiplex rate) of 1000Hz.

RAINFLOW METHOD AND DATA PRESENTATION

The technique of simplifying large amounts of single valued data by selecting "cells" (i.e. a range of values) and then constructing a histogram is well known.

For random, time-dependent data the (semi) continuous readings may be approximated by cycles each with an amplitude ("range") and a mean. These two parameters may then be used to construct a form of "two dimensional histogram" which represents the original dataset. The method is described in more detail in A.S.T.M. E1049-85 (ASTM 1990).

The RE49 has two internal programs. The first performs a simplification which ignores the mean of the cycles and records only the range (This is termed "without mean"); the second includes both pieces of data (and is termed, not surprisingly, "with mean"). The first approach, while ignoring some data, permits a simple two dimension representation of the information. The user must select whichever form is felt to be the more appropriate to the task at hand. Examples of both forms are given in Figs 1 and 2.

CALIBRATION OF DEVICE

The RE49 records values of voltage which are converted to strain by use of standard electronic bridge circuits and gage factors. These may then be converted to stresses by using an (assumed) value of Youngs Modulus. The device is supplied with a calibration device (a small steel beam loaded with known weights) and this was preferred over dependence on calculations for this study.

To verify the results obtained on the bridge, the results of a test loading on this bridge were used. In this prior Caltrans study trucks of known weights were placed at several locations on the bridge to serve as calibration loadings. Tests were conducted under both static and dynamic conditions.

A third estimation method, which will be studied in the future, is to use the results of computer calculations in combination with the known truck loadings.

RESULTS

Since this study is still in progress and space does not permit a full presentation of the data the results from only one gage will be discussed. This procedure will serve to demonstrate the procedures involved since the results from all gages are treated in similar fashion.

The gage selected is No.33 which is oriented longitudinally on the top surface of the bottom plate of the trapezoidal box. In the truck calibration tests the truck was placed at six equally spaced intervals across the 40m span (interval between locations 40/7 = 3.7 m). For a 26Mg truck the maximum stress in gage 33 was measured as 8MPa thus providing a calibration factor (for static conditions).

The time dependent history read by the RE49 is a function of many factors including

   1. Weights of vehicles
   2. Number of vehicles per hour
   3. Lane occupied by vehicle
   4. Speed of vehicles (recall RE49 multiplex rate of 1000Hz)
   5. Dynamic effects

With reasonable assumptions for factors 1,2,4 and 5 the means and ranges of the equivalent cycles of loading could be calculated and compared with the observed results (Fig 2). The agreement was reasonable. Ongoing research will attempt to improve knowledge of all of the above factors, and will be presented in a future publication.

SUMMARY

The device has proven its practicality as a lightweight, self powered machine for monitoring remote sites for long periods of time without the need to accumulate and handle large volumes of data. It should prove particularly useful in fatigue studies, and it is not limited to strain gage studies but can be used with any device which monitors parameters by means of electronic sensors.

REFERENCES

American Society for Testing and Materials (1990) Standard E101049-85 (renewed 1990). A Method for Reducing Fatigue Data by the Rainflow Method ASTM Philadelphia Pa. USA

Taylor et al. "Evaluation of a New Strain Monitoring Device: A Study at Bryte Bend Bridge" Report to Caltrans Division of Highways, Engineering Service Center 1995. 2 volumes.

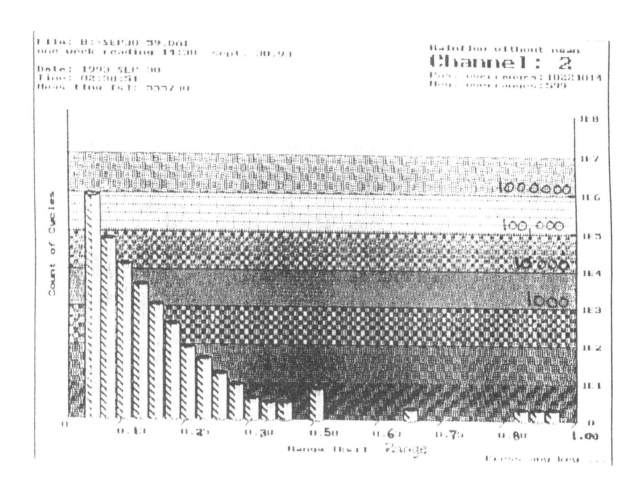

FIGURE 1 : RE49 output ("Without Mean")

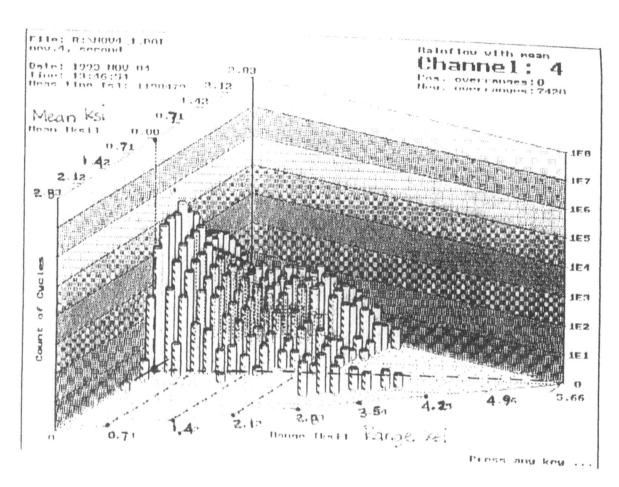

FIGURE 2 : RE49 output ("With Mean")
Vertical Axis = number of counts
+ is tensile

Emergency Monitoring of the Williamsburg Bridge, New York City

N. DINI

ABSTRACT

This paper presents the NDT methods and instrumentation used for monitoring of a major crack that developed in the masonry pedestal of a tower of the Williamsburg Bridge. The paper will explain the factors which contributed to the cracking of the pedestal, as well as the reason for monitoring the crack and the tower.

Williamsburg Bridge is a suspension bridge located in New York City. It is a major crossing of the East River and it is vital to the traffic flow in and out of Manhattan.

Due to poor maintenance, the "pinned" link at the top of one of the side span towers "froze", becoming a rigid connection between the tower and the deck. Consequently the longitudinal movements of the bridge deck induced significant horizontal forces at the top of the tower. As a result during the summer months of 1994 a wide crack developed on the entire west face of the tower's masonry pedestal. The crack's width varied in accordance with the horizontal movement of the bridge deck, due to the temperature variation of the bridge.

The bridge is a very important artery on which traffic must be maintained. Until a permanent repair or replacement of the "pinned" link on the top of the tower can be made, a monitoring program has been implemented.

The following are monitored:
- the variation of the crack width, the variation of the stress in the steel structure portion of the tower, and tower verticality.

The following methods (procedure, instruments) are used:
- electrical resistance strain gages for static and dynamic readings, dial gauge indicators, optical instrumentation

Nicolae Dini, Department of Transportation, 2 Rector Street, 4th Flr., New York, NY 10006

BRIDGE DESCRIPTION

Williamsburg Bridge (fig.1) is a suspension bridge located in New York City spanning the East River. It connects Brooklyn to Manhattan and was opened to traffic in 1903. It carries four traffic lanes in each direction and two subway tracks (one in each direction). The bridge is a major East River crossing and is vital to the traffic flow in and out of Manhattan.

The bridge deck is suspended from 4 main cables. At the top of the main towers the cables pass over saddles which are installed on roller nests. Consequently the saddles are free to move horizontally. The total deck's length, anchorage to anchorage is 2793 feet

There are no expansion joints (gaps) along the entire length of the deck. The entire deck is an uninterrupted (gapless) "unit" from Brooklyn anchorage to Manhattan anchorage made up by a string of deck sections joined (pinned) together. The central span (1600ft. between the main towers) is suspended from the main cables. The side spans, each 596ft 6in, are not suspended from the main cable. Each of the side spans rest on five supports: a) anchorage abutment, b) main tower and c) three intermediate towers. Of these five side span supports, three have roller nests (the abutment and the two adjacent intermediate towers) while the other two supports (the third intermediate tower PP20 and the main tower PP30) each have an upright link (fig.2).

The bridge deck has no fixed connections to abutments or towers for preventing bridge deck longitudinal movement. The centering of the span and maintaining its geometrical center fixed is achieved by the arrays of the main span wire rope suspenders which are not vertical but slightly inclined and symmetrical about the vertical $\mathcal{C}$ of the bridge. When the length of the deck changes due to temperature variation the center of the bridge remains fixed while the two halves of the deck expand/contract in opposite directions from the middle of the bridge.

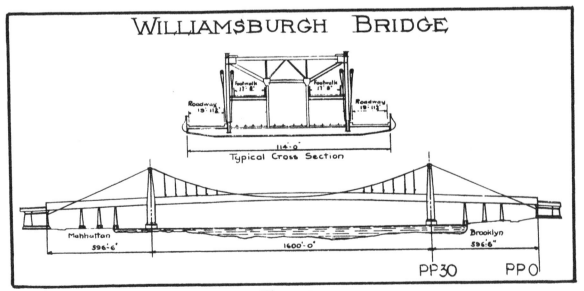

Figure 1

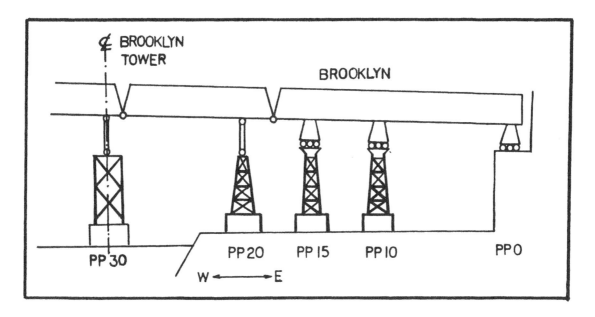

Figure 2

MAINTENANCE IMPORTANCE

The upright link on the intermediate tower PP20 of the side span is 9ft 6in tall. The link's lower end is pinned to the top of the tower while the upper end is pinned to the bridge deck. This pinned link which was intended to allow free horizontal movement of the deck for the thermal expansion/contraction cycle failed to function properly. Due to rust build up between the moving components and lack of lubrication, what were intended to be pinned connections (which would have allowed free rotation) became rigid connections. As a result any longitudinal movement of the deck due to thermal expansion/contraction, etc. induced a longitudinal force at the top of the intermediate tower which has the frozen link.

The longitudinal force generated a bending moment in the tower, which was large enough at its base to exceed the resistance opposed by the sum of the ultimate strength of the masonry in tension and the dead load. As a result a relatively wide, irregular crack developed in the tower's masonry pedestal on its entire west face. The crack was discovered in a matter of days after it appeared. The crack at the time when it was noticed did not in any obvious way endanger the safety or the integrity of the bridge. It was determined that the deck's trusses, by themselves, could not withstand the load in the case that the tower would not provide a proper support to the bridge deck. The situation had to be monitored and evaluated thoroughly and continuously, to properly assess if unrestricted traffic could be allowed on the bridge until the link is unfrozen or a permanent replacement of the link is provided. Due to continuous temperature changes the crack size varied and its adjacent surfaces were rubbing against each other. Masonry particles broken from these adjacent surfaces slowly started to fill the cracks so that the tower could not return to its original uncracked position when the temperature changed. Because of this slow but gradual deterioration of the pedestal of the tower and because the traffic on the bridge was not interrupted a program of continuous monitoring of the tower behavior was implemented. The purpose of the tower monitoring was to ensure the safety of the traveling public.

The following were monitored:
- the verticality of the tower
- the position of the link
- the position of the link in relationship to the tower
- the variation of the crack opening
- the variation of the stress in the tower's steel structure

The following methods/ instrumentation were used in the monitoring process:
- optical instrumentation
- dial gauge indicators (fig.3)
- in house custom made devices (fig.3)
- electrical resistance strain gages (fig.4)

Figure 3

Figure 4

To reduce as much as possible the damaging effects of the cyclical closing and opening of the crack an attempt to unfreeze the link was undertaken. By installing 8 hydraulic jacks of 60 tons each at the top of the tower, between the tower's structural members and the frozen link, the link was forced towards its normal position. This operation partially unfroze the link. The monitoring already in progress has showed that by this jacking operation the masonry gap was reduced from an average of 1" to an average of 3/8", and the top of the tower moved in longitudinal direction about 3". It must be mentioned that before the jacking operation there was no movement between the link and the tower. The link and the tower worked as a unit. After the jacking operation the "bond" between the link and the tower was partially broken and the dial gauge indicators showed intermittent movement between the link and the tower. The on going monitoring showed that after the jacking operation the crack range variation (closing/opening) due to temperature change narrowed. For monitoring the variation of the horizontal force at the top of the tower its effect on the tower's pedestal crack and the stresses induced into the steel structure, the tower was fitted with various brackets (fig.3) as well as 16 electrical resistance strain gages (fig.4).

THE MONITORING AND EVALUATION OF THE TOWER

The monitoring of the tower started on 14 of July 1994, the day on which the crack was discovered.

Optical instruments were first used to monitor the position and the movement of the tower. The optical measurements confirmed that the crack (gap) was developed due to uplifting and not to sinking of the foundation of the tower.

The dial gage indicators installed both at the top of the tower and at the masonry crack showed that:

     a) the link on the top of the tower was totally frozen before jacking operation

     b) the link was partially unfrozen after the jacking operation was performed and the link started to intermittently rotate.

The strain gages were the most accurate "sensors" of the stress/force variation into the tower. While the dial indicators read only when a movement occurred, that is only after enough force builds up to overcome the internal friction forces, the strain gages continuously showed the variation of the horizontal force at the top of the tower. The strain gages were very sensitive indicators of the force variation in the bridge deck.

Fig. 5 shows the variation of the strain/stress in the legs of the tower during 3 different days of October 1994.

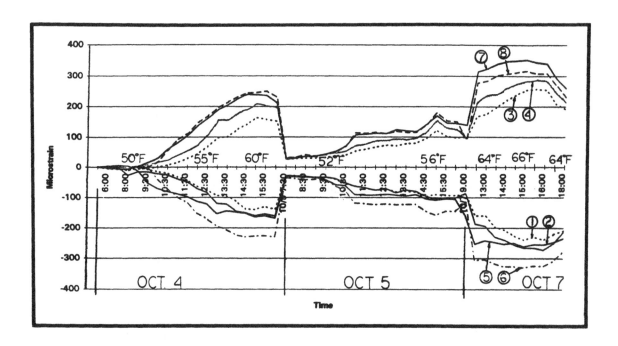

Figure 5

It is a graphical representation of the variation of the strain as a function of time/temperature into each of the 8 tower's legs. The readings were taken at intervals of between one to two hours during the days of 4,5 and 7th of October 1995. Strain gages No.1,2,5 and 6 were located on the east side legs of the tower while gages 3,4,7 and 8 were located on the west side legs of the tower. The value on the vertical axis is strain and is linearly proportional to the forces that develop in the tower due to thermal compression/shrinkage of the deck.

The long term monitoring which was performed with simple in house built brackets/instruments showed that during the winter of 1994-1995 the range in which the crack on the west face of the tower's pedestal opened-closed was 1/4 inch. It also showed that in the winter when the temperature dropped significantly, due to the shrinkage of the entire deck, a crack in the east side of the tower's pedestal opened while the west crack still remained open. Monitoring and evaluating the tower's position over long periods of time, the variation of the gap in the tower's masonry as well as the stress in the tower legs metal components, assured an unrestricted and safe traffic flow over the Williamsburg Bridge, proving at the same time that the traffic had no adverse impact on the damaged tower.

Quantitative Bridge Safety Assessment Utilizing Fracture Mechanics and Ultrasonic Stress Measurements*

A. V. CLARK AND T. A. ANDERSON

ABSTRACT

Fracture mechanics can be applied to assess cracked bridge members. If crack length and stresses are known the crack driving force (stress intensity factor, K) can be calculated. K was calculated for flange splices and hot-rolled beams, as a function of crack length. In both cases K eventually becomes negative indicating no further crack propagation. However a cracked girder will become compliant and "shed" load to uncracked neighboring members. Our calculations show that the changes in both compliance and load-carrying capacity of the cracked girder are small until the girder is deeply cracked. A finite-element analysis of a cracked girder showed that by determining the bending stresses about one beam depth from the crack it is possible to determine K. Measurement of these stresses was simulated in a field test. The method used small changes in sound speed to determine stress. The ultrasonic transducers used required no couplants and no surface preparation. They were also used to measure liveload stresses for fatigue monitoring and for determining stresses in an integral backwall bridge.

INTRODUCTION

Fracture mechanics provides a mechanism for quantifying the life of structural members. To apply fracture mechanics we must know the size and location of a crack, and the stresses acting upon it. For example the stress near a through-crack in a plate in remote tension S_0 is given by [Anderson, 1995].

$$S = S_0 \left(\pi \frac{a}{r}\right)^{0.5} = \frac{K_I}{\sqrt{r}} \tag{1}$$

A. V. Clark, Materials Reliability Division, National Institute of Standards and Technology, Boulder, CO 80303
T. A. Anderson, National Institute of Standards and Technology, Guest Researcher, on leave from Texas A&M University

where r is distance from crack tip and a is the crack length. K_I is the stress-intensity factor and gives the stress concentration at the crack tip; in this example $K_I = S_0(\pi a)^{0.5}$. K_I^2 is proportional to dU/da, where U is the potential energy in the cracked specimen [Anderson, 1995]. Hence K_I^2 is proportional to the crack-driving force. When K_I is greater than K_{Ic}, fracture occurs; K_{Ic} is the critical stress-intensity factor and is a material property. K_I will be more complicated than Eqn. 1 for more realistic structure. We calculated K_I for rectangular beams with vertical cracks using finite element analysis (FEA). We compared the FEA result with solutions in the literature [Anderson, 1995] for beams in 3- and 4-point bending. The result is shown in Fig. 1; good agreement is evident. Here we have scaled the moment M_c so that the same stress is applied at the crack location.

The reason for the agreement is as follows. The stress intensity can be calculated from [Anderson, 1995]

$$K_I = \int h(x)\ S(x)\ dx\ ,\tag{2}$$

where the weight function h(x) is the value of K_I for a concentrated stress at x. The stresses S are those which existed at the crack location before crack formation. Since the bending stresses for all three cases (uniform load: 3-and 4-point bending) are the same, K_I must also be the same.

APPLICATION OF FRACTURE MECHANICS TO BRIDGE GIRDERS

The above analysis pertains to rectangular beams. We can extend to the case of I-beams by scaling:

$$K_I = K_{Ic}\ \frac{I}{I_0}\ ,\tag{3}$$

where the subscript "0" indicates the values for rectangular beams and I is the area moment of inertia. The scaling follows

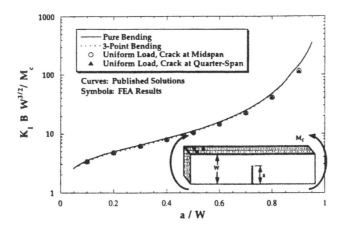

Fig. 1. Normalized stress intensity factor for rectangular beams; B is beam thickness.

from superposition, plus the fact that the bending stress is inversely proportional to I.

We now apply these principles to cases of more practical interest. Consider a retrofit to an I-beam with a vertical crack; a splice across the crack mouth (splice at the bottom of the flange). To model the splice we calculated the force necessary to close the mouth of the crack and also the associated stresses. The latter were input to the integral in Eqn 2 and the values of K_I as function of crack length calculated. The result appears in Fig. 2, along with values of K_I for the case of no splice. Comparing shows the beneficial effect of retrofitting. K_I for the no-splice case increases with crack length; a brittle fracture would occur. K_I for the splice reaches a maximum and then decreases; a brittle fracture of the entire beam will not occur. In the latter case K_I becomes negative. This has interesting consequences for fatigue crack propagation. If K_I is large enough in the negative sense, even fatigue crack propagation is not possible.

Suppose the beam is hot-rolled and has resulting residual stresses. Typically the flange will cool faster than the web (larger thermal inertia). The resulting mismatch in shrinkage causes the flange to have tensile stresses and the web compressive stresses; the stresses must be self-equilibrating. We assumed a parabolic distribution of compressive stresses in the web, and a corresponding distribution in the flange. The magnitudes are chosen to give zero net force. We used Eqn 2 to calculate K_I, as shown in Fig. 3. There we varied the peak residual stresses and superposed them with an assumed 69 MPa (10 ksi) deadload bending stress. For all cases K_I reaches a maximum and ultimately becomes negative; total fracture will not occur. A rolled beam with sufficient compression is a "smart structure" because it is fracture-proof.

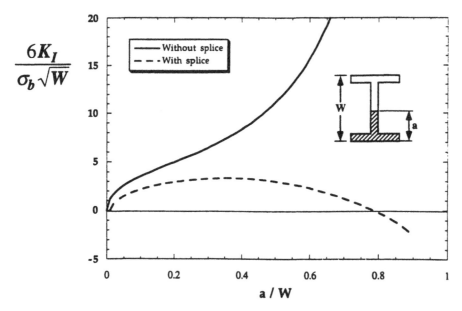

Fig. 2. Normalized stress intensity factors for beam with and without splice. σ_b is peak bending stress.

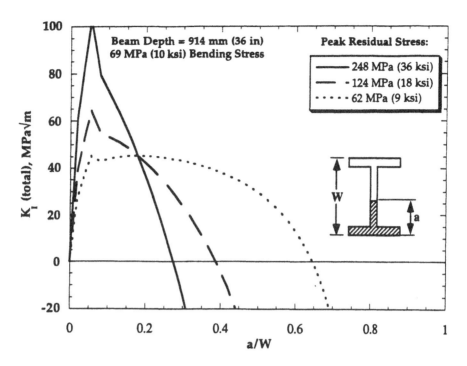

Fig. 3. Stress intensity factors for rolled I-beams as function of peak residual stress in flange. The beams have residual stresses plus a 69 MPa liveload stress.

Even if the beam does not totally fracture, it may still "fail" if it becomes too compliant (possibly causing damage to the deck, etc.) or if it loses load-carrying capacity and "sheds" load onto neighboring structural elements. We have considered both of these effects. To model loss of stiffness we use a scheme discussed in [Clark and Anderson, unpublished].

As an example we show the relative stiffness change $\Delta k/k_0$ in Fig. 4 for a crack at the weld of a cover plate. Two cases are identified: the small-scale case is the specimen size used in [Fisher et al., 1970] to generate S-N curves for fatigue life of coverplate details. The full-scale case is taken from an actual bridge (length = 26 m). There is little stiffness change with crack depth, even for the large scale beam. Similar calculations performed for cracks which have penetrated into the web show that stiffness changes of a few percent occur until the beam is about 90% fractured [Clark and Anderson, unpublished].

As the crack deepens, some of the load carried by the cracked girder is "shed" to the other girders. We calculated this for a 3-girder bridge, with results shown in Fig. 5. Here α is the ratio of the bending stiffness of the deck to the "spring constant" of the girders [Clark and Anderson, unpublished]. When α is large we approach the rigid bridge condition (the deck does not deform). For long, narrow bridges (α of order unity or less) load shedding is negligible until the girder is about 80% cracked.

We now relate these results to the two examples given above: the retrofit, and the rolled beam with compression in the web.

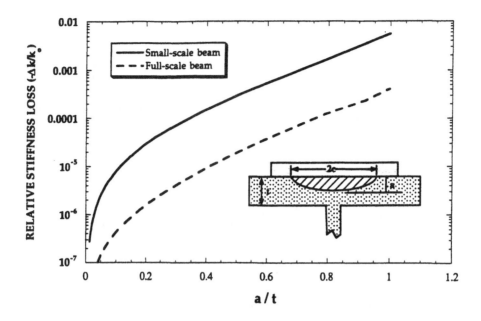

Fig. 4. Stiffness loss due to crack in coverplate weld; crack (shaded) has penetrated depth a into flange.

Any crack will arrest in both cases, since K_I eventually becomes negative (see Figs 2 and 3). Furthermore there will be little loss of stiffness and load capacity unless the crack is very deep (Figs 4 and 5). Hence we expect that the bridge will be "fracture-proof" for these cases.

APPLICATION OF ULTRASONIC STRESS MEASUREMENT TO FRACTURE AND FATIGUE

To this point we have considered only the effect of brittle fracture. Another significant failure mechanism is fatigue. The relation between the number of fatigue cycles N_f to failure and the stress range S_r is $N_f = AS_r^{-m}$ where A and m are constants. Their values are chosen based in part on the data of [Fisher et al., 1970]. S_r has conventionally been measured with strain gages, which require surface preparation. An alterative method is to use ultrasonics. For uniaxial stress (such as in a girder flange) the velocity V is related to the stress, S,: $V = V_0 + C_a S$ where V_0 is unstressed velocity, C_a is the "stress-acoustic constant." Hence measurement of velocity changes gives changes in stress. A description of this technique and its application to fatigue monitoring is given in [Fuchs, 1995]. There a noncontacting transducer was used which required no surface preparation. Unlike conventional (piezoelectric) transducers it required no couplants to transmit sound into the girder.

Ultrasonics could also be used to determine the crack-driving force. Recall that K_I is given by Eqn 2, where the stresses are those which existed at the crack before crack formation. We calculated the stress distribution in cracked

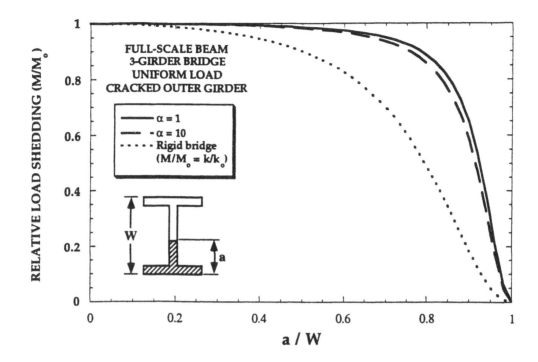

Fig. 5. Load shedding due to crack in outer girder of 3-girder bridge. M = moment carried by cracked beam; M_0 = moment carried in uncracked configuration.

beams using FEA and found that the stresses approached the uncracked values at a distance of approximately one beam depth from the crack [Clark and Anderson, unpublished]. Hence measurement of stress along a vertical scanline at this distance, and use of Eqn 2 should suffice to determine K_I. This has been simulated in a field test on a bridge in Virginia [Lozev, unpublished].

The method was also tested on an integral backwall bridge. There are no expansion joints in this bridge; the backwalls push against backfilled soil as the bridge expands. Because the bridge is on a skew the possibility exists of stress buildup in various girders. The bridge had already been instrumented with strain gages; because of anomalous readings it was suspected that some of the instrumentation amplifiers were malfunctioning. Ultrasonic measurements of bending stresses were made at backwalls on opposite sides of the bridge, both on bottom flanges (where strain gages were located) and on vertical scanlines. When the suspected amplifiers were subsequently replaced the strain gage data agreed to within about 60 MPa (9 ksi) with the ultrasonic data [Lozev, unpublished].

Another possible application of the ultrasonic method is characterization of stresses in pin and hanger assemblies. If corrosion causes one of the pins to "freeze" thermal expansion there will cause a bending moment in the hanger which is reacted by a torque in the frozen pin. Measurement of the shear stress $S_{r\theta}$ in the pin is difficult with ultrasonics. However the bending stress S_{xx} can be more easily measured with sound propagating through the hanger thickness. We can then use strength-of-materials relations to obtain

$$S_{r\theta} = 2 \frac{r}{h} \frac{I}{J} S_{xx} \qquad (6)$$

between peak stresses. r and h are pin radius and hanger width; I and J are moments of inertia of hanger and pin. Hence once S_{xx} is measured we can determine the stress in the pin.

REFERENCES

T.A. Anderson, Fracture Mechanics Fundamentals and Applications, CRC Press, Boca Raton, 1995.

A.V. Clark and T.A. Anderson, Nat'l. Inst. of Stand, and Tech., Gaithersburg, MD, to be published.

J.W. Fisher, et al., NCHRP, Rep. 102, Tram. Res. Bound Nat. Res. Council, Wash. DC, 1970.

P.A. Fuchs, A.V. Clark and S.R. Schaps, "Monitoring Bridge Fatigue Loads with Ultrasonic Transducers, Sensors, Vol. 12, No. 1, p. 20, 1995.

M.G. Lozev, A.V. Clark and P.A. Fuchs, to be published, Virginia Transportation Research Council, Charlottesville VA.

Computer Stereo Vision Method for NDE of Bridges

M. F. PETROU, W. S. JOINER, Y. J. CHAO, J. D. HELM AND M. A. SUTTON

ABSTRACT

The aging U. S. infrastructure has given rise to well-founded concerns for existing bridge structures; the number of bridges which require repair and/or additional monitoring is approaching 200,000. Advanced methodologies are needed to: (a) provide field data which can be used to quantitatively measure global bridge parameters (e.g. stiffness) in support of bridge management, (b) locate and assess local damage in existing bridge structures, (c) determine the durability of repairs and (d) assess the degradation of relatively new structures so that repairs can be implemented.

Since a bridge management system combines management, engineering and economic inputs, it is essential that accurate data on the current condition of each bridge be available to help the decision-makers make timely decisions relative to bridge repair/replacement. Currently, much of the data available to bridge management is based almost entirely upon routine visual inspections; these inspections are both subjective and limited in ability to detect hidden deterioration or flaws (i.e. such as early corrosion of the steel reinforcement or cracks).

In recent years, a completely computer-based optical method which measures the full, three-dimensional displacements and surface strains and uses normal white light (e.g. sunlight, incandescent light) illumination has been developed. The method is based on stereo vision principles and uses two cameras to record images digitally and store them in a portable computer. The images are analyzed at the convenience of the user to obtain the surface deformations and/or 3-D displacements at points of interest. Due to the character of this method, it has many advantages over previous optical methods including: (a) ability to eliminate effects of most vibrations, (b) ability to interrogate large or small regions; the system can be configured so that data for either global or local structural response can be quantified and (c) ability to be converted to a portable field device without affecting accuracy of the measurements.

This paper presents some preliminary results from an on-going research project on the monitoring of the performance of bridge structures using the computer stereo vision method. The computer stereo vision method and a DCDT were used to measure the vertical displacements of the steel girder frame of a small scale bridge structure. The preliminary results show very good agreement between the displacements obtained by the two methods.

M.F. Petrou and W.S. Joiner, Department of Civil and Environmental Engineering, University of South Carolina, Columbia, SC 29208
Y.J. Chao, J.D. Helm, and M.A. Sutton, Department of Mechanical Engineering, University of South Carolina, Columbia, SC 29208

BACKGROUND FOR COMPUTER STEREO VISION METHOD

The experimental technique was first developed at the University of South Carolina in the mid-1980's for two dimensional deformation measurement (Chu et al., 1985). In the early 1990's the method was extended to three dimensions in which two digital cameras (i.e., a stereo vision system) were used and more advanced computer software developed (Chao and Sutton, 1993, 1994; Helm, Sutton, and McNeill, 1994; Luo, Chao, and Sutton, 1993, 1994). Figure 1 shows a schematic of the stereo vision system with the typical two-camera arrangement. During the experiment, both cameras take images from the specimen surface before and after the specimen is loaded. Using a pin-hole model and including lens distortion correction, the software performs the calibration of the cameras (to obtain accurate camera parameters such as the focal length and the distance between the two cameras) and computes the three-dimensional displacement fields of the specimen surface. The accuracy obtained in laboratory testing of small specimens is approximately two to three microns.

The stereo vision system developed at the University of South Carolina has been shown to be a powerful and extremely versatile experimental method, capable of characterizing the deformation fields for three-dimensional bodies. The stereo vision system is fully computerized, non-contacting, full-field, accurate, and involves minimum of work during testing. Current development of this method is devoted to verifying the accuracy of the method for large structures, to accelerating the computation procedure, and to implementing a portable system which can be conveniently used in the field for large civil structures such as bridges.

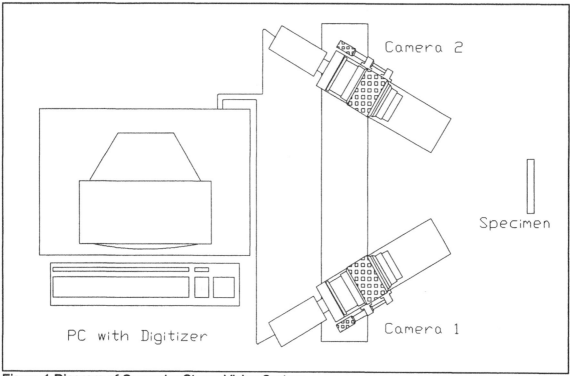

Figure 1 Diagram of Computer Stereo Vision System.

CURRENT STUDY

This project seeks to improve our ability to monitor existing and new bridges by developing and verifying a new NDE method based on stereo vision principles outlined in the previous section. Through analysis of the image data, long-term changes in bridge deformation can be correlated to increased damage and appropriate repairs initiated. The proposed system can be used to: (a) provide field data which can be used to quantitatively measure global bridge parameters (e.g. stiffness), (b) locate and assess local damage in existing bridge structures, (c) determine the durability of repairs, (d) assess the degradation of relatively new structures so that repairs can be implemented and (e) supply the structural analyst with experimental data for verification of deterioration models that will be the basis for improved life-cycle cost optimization. In addition, the proposed system could be used to (f) develop a data base for repaired structure to assess its long-term durability and (g) develop a data base for new bridges so that accurate comparisons can be made to assess the extent and severity of long-term damage.

The accuracy and utility of the new NDE system will be demonstrated through laboratory tests on a small scale model bridge. The laboratory tests will be performed on a 1/6.6 scale model of a complete bridge (see Figure 2), including reinforced concrete deck, steel girders and steel diaphragms between adjacent steel girders. The global structural response of the bridge, undergoing both static and long-term fatigue loading, will be recorded using the stereo vision-based NDE system. Specifically, (a) the three-dimensional deck deflection in a specific region will be measured to assess the degradation in the concrete and (b) the deformations near an artificially cracked steel girder will be recorded. These will be used to verify the ability of the system to acquire both local and global data for a bridge structure.

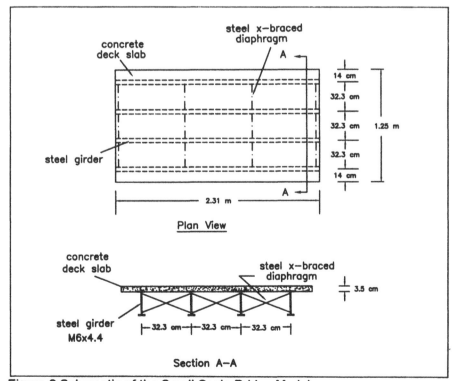

Figure 2 Schematic of the Small Scale Bridge Model.

An obvious symptom of long-term cumulative damage in a deck bridge structure is the change of the stiffness of the deck slabs (McCarten, 1991) and consequently the change of the deflection response of the structure under a constant load (known weight truck). Experimental studies (Perdikaris, Petrou, and Wang, 1993; Petrou, 1993) showed that the deflection increases slowly in the early stage of the fatigue process and then increases rapidly as the structure approaches incipient failure. It is crucial to the management of the highway bridge system that the onset of final failure be identified as early as possible so that the structure can be repaired. Clearly, the new method, which could accurately quantify the bridge displacements and deformations in general, has the potential to be an important tool in bridge structure safety assessment.

For local measurement purposes, one of the steel members will be artificially notched and the notch covered with paint to obscure its presence. During the application of various stationary loads, the stereo vision based NDE system will be used to record images and compute the 3-D displacements in the general region where the notch is located. The data obtained by the stereo vision system will be used to locate and determine the size of the flaw.

Initial tests were conducted on the steel girder frame (see Figure 2) of the bridge model to verify the accuracy and applicability of the method. The vertical displacement field obtained from the computer stereo vision method is compared in Figure 3 with deflection measurements obtained from a DCDT. The load was applied at the center of one of the steel girders and the DCDT placed 80 mm from the center of the same girder. The stereo vision method provided deflection data for a region extending 120 mm away from the centerline of the same girder. The girder centerline is along the negative sixty (-60mm) line in Figure 3. The difference in the deflection measured with the computer stereo vision method and a DCDT for the two loading cases (6.7kN & 13.3kN) varies between 2-4%. It is very important to note that the computer stereo vision method provides the deflection field of an area, reducing the possibility of a localized error.

REFERENCES

Chao, Y.J. and Sutton, M.A., "Accurate Measurement of Two and Three-Dimensional Surface Deformations for Fracture Specimens by Computer Vision," Chapter 3, *Experimental Techniques in Fracture, Vol III*, the Society for Experimental Mechanics, 1993.

Chao, Y.J. and Sutton, M.A., "Computer Vision Methods for Surface Deformation Measurements in Fracture Mechanics", *ASME AMD Novel Experimental Methods in Fracture*, 176, 123-133 (1994).

Chu, T.C., Ranson, W.F., Sutton, M.A., and Peters, W.H., "Applications of Digital Image Correlation Techniques to Experimental Mechanics," *Experimental mechanics*, 25(3), 232-245 (1985).

Helm, J.D., Sutton, M.A. and McNeill, S.R., "Three-Dimensional Image Correlation for Surface Displacement Measurement", *SPIE Videometrics III*, 32-45 (1994).

Luo, P.F., Chao, Y.J. and Sutton, M.A., "Accurate Measurement of 3-D Deformations in Deformable and Rigid Bodies by Computer Vision", *Experimental Mechanics*, 33 (2), 123- 133 (1993).

Luo, P.F., Chao, Y.J., and Sutton, M.A., "Application of Stereo Vision to 3D Deformation Analysis in Fracture Mechanics," *Optical Engineering*, March 1994.

McCarten, P.S., "Fatigue Life of RC Bridge Decks", *Proc. 8th Annual Intl Bridge Conf. in Pittsburgh*, Pa., Paper No. IBC-91-42 (1991).

Perdikaris, P.C., Petrou, M.F., and Wang, A., "Fatigue Strength and Stiffness of Reinforced Concrete Bridge Decks", *Final Report-FHWA/OH-93/016*, 1993.

Petrou, M.F., "Fatigue Performance of AASHTO and Ontario Design for Non-Composite Reinforced Concrete Bridge Decks", *Ph.D. Dissertation*, Case Western Reserve University, Cleveland, Ohio, August 1993.

Sutton, M.A., Chao, Y.J., and Lyons, J., "Computer Vision Methods for Surface Deformation Measurements in Fracture Mechanics," *ASME-AMD Novel Experimental Methods in Fracture*, 176, pp.123-133, 1993.

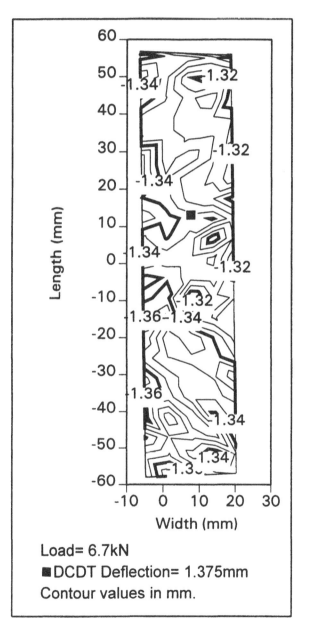

Load= 6.7kN
■DCDT Deflection= 1.375mm
Contour values in mm.

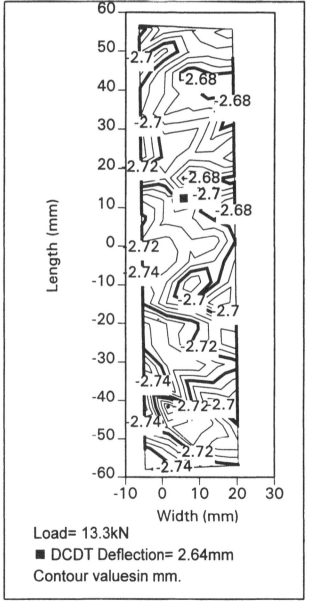

Load= 13.3kN
■ DCDT Deflection= 2.64mm
Contour valuesin mm.

Figure 3 Comparison of Deflection Measured by a DCDT and the Computer Stereo Vision System

Fracture Critical Inspection and Global Remote Monitoring of Michigan Street Lift Bridge, Sturgeon Bay, Wisconsin

P. E. FISH AND D. W. PRINE

ABSTRACT

This paper describes the serious cracks found during a Fracture Critical Inspection on a 65-year-old rolling bascule lift span and how a continuous Global Monitoring system monitors for changing conditions from a remote location. Discussion will include conditions found during several inspections on the bridge, major maintenance rehabilitation to keep the structure in service until it can be replaced at a future date, and Global Remote Monitoring to monitor the rolling bascule girders for changing conditions. With increased demands on inspection personnel and an aging infrastructure, new NDE systems need to be implemented which can help assure safe travel by the public.

HISTORY AND DESCRIPTION

The Michigan Street Lift Bridge was built in 1930. It crosses Sturgeon Bay, a major ship building area. It is one of two bridges that provide the only routes to a peninsula. The structure consists of thirteen spans for a total length of 1,413 feet. There are seven spans east of the lift span, five of which are combination steel-concrete girder and two are overhead steel truss. West of the lift span there are five overhead steel truss spans.

The overhead truss, rolling bascule lift span is 161'-6" in length. Each leaf of the span is 80'-9" in length. Roadway width is 24 feet with one sidewalk 8 feet wide. The two bascule girders are spaced 26'-6" apart. The bridge opens approximately 3,600 times a year for a total of 7,200 cycles.

Maintenance and repair history on the rolling bascule span is lengthy and important in determining the causes of conditions discussed. A summary of the history is as follows:

- 1930 New Structure
- 1959 New Steel Grid Floor
- 1960 Extensive Damage Repairs (Ship Hit Bridge)
- 1965 Estimated Date for Field Welds Placed on Bascule Girder Rivets and Rolling Plate
- 1970 Bridge Painted

Philip E. Fish, Specialized Bridge Inspection, Wisconsin Department of Transportation, 4802 Sheboygan Avenue, Madison, WI 53707-7915
David W. Prine, BIRL, Northwestern University, 1801 Maple Avenue, Evanston, IL 60201

- 1978 Fix Superstructure Deterioration
- 1984 Bridge Painted
- 1988 New Electrical Components on Drive System
- 1992 Rebuild Motors

The bascule girder is supported by two components, a track girder and a segmental girder. The track girder remains in a fixed position at all times and the segmental girder moves as the bridge opens and closes.

As shown in Figure 1, both the track girder and the segmental girder have two web plates spaced at 1'-2". On the track girder, a cast steel (A24 Material) track plate with teeth is attached to the top of the box girder with two single lines of rivets. The segmental girder has a rolled steel plate (A7 Material) with holes that match the teeth in the track plate. This is also attached in a similar manner with two single lines of rivets.

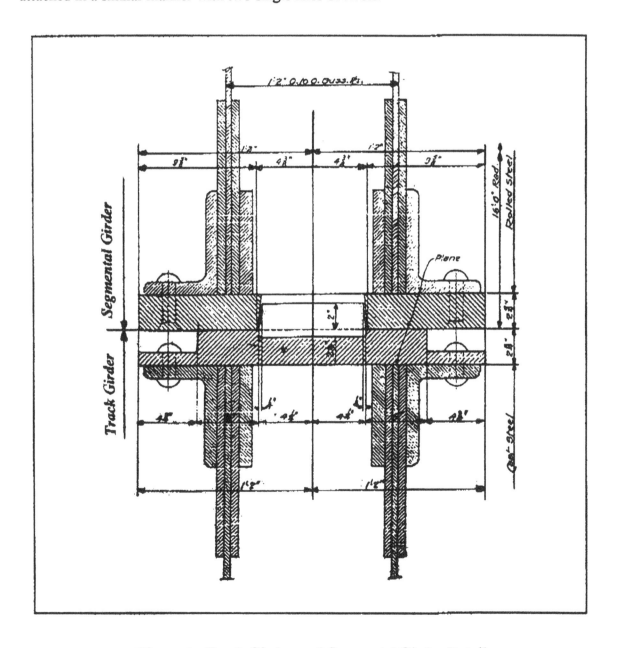

Figure 1. Track Girder and Segmental Girder Detail

Structural material in the bascule girders is (A7 Material). All mechanical equipment to operate the bridge is located above the roadway along with the counterweights for each leaf.

FRACTURE CRITICAL INSPECTION

June 1994 Initial fracture critical inspection was performed at the request of persons doing routine inspections on the bridge. They noted movement in the rolling plate where it attached to the segmental girder each time the bridge was opened and closed.

This inspection found serious cracking throughout the segmental girder. A majority of the cracks initiated from field repair welds that were placed in approximately 1965 although the date could never be clearly defined. Field welds were placed along the edge of the rolling plate and connection angles. Also, welds were placed around the rivet heads on the angles. Apparently this was done as an attempt to stop movement where the rolling plate was attached to the segmental girder.

The cracking was continuous along the rivet lines in the connection angles for several rivets. Cracks also migrated at random patterns from rivet heads. The connection angles were also cracked at the 90-degree corner of the angles from bending in several locations. Vertical bearing stiffeners were worn into the connection angle as much as 1/8". The field welds attaching the connection angles to the rolling plates were also cracked in many locations for several inches in length. All rivets connecting the rolling plate to the connection angles were ultrasonically inspected and determined to be failed. The only holding mechanism for keeping the rolling plate in place at several locations was corrosion in the rivet holes.

Emergency repairs were performed by placing bolts in failed rivet locations. Bridge was load posted and the number of openings was reduced as much as possible. Bridge placed on monthly inspection cycles.

August 1994 Inspection found increased cracking from field welds on rivet heads and field welds attaching the connection angle to rolling plate. New cracks were found in the rolling plate. These were located at the corners of the holes where the track tooth engages. The cracks were diagonal and varied in length.

Cracks were found in the 90-degree corner of the connection angles attaching the cast steel track plate to the box track girder. This was due to flex as the bascule span operated. Movement was visual.

Design started on rehabilitation plans and work was scheduled for February 1994, the minimal usage time for both vehicle traffic and ship traffic.

September 1994 Cracking continued, but the growth rate had decreased. A new area of concern was located. The bascule span had settled with the many years of use and the pinion gear (drive gear) teeth were bearing on the rack teeth, thus assuming load that was not designed for. The original design gap was 1/2".

October, November, December 1994, January 1995 Conditions continued to change, but at a much slower rate due to considerable decrease in operating cycles in the winter months.

Design plans were completed with several meetings to work out details for field conditions. A contractor was selected and material was ordered. All bolts were custom fabricated to match details of existing rivets.

MAINTENANCE REHABILITATION DESIGN

The design for rehabilitation took into consideration that the bridge was to remain in service a limited number of years until a new bridge can be constructed. The bridge would be permanently closed during the rehabilitation contract. It was decided to leave us much of the existing material in place as possible to achieve alignment when putting the segmental and track girders back in place after repairs.

To accomplish this, the severely cracked connection angles connecting the rolling plates to the segmental girders and the connection plates connecting the cast steel track plate to the box girder were reinforced by adding an additional plate on the exterior side. This provided a clamping force on the angles. Also, all cracks were to have holes drilled at the ends. The cracks in the 90-degree corners of the angles were to have holes drilled at the ends, with no other repair.

All vertical bearing stiffeners were to be attached in a similar manner by clamping new material on the exterior surfaces.

A325 high-strength bolts were specified. These had to be custom fabricated to meet the conditions of existing rivet details.

The pinion gear teeth were to be machined to provide clearance with rack teeth.

It was determined that the bridge would remain load posted after the repairs were complete.

MAINTENANCE REHABILITATION

Rehabilitation started in February 1995. The bascule girders were opened to their maximum open position and blocked to remain permanently open during the first part of the repairs.

The rolling plates were removed from the segmental girder and the track plates were removed from the box track girder. Both were marked clearly so they could be placed exactly back at their original location after repairs were accomplished.

Unexpected conditions found Deterioration was more severe than expected. The cast steel track plates had considerable cracking on the bottom side throughout. Where the cast steel track plates connected to the angles, there was heavy wear up to 1/8", and severe corrosion with deep pitting.

On the rolling plates, wear had taken place where the web plates of the segmental girder bear on the rolling plates. In this area, up to 1/8" had worn away. At the connection angles there was also wear and deep pitting corrosion. Additional cracking was found in the connection angles along the rivet lines. The cracks had migrated out from the one single crack that was visible on the top side.

Additional rehabilitation repairs After several contacts, it was determined that it would take several weeks to fabricate new cast steel track plates, thus delaying the contract time a considerable amount and delay re-opening of the bridge. To accommodate contract time and the traveling public, it was decided to repair existing plates.

Cast steel track plates were taken to a fabrication shop and weld repaired under close supervision. Weld material and welding processes were selected that would be most compatible with the existing material. The process included restraints to minimize distortion, preheat, and post-heat to minimize cracking and internal stress. The cast steel track plates were cleaned to remove deep pitting corrosion and primed with paint.

Deterioration from wear and original casting flaws made it impossible to completely weld repair two cast steel track castings. Mr. David W. Prine at BIRL was contacted and requested to develop a remote strain gauging system to monitor these plates from a remote location. Strain gauges were installed on the two plates before installation on the bridge. The gauges were located between the two web plates of the box girder which is inaccessible when the cast steel track plates are in position on the bridge.

Rolled steel plates on segmental girders were also taken to a fabrication shop for cleaning to remove deep pitting corrosion. Holes were drilled at ends of diagonal cracks in corners of tooth sockets. A prime coat of paint was applied.

To provide complete bearing surfaces when re-installing the cast steel track plate and rolled steel plates, titanium putty with a compressive strength of approximately 18 ksi was selected. It would be applied immediately before final bolting to all bearing surfaces to assure that would remain in a liquid form until bolts were tightened.

After adjustments for unexpected deterioration, the rehabilitation work progressed on schedule. The cast steel track plates and rolling plates on segmental girders were re-assembled with new clamping plates, titanium putty and bolts. New bearing stiffeners were placed by clamping with new materials and bolts. Teeth on the pinion gears and rack castings were machined to provide a clearance on 1/8".

The initial closing cycle of each bascule leaf was monitored closely for any areas of interference. After minor adjustments, the bridge was placed back in service.

GLOBAL REMOTE MONITORING

The serious deterioration in the track castings detected by the Wisconsin DOT inspection team and discussed above necessitates close monitoring of the bridge to maintain safe operation. The use of human inspectors to accomplish the desired level of monitoring would be prohibitive both from a cost standpoint and the increased drain on resources. Additionally, visual inspection of the accessible portions of the bridge is not likely to detect deterioration in a sufficiently timely manner. For these reasons, Wisconsin DOT decided to apply remote global monitoring techniques to this structure.

Northwestern University's Industrial Research Laboratory, BIRL under contract to Wisconsin DOT, conducted extensive tests on this structure and a remote global bridge monitoring system has been installed which has continuously and reliably monitored bridge condition since July of 1995. The monitoring system uses technology that was originally developed for the automotive industry. The suitability and practicality of applying this technology to bridges was verified on several bridges in Wisconsin, Indiana, Illinois, Kentucky, and California over the past two years under experimental efforts that were sponsored by the Infrastructure Technology Institute (ITI) of Northwestern University. Two rugged field computers mounted near the critical areas of the bridge on opposite sides of the shipping channel are networked by means of a spread spectrum radio link to a host computer that is installed in the bridge tenders office. The host computer is connected to a dedicated telephone line to allow remote access. The field computers and their radio links along with the necessary software were manufactured by Somat Corporation of Champaign, IL. The deteriorated areas in the track castings are being monitored with strain gages and the response of the bridge piers are being monitored with clinometers to detect tilt in the piers during lift cycles. A total of six sensors are being utilized at each test site. The data are continuously recorded and analyzed in several different analysis modes and the results are stored on the host computer and made available for later plotting and display. Additionally, real time data monitoring is provided to allow assessment of individual lift cycles. The latter mode which includes simultaneous data from any 4 sensors in a particular test site, can be observed using a computer equipped with a fast modem (14.4 KB) from any remote site equipped with a telephone connection.

This installation, which uses off-the-shelf technology is a successful field proven prototype for remote global bridge monitoring. The application to bridges has been proven and is ready for application to other bridges that require close monitoring to insure safe operation during life extension.

Session 3

C3—ULTRASONIC MEASUREMENT TECHNIQUES

S3—NDE APPLICATION

C3A—IMPACT-ECHO OF CONCRETE

S3A—INFRASTRUCTURE ASSESSMENT
AND MANAGEMENT

C3B—GLOBAL MONITORING OF STRUCTURES

S3B—INFRASTRUCTURE ASSESSMENT
AND MANAGEMENT

C3C—FIBERSCOPE AND NUCLEAR GAGE INSPECTION

S3C—NEW NDT TECHNIQUES

Nondestructive Evaluation of Timber Pile Length

R. W. ANTHONY

ABSTRACT

Knowledge of the pile length is a vital component in calculating the scour resistance of bridges. A nondestructive evaluation technique based on longitudinal stress wave propagation has been developed to determine the length of timber piles. A total of 33 piles from different bridge sites were selected for equipment evaluation and verification testing of the pile length technique. The piles were evaluated by the stress wave technique, then compared to construction records for verification of length. The stress wave technique has been proven to reliably estimate pile lengths between 20 and 60 ft with an accuracy of ±15 percent.

BACKGROUND

Knowledge of pile length is a vital component in calculating the scour resistance of a bridge. However, records of timber pile lengths may, in many cases, be nonexistent or incomplete due to the construction practices for timber piles. Piles are typically driven until they reach a predetermined resistance, then trimmed at the end to provide a solid, level surface for substructure construction. Records that specify initial pile length, depth of driven pile or length of pile trim are often unavailable, especially in older structures. Thus, it is difficult to obtain timber pile length data for scour evaluations.

The timber pile length determination technique is based on longitudinal stress wave propagation (Engineering Data Management, 1992). Stress waves, produced from a hammer impact, travel along the length of a pile and are reflected at boundaries until dissipation of the impact energy. The stress waves travel at a velocity that is dependent on the pile density, moisture content and material quality.

Pile length can be evaluated by measuring the time required for the stress wave to travel to the base of the pile and return. The reflection time is related to the resonant frequency of the pile. The measurement of reflection time or resonant frequency and stress wave velocity enable the calculation of pile length.

The three components of the data collection process for the pile length technique are the excitation source for inducing the stress wave, sensors for measuring the pile response and a

Ronald W. Anthony, Engineering Data Management, Inc., 2301 Research Boulevard, Suite 110, Fort Collins, CO 80526

data acquisition system for recording and processing the stress wave data. For the determination of pile length it is crucial that the induced stress wave propagates primarily along the longitudinal axis of the pile, avoiding transverse excitation of the pile as much as possible. Moreover, the induced stress wave must be of sufficient energy to travel the length of the pile and back.

After evaluating several approaches, it was determined that a direct impact to the pile induces the most desirable stress wave. However, since the end of the pile is generally not accessible, the impact must occur below the pile cap through an attachment on the side of the pile. The attachment (a lag screw inserted at 45 degrees) is positioned such that the operator can easily swing a hammer without the interference of the bridge stringers or other components near the pile cap.

Stress waves induced from the hammer impact travel through the pile and are reflected at the ends of the pile. From the information recorded by the two sensors, located at a fixed distance apart, an estimate of stress wave velocity is obtained. Pile length is calculated from stress wave velocity and the reflection time. Reflection time is the time taken for the stress wave to travel from the impact point to the bottom of the pile and back.

The resonant frequency of the pile is the inverse of the reflection time required for the stress wave to travel twice the length of the pile. The resonant frequency of a pile is related to pile length and stress wave velocity. Thus, the length of the pile can also be estimated using the resonant frequency of the piles

EVALUATION OF THE PILE LENGTH NDE TECHNIQUE

A total of 33 piles from different bridge sites were selected for equipment evaluation and verification testing of the pile length technique. The bridge sites were in four different states. Table I lists the number of piles tested, the age and condition of the piles for the different sites. The Colorado piles were used during the initial stages of the project to evaluate multiple sensor configurations and hammer impact methods. The Louisiana, Tennessee and Minnesota bridge sites each contained piles of various lengths which had been recorded during construction. The piles were evaluated by the stress wave technique, then compared to construction records for verification of length.

TABLE I – TEST SITES

Location of Test Site	Number of Piles	Age of Piles (Yrs.)	Condition of Piles
Colorado	13	46 to 59	Decayed to Good
Louisiana	9	0 or Unknown	Good to New
Tennessee	7	0	New
Minnesota	4	Unknown	Good

Pile lengths from Colorado and Louisiana were estimated using only the resonant frequency method whereas piles from Tennessee and Minnesota were evaluated using both the resonant frequency method and the reflection time method.

The advantage of using both the resonant frequency and the reflection time is that it is possible to evaluate and predict lengths for a greater percentage of piles tested. There are some piles for which an estimate of the length cannot be obtained using resonant frequency because of difficulties in interpreting the frequency record. In many such cases, it is possible to obtain an estimate of the length using the reflection time. In general, reflection time method provides a more accurate estimate for the length of piles.

As an example, a frequency record that is difficult to interpret is shown in Figure 1. This frequency record has only one dominant peak (at approximately 500 Hz). It is difficult to obtain an estimate of the resonant frequency from a single peak because spacing between peaks is used as a measure of the resonant frequency. For the same pile, a time domain record of the stress wave and its representation after filtering are shown in Figure 2. From Figure 2(b), an estimate of the reflection time is obtained as the time between the two marked points. Using reflection time, the length of this pile was estimated to be 30 ft. The actual length, obtained from the construction records, is 31 ft.

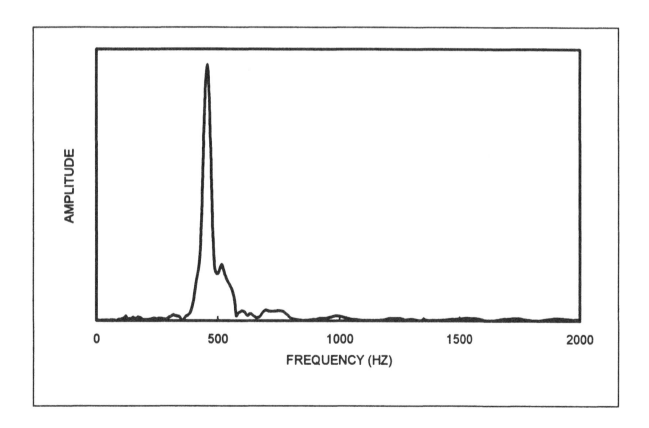

Figure 1.   Frequency Record for a 30 Ft. Pile.

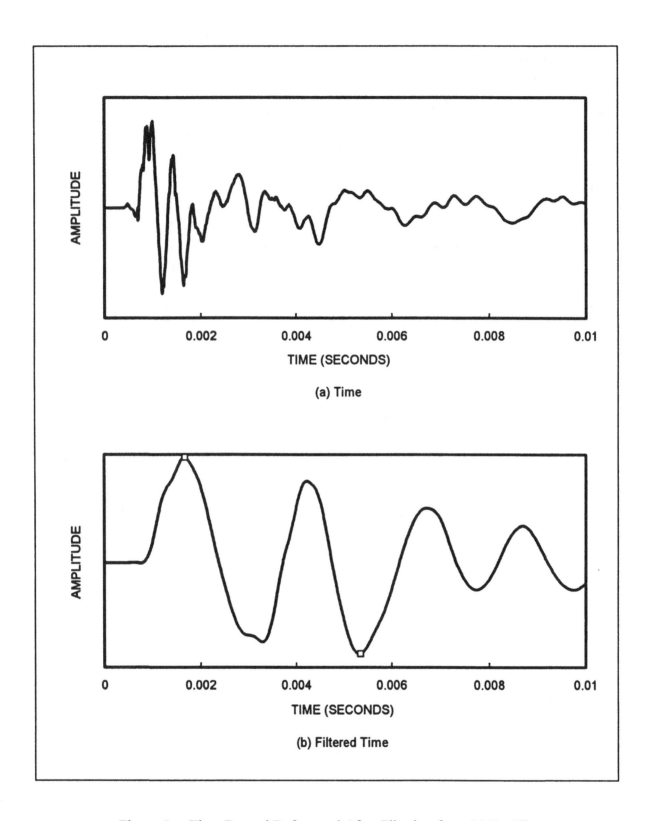

Figure 2. Time Record Before and After Filtering for a 30 Ft. Pile.

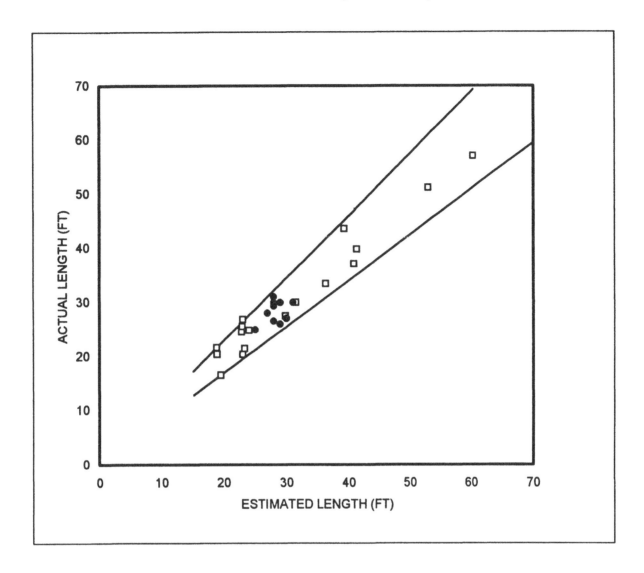

Figure 3. Plot of Actual and Estimated Pile Lengths for 33 Piles of Known Length in Table I.

Figure 3 depicts the relationship between the actual and estimated pile length for the 33 piles in Table 1. Points on the figure marked with a filled circle represents the piles from Tennessee and Minnesota. These piles were evaluated using both resonant frequency and reflection time. As shown in Figure 3, the range of pile length estimates fall within ± 15 percent of the actual pile length as defined by the two lines. For cases requiring conservative length estimates, pile length can be adjusted by reducing the estimated length up to 15 percent.

In addition to the piles discussed above, tests have been performed on a bridge scheduled for demolition in Colorado. NDE data on 60 piles were collected for predicting unknown pile lengths. Using the pile length technique, it was possible to obtain an estimate of length for most of the piles tested. These data will be used to verify the accuracy of the technique. Results from this Colorado project will be available in 1996.

SUMMARY

A reliable NDE method to evaluate the length of timber piles has been developed. Field-rugged impact methods and sensor attachments are available which are simple to install and use. The stress wave technique has been proven to reliably estimate pile lengths between 20 and 60 ft with an accuracy of $\pm$ 15 percent. Engineers and maintenance personnel who require knowledge of pile length to evaluate the effects of scour on pile capacity will find this technique invaluable for acquiring previously unavailable information.

REFERENCES

Engineering Data Management, Inc. 1992. "Determination of Timber Pile Length Using Stress Waves." Report prepared for the Timber Bridge Initiative Special Projects Program, Timber Bridge Information Resource Center, Morgantown, West Virginia.

Impulse Response Testing to Detect Cracks in Bridge Eyebars

M. RAGBAVENDRACHAR

ABSTRACT

This paper discusses a nondestructive procedure that is applied to detect a crack in bridge eyebar specimens. The nondestructive testing procedure that is used is known as the multireference impact testing method and involves studying changes in the modal characteristic of the test specimens due to cracking. Four identical eyebars specimens are tested before and after a crack has been introduced in the test specimens. The resulting reposes obtained from impact testing have been analyzed to identify the presence of the crack in these specimens.

INTRODUCTION

Since the collapse of the Silver Bridge (West Virginia) in 1967, the Federal Highway Administration has established more rigorous guidelines for bridge inspection. The collapse of the Silver Bridge, a truss-type bridge, has been attributed to a small crack in one of the eyebars which grew to a critical length leading to a brittle failure of the eyebar. Due to a lack of redundancy in load path, failure of an eyebar resulted in the collapse of the Silver bridge

In California as well as in other states, there are a large number of truss-type bridges comprised of eyebars. In some of these bridges which are non-redundant, the existing eyebars have been supplemented with additional eyebars to increase the redundancy of the eyebar chain/group. Typically, these eyebars are visually inspected for corrosion as well as for the presence of cracks. However, because of the presence of debris, accessibility problems and the "sandwich" type arrangement of multiple eyebars, visual inspection alone cannot reliably identify the presence of cracks in an eyebar.

In addition to visual inspection, other nondestructive techniques may also used for eyebar inspection. These techniques include ultrasonic testing, radiographic testing and acoustic emission techniques. Some of the drawbacks in these techniques include difficulties in mounting transducers on curved surfaces, difficulties in differentiating between a fatigue crack and the more innocuous corrosion pits as well as in interpreting

Madwesh Ragbavendrachar, Bridge Engineer, Division of Structures (OSMI), Caltrans, P.O. Box 942874, Sacramento, CA 94274-0001 results that may be affected by corrosion debris in the eye of the eyebar. In order to overcome these drawbacks and facilitate more reliable eyebar inspection, this research was developed to investigate the applicability of impact testing as a tool to detect the presence of a crack in an eyebar. A pilot research was conducted on 1/3 scale eyebar specimens in the Materials Testing Laboratory at the Division of New Technology, Materials and Research in Caltrans.

OBJECTIVES

The primary objective of this research was to determine if a comparison of signatures of specimen eyebars obtained before and after the introduction of a crack would reveal the induced damage. Signatures in the form of Frequency Response Functions (FRF) obtained by multireference impact testing were compared in this study. In addition, the study also addressed the following issues:

1. Whether undamaged eyebars which are geometrically identical (within the levels of acceptable tolerances), would exhibit identical modal characteristics (FRF). This study would establish the baseline for variations in FRF that can be expected when identical specimens are tested at different times.

2. Whether the effects of a crack on FRF would be different from those due to corrosion pits, in order that this method can be used to effectively distinguish between a corrosion pit and a crack in an eyebar.

RESEARCH METHODOLOGY

Multireference Impact Testing: In multireference impact testing, an impulse force is applied at a specified location of the test specimen with an instrumented hammer. The resulting acceleration is measured at two or more locations (measurement locations). The applied force and the measured acceleration response are recorded by a signal analyzer and transformed into FRFs. An FRF is a ratio of the measured response to the force applied. FRFs are regarded as signatures of the test specimen since they can be processed to yield modal characteristics (frequencies, damping and mode shapes) of the specimen.

Test Specimen: Four identical test specimens (within acceptable deviation in geometry) labeled as specimen 'A', 'B', 'C' and 'D' were fabricated in the laboratory. Each of these specimens, which was 14.625 in. long with 3 in. dia. heads at both ends, was fabricated from a single sheet of mild steel which was 7/16 in. thick. The diameter of the eye (to accommodate a pin) as well as the width of the "stem" of the eyebar was 1.2 in. Each specimen was discretized into 13 points. A crack was introduced in the eye of the eyebar using the Electrically Discharged Machining (EDM) process. The crack thickness was 0.005 in. for all the tests while its length varied from 1/32 in. to 1/8 in.

Test Hardware and Software: The test hardware included a PCB 086B03 impulse hammer, PCB303A02 accelerometers and a HP 3566A signal analyzer and data acquisition system. The analyzer was interfaced with a DOS based PC and the data acquisition was controlled through HP data acquisition software that operates in the Windows environment.

Test Set-up: The test specimen was mounted on a MTS testing machine in the laboratory and a 5000 lb axial tensile force was applied and maintained for the duration

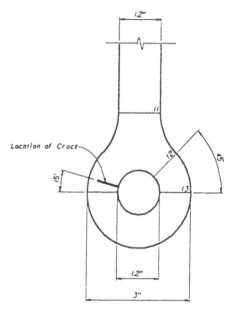

Figure 1. Test Set-up

of the test. The acceleration responses due to applied impact were measured at three locations by mounting accelerometers at locations 13, 11 and 8. Fig. 1 shows the test set-up with the inset showing the details of the crack location.

TEST PROCEDURE

The first series of impact tests were conducted on undamaged test specimens. Impact was first applied at location #1 and both the impulse force and the resulting acceleration responses measured at locations #13, #11and #8 were recorded by the HP data acquisition system and transformed to FRF. This procedure was repeated by applying impacts at all the remaining measurement locations. In order to minimize random errors, an average of six FRF at each location was used in this study. These sets of FRFs were termed as "Reference FRF".

Reference FRFs were obtained for each specimen and mutually compared to determine the minimum deviation in FRF characteristics that may be expected when FRFs of identical specimens are compared. The comparison primarily focused on the location of the peaks of the imaginary component of the FRF. These peaks correspond to the natural frequency of the test specimen.

Fig. 2 shows a typical comparison of reference FRFs obtained from response at location #11 due to impulse applied at location #11 for specimens A, B and D.

The second series of impact tests were conducted after the introduction of a 1/32 in. long crack in the eye of the eyebar specimen B. After obtaining the FRFs, the length of the crack was increased to 1/16 in. and a third series of impact tests was conducted on specimen B. Fig 3(a) shows a comparison of reference FRF and FRF of the cracked specimen B (1/32 in. 1/16 in. cracks) obtained from response at location #11 due to impact at the same location. Fig. 3(b) shows a similar comparison for FRF obtained from response at #11 due to impact at #13.

The fourth series of tests were conducted on Specimen D in which pitting damage was introduced in the eyebar. To simulate corrosion pitting, several small chisel marks were introduced with a round-tip chisel. Fig. 4 shows a typical comparisons of measured FRF of such a specimen.

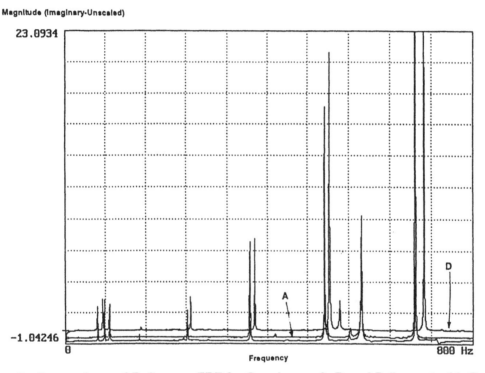

Figure 2. Comparison of Reference FRF for Specimens A, B and D (Impact - 11; Resp - 11)

RESULTS AND DISCUSSION

Analysis of Fig. 2 reveals that the peaks of undamaged specimens A and B occur at nearly identical frequencies. The peaks in the FRF for specimen D occur at a slightly higher frequency with reference to the corresponding peaks for specimens A and B. It may be noted that this difference in the frequency of the peaks for specimen D is consistent for all the modes of vibration. Therefore, it is clear that the undamaged specimens A, C and D exhibit nearly identical modal characteristics in the undamaged condition.

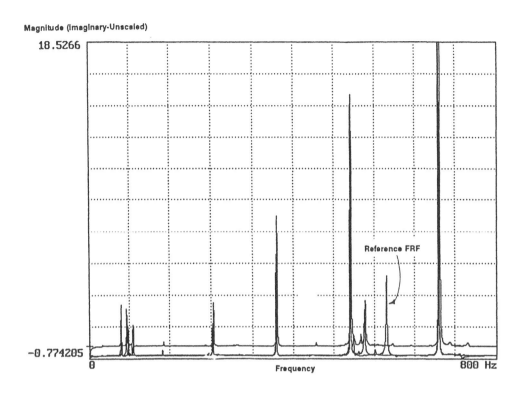

(a) Impact at 11, Response at 11

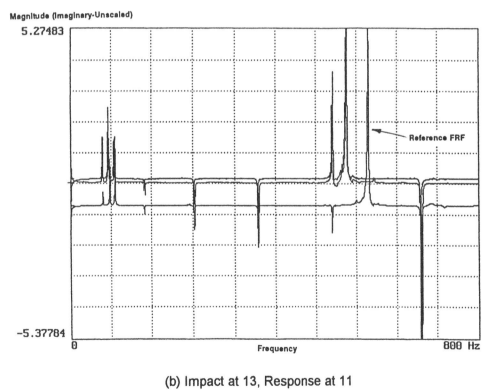

(b) Impact at 13, Response at 11

Figure 3. Comparison of Reference FRF and FRF of Cracked Spec. B (Impact- 11, Resp- 11)

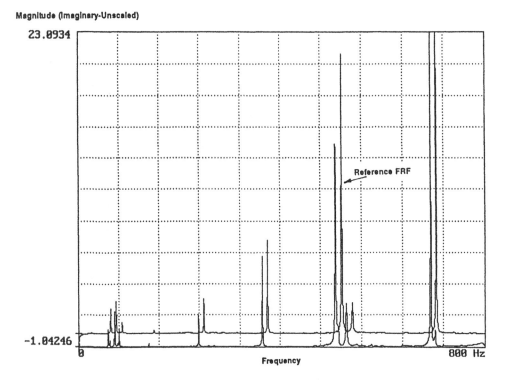

Figure 4. Comparison of Reference FRF and FRF with Pitting (Spec C; Impact 11; Resp- 11)

A study of Fig. 3(a) reveals that the introduction of a crack resulted in an appreciable shift in the peak at 580 Hz. However, there was no discernible shift in the frequencies of other peaks in the 0-800 Hz frequency span. An identical trend is observed in Fig. 3(b). This clearly illustrates that the introduction of a crack affects only a specific
mode without significantly affecting other modes of vibration. It may also be observed that the crack size did not have a significant effect on the peaks of the FRF. It is noted that the FRF obtained from responses measured at location #11 were more sensitive to the induced damage, particularly with reference to impact applied at location 10, 11 and 13. In general, FRFs obtained from responses at location #8 were not sensitive to damage.

Fig 4 shows a comparison of reference FRF and FRF for specimen C with pitting damage. It may be seen that the introduction of pitting uniformly affected all the modes of vibration in contrast to the results of introducing a crack.

CONCLUSIONS

This research clearly demonstrates that impact testing was successful in identifying a crack in a specimen eyebar. Furthermore, the results of this study reveal that the effects of a crack on FRF are significantly different from those due to pitting. Therefore, this method may also be used to distinguish between a crack and a corrosion pit. Hence, multireference impact testing has the potential to be developed into a tool for a rigorous eyebar inspection.

Ultrasonic Measurement of Applied Stress in Steel Highway Bridges

E. A. MANDRACCHIA

ABSTRACT

Many of the nation's steel highway bridges have been classified as structurally deficient. To ensure their safe and reliable operation an accurate and cost effective measurement technology is required. This paper examines the applicability of an alternative ultrasonic measurement technique. Laboratory testing on mild steel structural members verifies linearity of the acoustic measurement technique relative to surface strain as measured by a strain gauge ($r^2 = 0.9984$). Field tests conducted on three highway bridges support the results obtained during laboratory testing. Good correlation was achieved between surface stress and acoustic measurements for the three bridges examined in this study. Furthermore, measurement speed and resolution proved to be adequate for monitoring bridge loads on rough, pitted, and painted surfaces. The ultrasonic technique examined in this study is comparable in performance to the traditional strain gauge measurement technology, and offers a portable, cost effective method for monitoring bridge loads and other types of structural loading.

BACKGROUND

Currently, fatigue and load monitoring is accomplished by bonding an array of strain gauges to a structural member. This technique is inherently labor intensive, time consuming, and inadequate for long-term or continuous monitoring. Other nondestructive measurement techniques include x-ray and neutron diffraction, both lack portability and are limited to the measurement of stress at or near the structural member's surface. Ultrasonic measurement of stress offers the ability to characterize near surface, as well as bulk material properties in a portable, inexpensive manner.

The basis of the ultrasonic measurement technology examined in this paper originates in the application of acoustic physics to the measurement of stress. This phenomenon is known as the acoustoelastic effect and applies to all acoustic waves traveling within a material. A detailed theoretical description of the acoustoelastic effect is outside the scope of this presentation (Hughes and Kelly, 1953).

E.A. Mandracchia, Sonic Force Corporation, Burlingame, CA 94010

Acoustoelasticity defines the relationship between sound velocity, V, and stress, σ, to be of the form (Buck, 1990)

$$V = V_0 + S\sigma \tag{1}$$

where V_0 is the stress free velocity of sound and S dependents on a combination of both second and third order elastic constants.

The measurement technique examined here involves precisely determining the Time-of-Flight (TOF) of an acoustic wave between a pair (transmitter and receiver) of transducers. The acoustic wave mode chosen for this study is the Rayleigh wave (RW). Rayleigh waves are primarily a surface phenomena and are therefore geometrically bounded near the measurement surface. This wave mode allows for the direct comparison between the acoustic and traditional strain gauge based measurement techniques. Rayleigh waves can also be generated and detected through paint as well as operate on rough, pitted and rusted surfaces. Furthermore, changes in the TOF due to acoustoelasticity can be enhanced by allowing the transducers to move freely relative to each other on the measurement surface. Straining of the specimen changes the distance between the two acoustic transducers affecting the acoustic waves TOF. This "strain" effect enhances the sensitivity of the measurement (Fuchs, P.A. and A.V. Clark, 1995).

MEASUREMENT STRATEGY

The induction of acoustic energy into the structural member is accomplished using ElectroMagnetic Acoustic Transducers (EMATs). EMAT operation is based on direct electromagnetic coupling of acoustic energy into the structural member under examination. Essentially, EMATs utilize magnetic coupling between an external coil and induced eddy current within the surface layer of a metal. Within this surface layer electrons are constantly in collision with the ions and consequently transferring momentum to the atomic lattice via electron-to-ion collisions. In the presence of a static "biasing" magnetic field the motion of the eddy current and resulting electron-to-ion collisions are determined by the Lorentz Force. This body force generates ultrasonic energy that propagates equally in all directions. Consequently, the direction of both the induced eddy currents and the biasing magnetic field determine the possible elastic modes that can be generated including longitudinal, bulk-shear, Rayleigh surface, and plate waves. In the case of this study all testing was conducted using 750kHz Rayleigh wave EMATs.

Basic operation of the measurement system is managed by an internal microcontroller which accepts operational instructions via an external computer-based control system. An individual measurement is initiated when the Microcontroller issues a start pulse to the Signal Processor. The Signal Processor sends an ultrasonic tone burst to the Transmit EMAT and simultaneously starts a Counter/Timer to begin the Time-of-Flight measurement. The induced acoustic wave traverses the structural member where it is detected by the receive EMAT. The ultrasonic burst is then amplified and sent to the Signal Processor terminating the Time-of-Flight measurement.

Individual TOF measurements are summed and time averaged. The number of TOF measurements averaged improves the effective Signal-to-Noise ratio of the measurement by a factor of $(N)^{1/2}$, where N is the number of time averaged measurements. For the data presented in this paper the number of time averaged measurements are N = 300 at a measurement rate of 1200 Hz. This results in 4 complete measurements per second, or an overall measurement rate of 4 Hz. The basic tradeoff made when increasing the time averaging sample size is measurement speed. Once an adequate number of TOF samples have been accumulated the average is read by the Microcontroller, transfered to a laptop PC via an RS-232 interface, displayed in real-time, and magnetically stored on diskette in ASCII format for post-processing.

LABORATORY TEST RESULTS

Laboratory tests were conducted on various structural members to verify linearity of the acoustic measurement technique. A variety of structural specimens (Plate, Bar, I-Beam, L-Beam, etc.) were placed in an Instron force application instrument capable of applying 66,000 lbs and subjected to bending, compressive and tensile stresses. Strain gauges were placed on the surface of each test specimen allowing direct comparison of the two measurement techniques. Figure 1 illustrates the measured variation of the Rayleigh wave's TOF for various applied loads on a L6 x 4 mild steel angle beam.

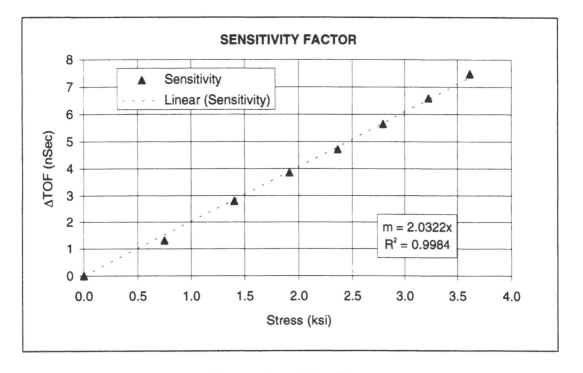

Figure 1: Sensitivity Factor

As shown, the data exhibits a high degree of correlation ($r^2 = 0.9984$) between the acoustic wave's TOF and the surface stress as measured by the strain gauge. Additionally, if we convert the measured strain to surface stress a sensitivity factor, m, can be defined by the expression:

$$m = \Delta(TOF) / \Delta\sigma \qquad (2)$$

where $\Delta(TOF)$ is the change in the acoustic wave's Time-of-Flight and $\Delta\sigma$ is the change in surface stress. This relationship defines the sensitivity factor used to calibrate the acoustic measurement. For the laboratory data illustrated above the sensitivity factor is determined by the slope of the best fit trendline, m = 2.0322 nSec/ksi.

FIELD TEST RESULTS

With the cooperation of Caltrans a field test program was conducted on the Pioneer bridge located on I-5 near Sacramento, CA. The superstructure of the bridge consists of a reinforced concrete deck supported by four steel plate girders. Transverse diaphragm details include intermediate cross frames and lateral bracing.

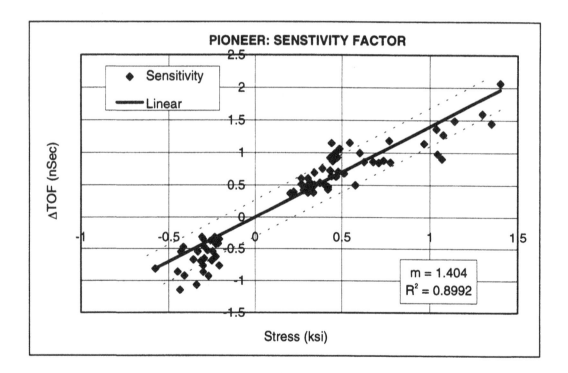

Figure 2: Pioneer Bridge Sensitivity Factor

The acoustic measurements were recorded by placing transmit and receive EMATs on either side of a strain gauge that was installed by SonicForce personnel on a simply supported A-36 I-Beam girder. EMATs recorded acoustic measurements on the tension (bottom) side of the I-beam at midspan, and were aligned in an axial direction parallel to the direction of maximum stress. Both acoustic TOF and strain gauge measurements were displayed and stored simultaneously during typical mid-day traffic. Figure 2 illustrates the relationship between the acoustic TOF and strain measurements over a 10 minute period. The sensitivity factor for this data can be determined by using the same best fit trendline technique as use on the laboratory test data. As shown, the sensitivity factor for this 10 minute period is 1.37 nSec/ksi.

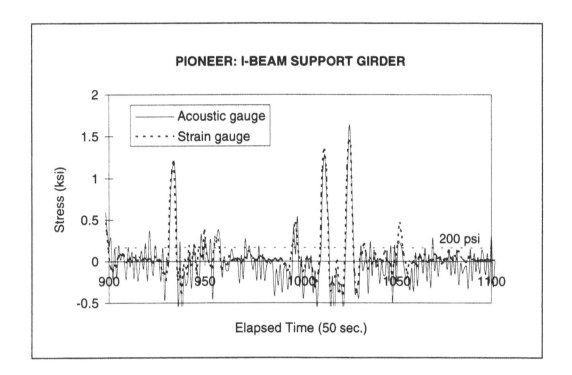

Figure 3: Pioneer Bridge

Illustrated in Figure 3 is the resulting loading history when the sensitivity factor of 1.37 nSec/ksi is used to calibrate the acoustic TOF measurement. As shown, three substantial loading impulses occurred during the 50 second period. The magnitude of these events were approximately 1.3, 1.5, and 1.6 ksi, respectively. The sensitivity factor of 1.37 nSec/ksi yields an acoustic-to-strain measurement error of approximately 200 psi.

Additional field tests were conducted on the Bear Creek (36C-0084) and Rodeo Gulch (36C-0042) bridges with the cooperation of the Santa Cruz County Department of Public Works. The Bear Creek bridge is located at Highway 9 over the San Lorenzo river less than a mile outside the town of Boulder Creek. The superstructure of the bridge consists of concrete deck on continuous steel I-beam girders. The Rodeo Gulch bridge is located approximately 3 miles Northeast of Capitola on Soquel Ave. The superstructure is similar to that of the Bear Creek bridge.

The test procedure and data analysis were identical to that used during the Pioneer bridge test program. The acoustic-to-strain best fit trendline was used to determine the sensitivity factors for each test site. These sensitivity factors were then applied to the recorded data for both locations. Illustrated in Figure 4 is the resulting load history for the Bear Creek bridge comparing both strain and acoustic measurement techniques. The best fit trendline analysis for both locations yielded sensitivity factors of 1.25 nSec/ksi and 1.75 nSec/ksi, respectively.

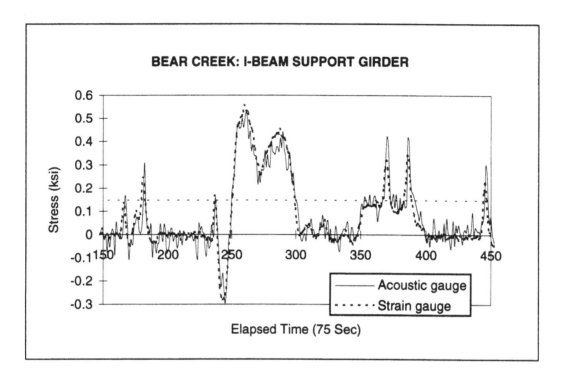

Figure 4: Bear Creek

CONCLUSION

A new ultrasonic measurement technique for measuring stress in structural members has been successfully demonstrated. The results of laboratory testing verify the linearity of the acoustic measurement technique. Field testing conducted at three highway bridge test sites supports results obtained during laboratory testing. Good correlation was achieved between surface stress and acoustic measurements for the three bridges examined. Furthermore, measurement speed and resolution proved to be adequate for monitoring bridge loads on rough, pitted, and painted surfaces. In summary, the ultrasonic technique examined in this paper appears to offer cost effectiveness and portability with sufficient measurement speed and resolution.

REFERENCES

Buck, O. , 1990. "Nonlinear Acoustic Properties of Structural Materials - A Review," *Review of Progess in Quantitative Nondestructive Evaluation*, Vol. 9, pp. 1677-1684.

Fuchs, P.A. and A.V. Clark, 1995. "Monitoring Bridge Fatique Loads with Ultrasonic Transducers," *Sensors*, Vol. 12.

Hughes, D.S. and J. Kelly, 1953. *Phys. Review*, Vol. 92, pp. 1145.

Privatization versus In-House State Forces for Structural Inspections

R. J. SCANCELLA

ABSTRACT

This paper presents the findings of the New Jersey Department of Transportation (NJDOT), a State agency, which has been reviewed for its efficiency and cost/benefits as compared to private inspection firms that perform the same tasks. This paper not only compares the cost of inspection services and testing operations, but also considers the responsibility of State agencies to ensure a quality built infrastructure.

INTRODUCTION

Because public scrutinizing of State agencies has become commonplace, it is imperative that high costs be avoided. Private industry has boasted about their cost effective operations, but is this claim valid when compared to the State's cost for the same operation? Should the government set cost effectiveness as part of the standard to which public services are measured?

The State of New Jersey, to promote cost benefits to the State's taxpayers, has initiated a program entitled " The Government that Works". A committee comprised of management level personnel from various State agencies continue to review all aspects of New Jersey government in an attempt to streamline departments and save money. This committee posed the question of whether to privatize material inspection and testing.

It is noted that although all in-house material testing and inspection was compared to privatization, this paper is only concerned with the Inspection Unit of the Bureau of Materials Engineering and Testing of the New Jersey Department of Transportation. Since the Bureau of Materials Engineering and Testing is comprised of two distinct areas (Testing and Inspection/Engineering) the original review was performed in two parts. The Inspection Unit performs the non-destructive testing for the approval of newly fabricated ferrous and non-ferrous structural items.

Robert J. Scancella, PE, New Jersey Department of Transportation, Project Engineer Materials, 1035 Parkway Ave., CN-600, Trenton, NJ 08625

This paper will briefly explore the elements that can only be obtained from an in-house unit. An example of these elements is a conference such as the NDT conference. Private Inspection/ Consultant companies attend and participate in these conferences, but the companies are not organizationally set up to perform various extra tasks.

GOAL

The goal of the New Jersey Department of Transportation, Bureau of Materials Engineering and Testing is to provide inspection, testing, and quality assurance of all materials associated with its roads and structures. This goal is to be accomplished at the lowest possible cost with the maximum benefit to the taxpayer based on the specifications set by the Governor's committee.

Private inspection/testing services have claimed that they can save money for both the taxpayer and the State because they can perform these tasks more efficiently. Optimum inspection is defined in terms of quality inspection and inspection forces.

SOLUTION

In order to achieve the required goal, it is necessary to determine the cost of the individual inspections compared to the actual cost of the Department of Transportation inspection activities. A simple comparison of inspection costs would fulfill only the minimum, therefore, further investigation is needed for optimal benefit. In addition, the inspection ability of the personnel performing the inspection must also be explored. The "Biggest Bang for the Smallest Buck" technique will be the final criteria in determining how taxpayer money is spent.

COST ANALYSIS

The following tables compare the actual cost of the New Jersey Department of Transportation inspection forces in performing inspection of only structural metals and concrete items. The final report to the "Government that Works" committee included all inspection tasks. The numbers have been rounded to the nearest $0.25 and the in-house costs also include a 7% increase for increased Social Security costs as of July 1995. The tables of private company costs show the minimum, average and maximum costs of ten companies that tendered bids from 1989 and 1992 on inspection services to State and non-State agencies in New Jersey, New York and Pennsylvania. Non-State agencies are considered to be organizations where user pay is the main funding versus a State agency where tax dollars are budgeted by the government.

NEW JERSEY DEPARTMENT OF TRANSPORTATION INSPECTION FORCES

Inspection Item	Minimum Day	Daily Rate	Overtime Rate	Minimum Sat, Sun, Hol	Daily Sat, Sun, Holiday
Plant Inspection Prestressed Concrete	$108.25	$213.50	$34.00	$139.25	$275.25
Plant Verification Prestress Concrete	$108.25	$213.50	$34.00	$139.25	$275.25
Inspection of Structural Steel, Aluminum and other Metals	$107.50	$214.75	$34.75	$139.00	$277.75
Field X-Ray	$320.50	$53.75/HR			
Field Radiographic (E-99 Cast.)	$320.50	$53.75/HR			
Field Ultrasonic	$142.50	$27.00			
Field Dye Penetrant	$158.75	$27.00			
Field Mag Particle	$182.50	$27.00			
Travel Reimburse-ments Mileage Per Diem	$0.25/mile In accordance with Federal Register				

PRIVATE COMPANIES

Inspection Item	Minimum Day	Daily Rate	Overtime Rate	Minimum Sat, Sun, Hol	Daily Sat, Sun, Holiday	Low High Avg. Cost
Plant Inspection Prestressed Concrete	$150.00 $170.00 $156.00	$200.00 $250.00 $221.00	$35.00 $35.00 $35.00	$160.00 $190.00 $177.75	$240.00 $300.00 $288.50	Low High Avg. Cost
Plant Verification Prestress Concrete	$140.00 $170.00 $156.25	$200.00 $320.00 $245.75	$30.00 $40.00 $38.25	$170.00 $190.00 $176.75	$240.00 $300.00 $288.50	Low High Avg. Cost
Inspection of Structural Steel, Aluminum and other Metals	$140.00 $200.00 $171.00	$200.00 $300.00 $246.75	$35.00 $40.00 $38.00	$160.00 $200.00 $187.50	$240.00 $330.00 $303.25	Low High Avg. Cost
Field X-Ray	$360.00 $700.00 $502.25	$75.00/HR $100.00/HR $87.75/HR				Low High Avg. Cost
Field Radiographic (E-99 Cast.)	$360.00 $700.00 $502.25	$75.00/HR $100.00/HR $87.75/HR				Low High Avg. Cost
Field Ultrasonic	$200.00 $350.00 $309.25	$35.00/HR $84.00/HR $46.00/HR				Low High Avg. Cost

Inspection Item	Minimum Day	Daily Rate	Overtime Rate	Minimum Sat, Sun, Hol	Daily Sat, Sun, Holiday	
Field Dye Penetrant	$200.00 $350.00 $281.50	$35.00/HR $50.00/HR $45.00/HR				Low High Avg. Cost
Field Mag Particle	$200.00 $350.00 $281.50	$35.00/HR $50.00/HR $45.00/HR				Low High Avg. Cost

Note: There is no comparison shown for the reimbursement of travel expenses because State employees must follow regulations set forth by the State governing agency.

INSPECTION CAPABILITIES

All NJDOT inspectors are required to be certified in many aspects of construction inspection before they can perform any inspection tasks. Once the inspectors have been trained and certified to the minimum requirements they must continue to maintain and increase their knowledge in order to perform more intricate inspection tasks independently. The inspectors must ultimately obtain the following certifications in order to be assigned to a project as a lead inspector: ASNT Level II for Dye Penetrant, Magnetic Particle, Ultrasonic Testing, and Radiographic Testing. In addition CWI certification in the structural metal area and ACI, PCI, and ASNT Level I in all areas for the Prestress/ Post Tension Concrete area are needed. The NJDOT also encourages the inspectors to obtain continuing education certifications. The Department has set a criteria for obtaining these certifications based upon industry standards, classroom instruction, experience time, and written/practical tests. To remain certified the inspector must meet the re-certification requirements of time, formal training, and testing together with a recommendation from the immediate supervisor. The re-certification is completed every three years.

The minimum level requirements for private companies were not able to be established. Private companies place emphases on the initial certifications and little or no emphases on recertification or continued education. Private companies employ only a minimum number of inspectors to cover the fabricator's work load.

CONCLUSION

Most societies, such as the American Concrete Institute and American Society for Non-destructive Testing, set the industry certification standards. However, the standards often reflect a combination of society requirements and private industry requirements which makes it difficult to compare State government and private industry.

Independent companies have less competition for positions/titles and their requirements for inspectors are more likely to be less stringent. Since the State also encourages training in other disciplines, the State inspector usually has a better working knowledge of materials in addition to the testing procedure. Although private companies maintain a minimum staff that meet the basic requirements of education/certification, additional staff are usually composed of short-termed employees possessing less experience. Generally, State employees exhibit more experience, training, and technical degrees than their private industry counterparts. The exception would be in private industry where research is a main objective.

Another aspect to be addressed when considering privatization is the inspector's ability to remain neutral while performing job tasks. Because the State is a non-profit organization and the wages of an employee is based solely on time worked as opposed to a private company's money making objective, it is easier for the State agencies to remain neutral regarding the inspection tasks. In addition, a conflict of interest may occur because some inspection companies are subsidiaries of other engineering or fabricating companies.

State inspectors at the job site strictly comply with the specifications and ascertain that fabricators adhere to any design or procedural change. Conversely, private companies are more likely to approve "minor" changes without design approval to maintain a smooth, working relationship with the fabricator. Therefore, structures built under the supervision of State personnel, meet the goal of a superior product requiring little or no maintenance.

Review of the cost analysis between the NJDOT and private inspection companies revealed several main points:

1.) If the NJDOT allowed private companies to perform their structural inspections, a core staff would be required for supervising private industry. The amount needed for this supervision would be approximately 22% of the inspection costs. This figure is the supervision costs for the present state inspection staff. It is anticipated that the core staff cost for supervising the privatized companies would be the same. An increase in staff may also be needed to perform random quality assurance inspections of the private companies.

2.) The cost for each of the inspection items performed by the NJDOT in 1995 were lower than all the private companies' costs for 1989 and 1992. The previous tables prove that the inspection by the NJDOT was not only lower than the average cost but was significantly lower than the lowest cost for an inspection item by a private company.

In conclusion, it is noteworthy to remember that maintaining the integrity of our structures and saving taxpayer money can best be accomplished through the use of State inspectors as the primary inspection unit for structural items. This is the best way to achieve the "Biggest Bang for the Buck"!

REFERENCES

New Jersey Department of Transportation
New Jersey Turnpike Authority
PennDOT
New York Department of Transportation
Delaware Department of Transportation
Texas Department of Transportation

SPECIAL THANKS

Henry Justus, NJDOT, Chief Bureau of Material Engineering and Testing for his continuing help, support and advice.

Rich Smetanka, Robert Skalla, Fred Lovett, Warren Cummings, and Victor Scarpinato of the NJDOT, Bureau of Material Engineering and Testing for their research on the in-house inspection costs.

Art Egan and Ric Barros, NJDOT, Bureau of Material Engineering and Testing for their review of the cost analysis charts.

Analysis–Reality and How the Two Compare

D. BRIERLEY-GREEN

ABSTRACT

Non-destructive techniques are useful in obtaining confirmation of analytical predictions. Present-day analysis using high speed computers leaves little in reserve. Never before has there been so much pressure placed on the engineer to confirm analytical behavior. The dynamic response of a structure can reveal important information both in confirmation and as a diagnostic tool when results are compared over time. This technique is a very powerful non-destructive testing method and the results of three applications on complex bridge structures will be discussed.

INTRODUCTION

On completion of their formal education and training most engineers are confident they can solve just about every analytical problem that could be encountered. This is even more true today with the increased use of fast, powerful desk top computers. The computer program produces an abundance of information and the engineer has become more remote from the problem. Never before has the engineer been better equipped to calculate the forces and stress levels within structural systems and never before has there been greater pressure on the engineer to justify his predicted behavior of a structural system. It should never be forgotten that the accuracy of the analytical result is purely dependent on the accuracy of the mathematical model and how that model captures the behavior of the real structure. The accuracy of the analysis should always be checked when dealing with the rehabilitation of large structures or when dealing with the more complex forms of construction regardless of the engineers' confidence level. Not only have computers allowed greater analytical sophistication, their development has also lead to the development of non-destructive testing equipment that is economical, less cumbersome, and easy to use. I would now like to describe field testing of three bridge structures using dynamic response evaluation to confirm predicted behavior.

David Brierley-Green, Luscombe-Green Associates 14406 Knowles road, Tenino, WA 98589

DYNAMIC RESPONSE INVESTIGATION

Dynamic response evaluation can be used not only for verification of modeling but also to determine internal changes in structural behavior possibly due to deck slab deterioration, malfunctioning of major components, or change of cable tensions in the case of cable supported structures. Evaluation records include the measurement of the natural frequencies, mode shapes, damping ratios, and impact factors. These measured values can be considered as the dynamic signature of the structural system.

Scope of work for a field investigation would be:

Task 1: Analytical Study

* Develop an analytical model using finite element modeling techniques.
* Analytically estimate the natural frequencies and associated mode shapes.
    Identify optimum locations for positioning accelerometers to best capture mode shape to be verified.

Task 2: Field Measurements
* Perform a field investigation using a spectral analyzer and an impact hammer to derive natural frequencies, associated mode shapes and estimated damping values.
* Record the response of the structure under wind loading.
* Record the response of the structure under moving truck loading.
* In a cable supported structure measure cable tensions by exciting the cables and measure their responses.

Task 3: Data Analysis and Comparison
* Compare field measured model parameters and analytical values.

    Correlation of the two sets of data may require refinement of one or both procedures. Potential dissimilarities discovered will be used as a diagnostic aid toward obtaining a credible mathematical simulation. A correlation may be established if the critical frequencies associated with analogous mode shapes are similar. A correlation of such results within 10% would be considered excellent.

Equipment

    What is needed to undertake such field investigations.? First, and probably most important, will be an abundance of confidence, a large measure of imagination, and complete commitment. The more tangible requirements are:

* Four channel spectral analyzer.
* At least four force balance accelerometer.
* Power supply.
* Lap top computer.

A schematic layout of the equipment is given in Fig. 1 which shows the setup using the accelerometers.   Fig. 2 shows the setup when using an impulse hammer.

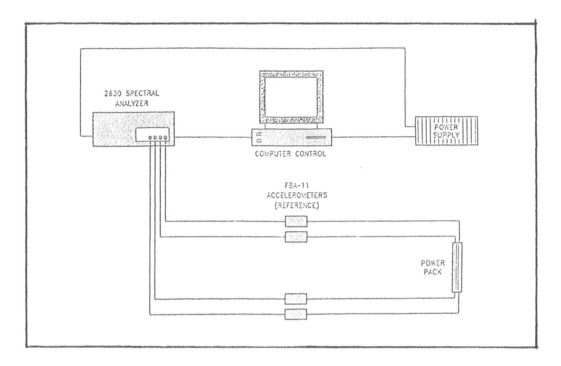

Figure 1

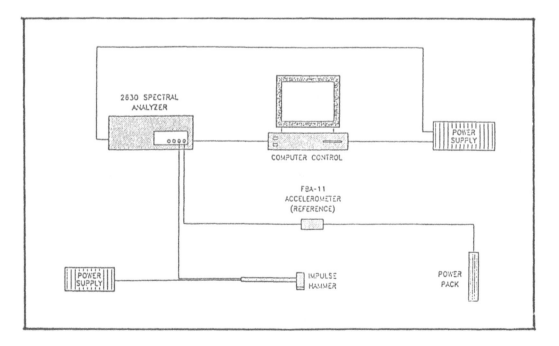

Figure 2

APPLICATION

I have used the above techniques and equipment on three bridge retrofit and seismic retrofit projects in recent years:

CAPTAIN WILLIAM MOORE CREEK BRIDGE - 1993

The Captain William Moore Creek Bridge was built in 1975. It is located on the Skagway-Klondike highway and is a very important route used predominately by tourist buses and heavy trucks transporting ore from the mines in Yukon to the Port of Skagway. The bridge, which crosses a deep rocky ravine, is comprised of two 300 feet long steel edge box girders, floor beams, timber deck, a set of cable stays, and an inclined steel pylon. The structure was highly susceptible to dynamic excitations under moderate winds and light traffic, and exhibited poor damping. As an attempt to remedy its noted weaknesses and to increase the load carrying capacity, a major bridge rehabilitation program was undertaken in the Fall of 1992.

Analytical Study

In order to obtain a quantifiable measurement of the bridge's dynamic character a dynamic response investigation program was initiated. The field investigation took place in June, 1993, just prior to the start of the rehabilitation. The analytical study comprised 3D finite element modeling to determine the natural frequencies and associated mode shapes, and the optimum location for the placing of testing equipment.

Field Work

For development of vertical mode shapes, field measurements were taken by placing an accelerometer at a reference point and impacting specified locations with the impulse hammer. Excitations of the deck due to a moderate gusty wind were also recorded to determine the lateral response of the deck to low frequency excitations, and backstay and forestay cable tensions were measured using wave velocity techniques. Ore trucks were also used to estimate the impact factors at three controlled speeds of 10, 15 and 20 miles per hour.

Results and Comparisons

Analytically derived and field measured natural frequencies and associated mode shapes within the frequency range of 1.5 to 20 Hz was found to be reasonable. The frequencies corresponding to similar mode shapes compare within 10% on average over this range. Live load impact factors derived from ore truck loading were found to be 20% for a vehicle speed of 20 MPH. This is in excess of the AASHTO designated impact factor of 12% used for the pylon and main girders design. It was clear something needed to be done to reduce impact forces. Average damping factors of 2.6% and 0.5% were derived over a frequency range of 3 to 10 Hz and 10 to 25 Hz respectively. I believe that the damping value of 2.6% is high in the lower frequency range, and a value less than 2.0% would be more appropriate. Stay cable tensions were estimated using both the measured fundamental frequencies and reflected wave velocities of the cables. These tensions were found to be in good agreement with calculated values. Wind excitations provided three torsional, one coupled bending/torsional, and two bending mode shapes within the frequency range of 1 to 6 Hz. The lateral responses compared well with calculated values and the vertical responses compared well with those derived using the impulse hammer.

TACOMA NARROWS SUSPENSION BRIDGE -1994

The second structure to be field tested was the Tacoma Narrows Suspension Bridge. A seismic vulnerability study was performed to identify seismically vulnerable components. I had already developed a 3D model for the bridge for a previous investigation. This model had been confirmed for static loading of the structure by measuring chord stress under predetermined truck loading. However, this 3D model was now required to undertake a dynamic analysis for identification of vulnerable components. In keeping with previous philosophy I felt it necessary to confirm dynamic characteristics of the bridge before confidence could be gained in any analytical predictions.

Field Tests

Significant mode shapes and associated frequencies would be determined together with an estimate of structural damping of the system when operating within the elastic range of the construction materials.

Field Measurements

Field measurements of transient vibrations commenced on the night of September 18, 1994 with measurements being taken between midnight and 5.00am Sunday morning. All measurements were taken with the bridge closed to traffic. The bridge was excited in the vertical direction by running the rear wheels of a 45 kip fully loaded gravel truck over a 4.75 inch high timber platform. This resulted in a significant bump causing the bridge to vibrate vertically. Excitation in the longitudinal direction was produced by two 45 kip gravel trucks traveling at 15 MPH and braking suddenly.

Evaluation of Transient Field Measurements

The first mode of vibration captured had a value of 0.074 Hz. This is in close agreement with the analytically derived value of 0.068 Hz. The second mode of vibration was not captured. The third mode of vibration, a predominantly vertical mode, was captured both in the side span and the center span. Both measurements show good agreement with analytical values. The fourth mode of vibration, a center span lateral mode could not be identified on the field records. The fifth and sixth modes of vibration were captured in the center span in the vertical and longitudinal directions with good agreement between field and analytical values.

Damping

The Half Power Band Width method was applied to eight frequency peaks from data collected in one side span and one center span bump test. Only peaks between zero and 2 Hz were selected as the structural damping ratio varies with frequency, and we were primarily interested in the lower frequency range. The damping ratios fluctuated between 0.3% and 3.5%. The average value over all readings was found to be 1.3%. This is a reasonable value and is probably close to the true value for a suspension bridge of this size operating within the elastic range.

LAKE KOOCANUSA BRIDGE - 1994

The third structure tested was Lake Koocanusa Bridge in Eureka, Montana. This steel truss bridge carries two lanes of traffic, has 2 central spans of 500 ft. with adjacent 400 ft. spans and end spans of approximately 300 ft.. A complete rehabilitation was undertaken to replace the lightweight concrete deck, and carry out a seismic retrofit of the structure. After building the 3D model and deriving respective dynamic responses, field testing was performed to verify the predictions. In a similar manner to the Tacoma Narrows Bridge the first, lateral, vertical and longitudinal frequencies were to be confirmed. The first vertical frequency was excited by running the front wheels of a car over a 4.75 high timber. This test was performed at mid point of the 500 ft. span. The first lateral frequency was excited by wind loading and the longitudinal frequency was excited by a fully loaded logging truck applying full braking power when traveling at 15 MPH. All field measured values compared very well with analytical derived values.

CONCLUSION

It can be said with confidence that I have learned a great deal about the structural behavior of bridges since undertaking this verification process and have gained continued confidence in the method developed for modeling large complex structures. Unless such tests are undertaken and confirmation obtained the engineer is indulging in pure guess work. There is no valid reason for not performing these tests for special structures as the equipment used is relatively inexpensive and easy to use.

A Passive Structural Health Monitoring System for Bridges

B. WESTERMO

ABSTRACT

The paper discusses the development and testing of a passive, peak strain sensor technology for use in damage assessment and health monitoring of bridges. The technology is based on the irreversible magnetic property changes that occur in a class of steel alloys when strained. This feature provides for sensors which can passively monitor the peak applied strain (or deflection) at a specific location within a structure. Results of past and ongoing test applications as well as a discussion on future instrumentation sites are presented.

THE TECHNOLOGY

There are a variety of active sensors available for structural monitoring, such as bonded strain gages, displacement sensors, accelerometers, and fiberoptic-based types. These devices are "active" in the sense that they require *constant* monitoring in order to measure the necessary features of the strain (or deflection) history required for damage assessment (especially in seismic hazard applications). A passive approach, as possible with this technology, would require fewer components and would be more reliable as a simple means of routine safety and operational inspection of highway structures. The system types discussed here are those which can be installed, as either a retrofit or into new construction, in their monitoring location and would measure changes in the response from the time of installation. Damage to a structure can take on different forms for different systems, but the final goal of the instrumentation system is to establish the structure's degree of usability, its capacity to withstand further loading, and the location of any damage detected.

The new technology for non-destructive evaluation of materials, which has been developed by Strain Monitor Systems, Inc., is based on a series of steel alloys which irreversibly transform from non-ferromagnetic to ferromagnetic behavior as a function of the applied peak strain. When employed as elements within a strain sensor they provide for a passive means of monitoring damage which is economical, simple, and reliable. Although only the peak values are assessable by this approach, this information is significant and oftentimes sufficient to define the damage state of a structure when employed in a thoughtfully designed instrumentation system. The technology is equally applicable to any situation where prevention of a load or deformation-induced failure is the concern.

A common variety of these phase transforming materials are the TRIP (**TR**ansformation **I**nduced **P**lasticity) steels (Westermo and Thompson; 1994a, 1994b, 1995). These metastable alloys (originally developed as high-strength, high-ductility steels) irreversibly transform from face-centered-cubic, austenite (non-ferromagnetic) to the body-

Bruce Westermo, Strain Monitor Systems, Inc., 5151 A Long Branch Ave., San Diego, CA 92107

centered-cubic, martensite (ferromagnetic) phase in proportion to the applied peak strain. Measurement of the relative amount of ferromagnetic material in the element, accomplished by several viable means, is used to infer the peak strain applied to the element. In this manner the material serves as a peak strain (or displacement) sensor. These alloys typically have a high ductility, thus allowing for a large measurement range of strain. The alloy composition and the material's thermomechanical processing control the strain vs. martensite content behavior and, hence, dictate the range and sensitivity of a particular alloy.

The design approach is to use sensors based on a small element of the alloy placed within mounting hardware and attached to the monitoring sites. Figure 1 shows a simple schematic for this approach. The deformations create a tensile strain in the TRIP alloy which, in turn, progressively transforms to ferromagnetic material. A small transducer (a Hall-type sensor is shown in Figure 1) attached to the element outputs a signal voltage in direct proportion to the amount of ferromagnetic material, and hence, the peak applied strain. When used in an attached gauge as a strain sensor, the measurable strain range and sensitivity can be tailored to meet the need through the attachment hardware and mounting configuration (as well as through the selection of the alloy employed in the element).

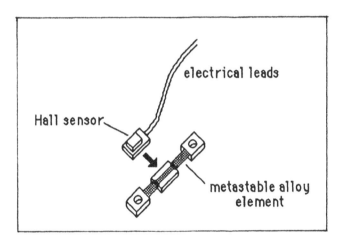

Figure 1. Schematic of Tensile Strain Sensor Components.

A variety of sensors can be developed from the use of the sensor element configuration. One such sensor type is the surface-mounted gauge, used to monitor tensile strains or deflections on planar, non-ferromagnetic surfaces, such as concrete, composites, and aluminum. A schematic of this is shown in Figure 2. The elements are epoxied, at each end, to the structural surface and monitor the peak, average strain over the element length. The Hall transducer, attached to the sensor strip, has electrical wires which are connected into an output junction or into the network junction for the system. Alternately, the transducer can be omitted and the sensor interrogated by manually placing a portable magnetometer over the sensor to take a reading. In either case, the transducer output is then compared to the laboratory calibration curves to determine the effective peak strain.

As an externally attachable sensor, an element as in Figure 1 can be mounted within a housing unit which is then attached to the structural component. Figure 3 shows just such an application in the form of a peak, tensile strain sensor for concrete reinforcing bar (shown here installed onto a length of #16 bar). The housing shown in the background fits over the gauge and provides a tight seal against the poured concrete.

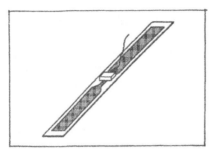

Figure 2. Surface Mount Tensile Sensor.

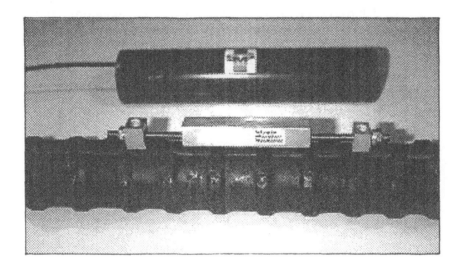

Figure 3. Reinforcing Bar Sensor and Housing (background).

As part of a series of tests, a sensor was placed on the tensile flange of a steel, box beam which was loaded in three-point bending to failure (i.e. severe flange and web buckling). Figure 4 shows the output voltage from a peak strain sensor (utilizing a Hall transducer) versus the strain at the same location measured from a bonded resistive strain gauge. The upswing in the curve at high strains is created by the geometry changes in the sensor operation due to the onset of the buckling in the beam walls.

EARTHQUAKE DAMAGE ASSESSMENT

The National Science Foundation is participating in a project directed at the monitoring of bridges and buildings for post-earthquake damage assessment. It may currently take from hours to days before a suspect structure can be inspected following an earthquake. Use of that structure in the interim is risky, and there is a need for the ability to quickly inform people of its condition and to warn emergency relief support as to problems which need attention. Simulation experiments were conducted to determine the strain sensor response characteristics for excitations produced during an earthquake. An axial test machine was programmed to reproduce a numerically calculated dynamic response at a hypothetical gauge site on the new I-5 Gavin Canyon Bridge (Los Angeles, CA). A 0.13 in. gauge length sensor element was used with a 6 in. attachment distance where the strain monitor was considered to be attached to the vertical side of a column on the bridge at a point near the deck. Figure 5 shows the peak strain output from the strain monitor superimposed on the actual strain produced by the earthquake acceleration history. Note that there are five tensile peaks produced during the earthquake, the sensor output follows the increasing strain levels and is stable between the individual peaks. An active or semi-active monitoring system would record the entire strain history as shown in the figure whereas a passive monitoring system would indicate only the maximum peak tensile strain (occuring at about 19 secs.) during the earthquake, which would correspond to the most damaging strain to the structure.

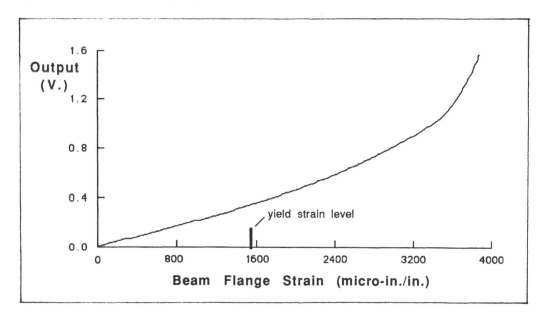

Figure 4. Sensor Transducer Output as a Function of the Peak Tensile Strain
for a Steel Beam Tested in Three-Point Bending.

As part of this ongoing program, it is planned that sensor networks will be installed onto three Caltrans bridges in 1996. These systems will have a series of peak sensors placed at critical locations on the bridges (e.g. at column connections, on reinforcing bar, and on composite wrapped retrofit columns) which will be digitally networked to an on-site microprocessor that monitors the sensors. These installations will have a cellular phone communication capability and will be able to immediately notify one of any significant changes in the peak strains or deflections as well as being open to interrogation at all times. The Federal Highway Administration is also about to begin a two year program (in conjunction with the University of Cincinnati Infrastructure Institute) for evaluation of the sensor technology for bridge health applications with installations to occur on several bridges in Ohio.

COMPOSITE MATERIALS TESTS

The technology has been employed in several load tests of composite and fiberglass reinforced columns. The use of carbon-fiber matrix and S-type fiberglass materials as retrofit reinforcing materials in civil structures is currently being explored and tested because of their potential benefits of being easy to install and their high-strength. Their long-term performance in these applications is not well understood and so a passive, peak sensing technology is ideal for their operational monitoring.

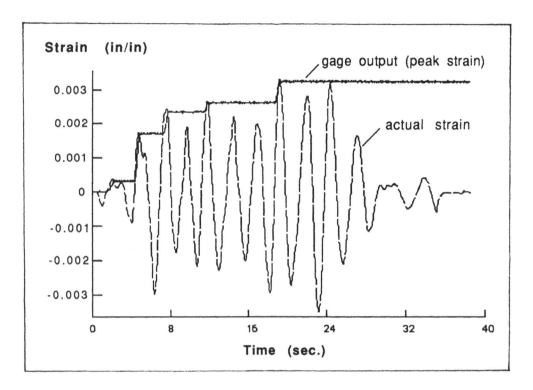

Figure 5. Strain History Vs. Peak Sensor Output for the Response of a Bridge Column to a
Design Earthquake.

A tensile sensor was mounted onto a 24 in. diameter, reinforced concrete column that
had been spirally wrapped with 7 layers of carbon fiber strand. A sensor element was
mounted near the base onto the outer surface of the wrap and measured the peak, tensile hoop
strains during the cyclical load testing. The calibrated sensor output is shown in Figure 6. At
each ductility level, the column was loaded through three cycles and then the peak
displacement was increased to the next ductility level. The change in the peak strain reading
within a set of loads that is present at ductility of 8 and above is due to a geometry change in
the early-type prototype gauge used in this test.

In another test, two sensor elements (of the type shown in Figure 2) were mounted
onto the outer surfaces of several concrete posts which were wrapped with fiberglass mat for
compressive load enhancement. Again the sensors monitored the hoop strain on the surface
of the composite. The compressive load was taken to failure (about 4 times the failure load of
a non-wrapped post). The peak strain measured in the wrap was 5,000 micro-in/in (or
0.5%). The failure strain for this material was about 3%.

The passive behavior of the sensor alloys implies that they can be embedded within
nonferromagnetic materials to witness the straining going on and would be interrogated via an
external magnetometer. This would provide a practical means of inspection for composites,
plastics, and concretes. Thin sensor material wires (0.01 in. diameter) were embedded in a
uni-directionally- reinforced carbon fiber-epoxy matrix, multilayer, laminated composite panel
for evaluation to detect and monitor tensile straining damage. One specimen was pulled to
failure to determine the failure strain which was about 4% in tension. Other identical
specimens were tested to a strain of 2% and 3%, respectively, at which point no evidence was
externally observed to indicate that the composite had been damaged. The ferromagnetic
responses were externally measured using a SQUID (Superconducting QUantum Interference
Device) magnetometer. Compared with control wires, the embedded wires that were tested in
tension displayed considerably higher output voltages (i.e. higher peak strains). Straining was

detected at many locations along the wires in the strained sample whereas the readings were very consistent for the unstrained control wires. These data are the first experimental evidence for a feasible, inexpensive technique to monitor composite material damage. Further research is required to perfect both the placement and detection methods; although it was possible to detect a change in ferromagnetic response using Hall sensors, the SQUID magnetometer is much more effective.

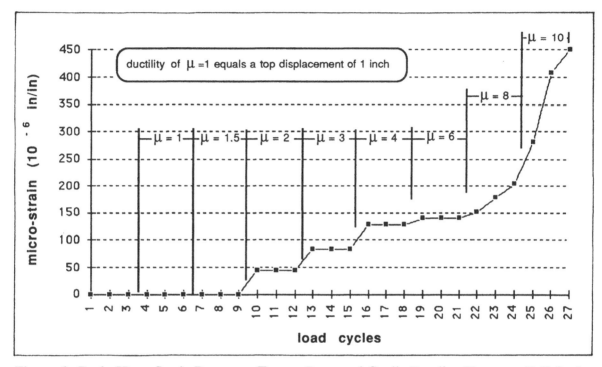

Figure 6. Peak, Hoop Strain Response From a Repeated Cyclic Bending Test on a Full-Scale Carbon-fiber Wrapped, Reinforced Concrete Column.

ACKNOWLEDGMENTS

The author wishes to thank the National Science Foundation, Caltrans, the Georgia Department of Transportation, the Federal Highway Adminstration, and the Advanced Projects Research Agency (ARPA) for their support on the projects discussed.

REFERENCES

B.D. Westermo and L.D. Thompson. 1994a. "Smart Structural Monitoring: A New Technology," SENSORS, 15-18.

L. D. Thompson and B. D. Westermo. 1994b. "Applications of a New Solid-State Structural Health Monitoring Technology," Proceedings of the Second European Conference on Smart Structures and Materials, Glasgow, Scotland.

B.D. Westermo and L. D. Thompson. 1995. "Passive Monitoring Systems for Structural Damage Assessment," Proceedings of the 1995 North American Conference on Smart Structures and Materials, San Diego, CA.

Applications of the Impact-Echo Method for Detecting Flaws in Highway Bridges

M. SANSALONE AND B. JAEGER

ABSTRACT

Engineers involved in the evaluation, repair, and rehabilitation of concrete bridges have long had a need for nondestructive test methods which can provide information about the internal condition of a structure without the need to drill holes into the structure. The impact-echo method -- a versatile and powerful technique that can detect internal flaws in a wide variety of reinforced and prestressed concrete bridges -- is rapidly becoming an extremely useful tool for engineers involved in condition assessment situations.

This paper provides a brief explanation of the impact-echo method and describes a variety of case studies where the method has been used effectively for condition assessment of bridges.

INTRODUCTION

The research that led to the development of the impact-echo method and field instrument was carried out from 1983-87 at the National Bureau of Standards (NBS) by the first author and Dr. Nicholas Carino [1], and from 1988 to the present at Cornell University by the first author and her graduate students [2]. This work led to the development of the method, its applications, and a portable field unit, which was patented by Cornell University and licensed to Germann Instruments, of Chicago and Denmark. The field instrument became available commercially in 1992. Three years later, about 45 impact-echo field instruments are being used in North America, Asia, Australia, and Europe.

THE IMPACT-ECHO METHOD AND ITS APPLICATIONS

Impact-echo is a technique based on the use of transient stress (sound) waves for nondestructive testing. The method is based on using a short-duration mechanical impact (produced by tapping a small steel ball against a concrete surface) to generate low frequency stress waves that propagate into the structure and are reflected by flaws and external surfaces. Surface displacements caused by the reflections of these waves are recorded by a transducer located

Mary Sansalone and B. Jaeger, Cornell University, Ithaca, New York 14853 adjacent to the impact, as shown in Figure 1. This signal is sent to a portable computer which contains a high-speed data-acquisition card and signal processing and display software. The recorded displacement versus time signals are then transformed into the frequency domain. Multiple reflections of waves between the impact surface, flaws, and/or other external surfaces give rise to transient resonance conditions which can be identified and used to determine structure thickness or the location of flaws. It is the patterns present in the displacement waveforms, and especially in the corresponding amplitude spectra, that provide information about the existence and location of flaws.

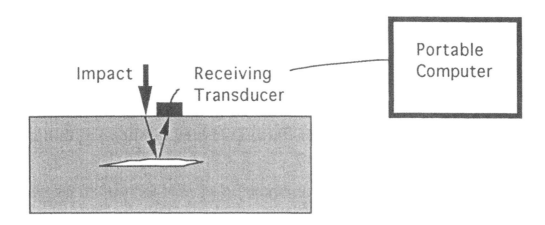

Figure 1. Schematic representation of an impact-echo test.

Current applications for the impact-echo method and field instrument include determining the location and extent of cracks, voids, delaminations, honeycombing, and debonding in plain, reinforced and post-tensioned concrete structures. The structural component geometries that can be tested currently, include plates (slabs, pavements, walls, and decks), layered plates (including decks with asphaltic concrete overlays), beams and columns (round, rectangular, and many I and T cross-sections), and hollow cylinders (pipes, tanks, and mine shaft and tunnel liners). Voids in post-tensioning ducts can be detected for a variety of structural component geometries and duct geometries. The presence of unbonded area indicating poor bond quality at an interface between two concretes can be detected.

In many applications, the wave speed in the concrete is determined from an impact-echo test on an area of known thickness. This wave speed is then used to test other areas of the structure. However, because the wave speed in concrete often varies in different parts of a structure, or in some cases, the thickness is unknown, an independent means of determining wave speed using direct P-wave speed measurements was developed. This allows the wave speed to be determined at any point in a structure. More precise wave speed information results in greater accuracy in determining depth using the impact-echo method.   The current field system now includes both the wave speed measurement technique and the impact-echo technique.   One of the common applications that requires this combined approach is measurement of the thickness of new, concrete highway pavements.

Readers interested in obtaining more information on the method and specific applications should contact the authors.   Over 40 papers and reports on the impact-echo and its applications have been published.

CASE STUDIES

The following case studies illustrate how the impact-echo method and instrument can be used as a condition assessment tool for an engineer involved in evaluation of concrete bridges.   (One or both of the authors were involved in each of the investigations discussed, often working with the consultant or agency responsible for structural evaluation and repair.)   Each of the cases presented includes a description of the structure and the problems to be diagnosed.   The role of impact-echo is discussed, and the results of impact-echo testing are summarized.   A statement about how the impact-echo results were verified is given.

Detecting Delaminations in a Concrete Bridge Deck with an Asphalt Overlay

Location:	Northeastern United States
Problem:	Detecting delaminations caused by corrosion of reinforcing steel in a 200-mm thick concrete bridge deck covered with a 100-mm asphalt overlay.
Results:	Impact-echo located delaminations at the top layer of reinforcing steel in the concrete deck.
Verification:	Cores
Outcome:	Concrete deck repaired. New asphalt overlay applied.

Detecting Cracking in a Reinforced Concrete Railroad Bridge

Location:	Europe
Problem:	Determine extent of structural cracking caused by alkali-aggregate reaction in a 1-m thick reinforced concrete railroad bridge deck.
Results:	Impact-echo detected the presence of a horizontal crack at mid-depth in deck across virtually the entire span. Figure 2 shows impact-echo tests being performed.
Verification:	Cores (and mode of failure during demolition).
Outcome:	Bridge was judged unsafe and was demolished. Upon impact with a wrecking ball during demolition, the lower half of the bridge deck separated from the upper half and collapsed.

Determining the Depth of Surface-Opening Cracks in a Segmental, Box-Girder Bridge

Location:	Europe
Problem:	Cracks in web of precast box-girder which originated from

post-tensioning anchorages.   Cracks open on the inside surface of the web, but not apparent on the outside of the web. The depth of the cracks into the web needed to be determined.

Results: Impact-Echo was used to determine the depth of the surface-opening crack (see Figure 3.).

Verification: Drilling holes.   Crack depths predicted by impact-echo were accurate to within ± 5 mm.

Outcome: Cracks were injected and sealed.

Locating Voids in the Grouted Ducts of a Post-tensioned Highway Bridge

Location: Northeastern United States

Problem: Determine of extent of grout in the grouted tendon ducts of post-tensioned bridge girders.

Results: 14 girders were tested in the initial phase of the work. Full or partial voids were detected in the tendon ducts of three of the girders using the impact-echo method [3].

Verification: Tendon ducts were opened up and inspected.   Figure 4 shows a photograph of a duct which was found to be empty.

Outcome: The consultant responsible for the structural investigation specified that impact-echo be used to determine the extent of grouting in tendon ducts in the bridge girders as part of a comprehensive evaluation and repair program.

Durability of Repairs in a Precast, Segmental, Box-Girder Rail Transit Bridge

Location: Southeastern United States

Problem: Large delaminations were observed in the web of several box-girder segments located over piers.   Prior to opening the bridges to traffic, the webs were repaired. Two years later impact-echo was used to assess the integrity of the repairs [4].

Results: Impact-echo test results showed that a significant portion of the repair patch on one segment had delaminated.   Test results also indicated several areas that exhibited signs of poor bond quality (high ratio of unbonded area at interface).

Verification: Cores

Outcome: The delaminated portion of the repair patch was removed and replaced with a properly compacted concrete repair material to restore the long-term durability of this bridge component.

SUMMARY

The impact-echo method is a powerful and useful method for detecting flaws in a wide range of applications involving concrete highway and rapid transit bridges.   It is slowly gaining acceptance for nondestructive testing of concrete structures.   There are a number of highly-qualified consultants in the U.S., Canada, Europe, and Asia using the method effectively as a tool for integrity testing of structures [4,5].   Most of these consultants are structural engineers involved in the maintenance, repair and rehabilitation of structures.   As the effectiveness of the method is demonstrated in more and more situations, as standards are developed, and as agencies adopt impact-echo as a method for testing, it is expected that the method will gain increasing acceptance and use.

REFERENCES

1. Sansalone, M., and Carino, N.J., "Impact-Echo: A Method for Flaw Detection in Concrete Using Transient Stress Waves," NBSIR 86-3452, National Bureau of Standards, Gaithersburg, Maryland, Sept., 1986, 222 pp.
2. Sansalone, M., and Streett, W.B., "Impact-Echo: Development and Technology Transfer," Proceedings of Research into Practice, National Science Foundation, May 1995.
3. Jaeger, B., Sansalone, M., & Poston, R., "Detecting Voids in the Grouted Tendon Ducts of Post-Tensioned Structures," ACI Structural Journal, scheduled for publication in 1996.
4. Poston, R., & Kesner, K., "Impact-Echo Strikes Home" Civil Engr., June 1995.
5. Hendrikson, C. "Impact-Echo," Concrete International, May 1995.

Figure 2.   Impact-Echo Testing of a Bridge Deck

Figure 3.   Impact-Echo Testing of a Post-Tensioned Box-Girder

Figure 4.   Photograph of a duct which impact-echo identified as empty.

Damage Assessment by Impact Testing

I. VILLEMURE, C. VENTURA AND R. A. SEXSMITH

ABSTRACT

Five models of a typical bridge pier at 0.45 scale were constructed in their original (1950's) as-built condition. Four were then retrofitted following several alternative schemes. In the structural laboratory at the University of British Columbia, each bent was subjected to sequences of lateral slow cyclic loading, which was increased for each sequence up to a high damage level. Between the sequences, corresponding to increasing levels of structural damage, forced vibration tests were performed on the bents with an instrumented hammer. Frequency domain analyses of the experimental data were performed to identify dynamic characteristics of each bent at each level of structural damage. The decreasing fundamental frequencies were related to the increasing ductility demands, the latter reflecting the structural damage of the bents. Sensitivity analyses were also performed to relate damping ratio changes to ductility levels imposed on the specimens.

INTRODUCTION

Decisions for pre-seismic strengthening and post-seismic repair depend on reliable predictions or estimates of existing and future damage. The seismic retrofit of Vancouver's Oak Street Bridge, which involved a program of retrofit tests (Anderson et al 1995), provided the opportunity to perform an additional program of damage assessment. Damage was estimated in terms of changes to frequency and damping as ductility demand was increased.

DESCRIPTION OF THE SPECIMENS

The model bents consisted of concrete cap beam supported by two columns. All specimens had overall scale dimensions of 3.53m in width by 2.75m in height (see Figure 1). The models replicate the upper half of the actual bents, above the midheight inflection point.

Isabelle Villemure, 914 Etienne Marchand, Boucherville, Quebec
Carlos Ventura and Robert Sexsmith, University of BC, 2324 Main Mall, Vancouver, BC

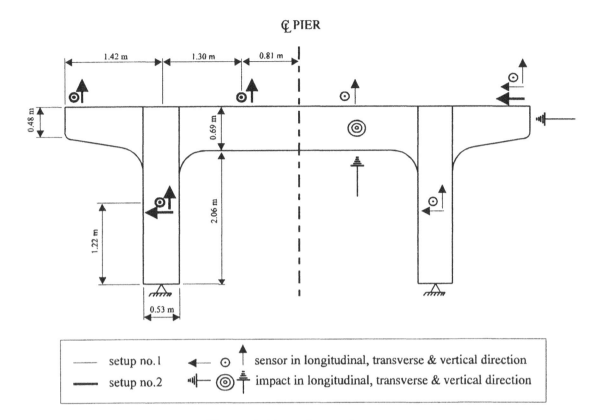

Figure 1. Experimental setup

The first specimen (S1) was the as-built model of the actual bent of the bridge. The retrofit of the second specimen (S2) consisted of coring the cap beam along its longitudinal axis and grouting two post-tensioned tendons in the cap beam. On specimen S3, the cap beam was retrofitted with a concrete underbeam. 1/4" steel jackets were added to the columns. Specimen S4 had vertical prestressing and longitudinal prestressing in the cap beam, and 1/4" steel jackets to retrofit the columns. Specimen S5 was given a fiberglass wrap in the regions of high shear in the cap beam and columns. This retrofit also included external prestressing on the longitudinal axis of the cap beam.

EXPERIMENTAL PROCEDURE

Vertical loads of 169 kN, representing scaled dead load of the superstructure, were imposed at each of five locations on the bent in order to simulate the superstructure dead load. Lateral loading was applied through a horizontal jack along a longitudinal axis at the deck level above the bent.

Each lateral load sequence, corresponding to increasing levels of damage, consisted of three complete cycles. Displacement ductility was used to quantify structural damage of the bents. Typical ductility values were 0.75, 1.0, 1.5, 2.0, 3.0, 4.0, 6.0,9.0 and 12.0. Figure 2 summarizes failure mode and maximum ductility level for each specimen.

Figure 2. Details on failure behaviour

Specimen	Failure Mode	Maximum Ductility Level
S1	shear failure in the cap beam	$\mu = 4.0$
S2	shear failure in the north column	$\mu = 6.0$
S3	failure by hinging in the beam/column joints	$\mu = 12.0$
S4	failure by hinging in the beam/column joints	$\mu = 12.0$
S5	high damage caused by spread cracking in the columns	$\mu = 12.0$

Two types of vibration tests were performed on the bents: impact testing and ambient vibration testing. The vibration tests were performed before and after each sequence of cyclic loading. Impacts were induced by an impulse hammer that includes an integral piezoelectric force sensor to measure the force applied. Hammer impacts were applied along the axis of the cap beam, and in the transverse and vertical directions in the middle part of the cap beam, as shown in Figure 1. Accelerometers were used to measure the induced vibrations on the bents at locations shown in Figure 1.

FREQUENCY ANALYSIS

Analysis of frequency was limited to the first longitudinal mode. Typical time histories induced in longitudinal direction are shown in Figure 3 for different ductility levels. The frequency response function (FRF) was used to determine the natural frequency at each damage level. In this study, the response of the longitudinal sensor on the cap beam was taken as the output. The hammer impulse in the longitudinal direction represented the input signal. Typical FRF plots are shown in Figure 4. FRF's were evaluated for all specimens at each ductility level (Villemure, 1995) using a program developed at U.B.C. by Horyna (1995). Figure 4 clearly shows the peak associated with the first longitudinal mode. The preliminary frequencies vary between 18.9 Hz (S5) and 22.3 Hz (S3) while the failure frequencies fluctuate between 12.9 Hz (S1) and 17.2 Hz (S5).

The frequency sensitivity to structural damage is shown in Figure 5. Generally, all specimens show decreasing frequency with increasing damage or ductility level. However, the rate of decay varies for each specimen. Specimens S2 and S5 present some irregularities.

DAMPING ANALYSIS

Two methods were used to assess damping sensitivity to structural damage. Specimens were first assumed to behave with viscous damping. High frequencies of typical response signals were filtered to remove contributions of higher modes. Damping ratios were then

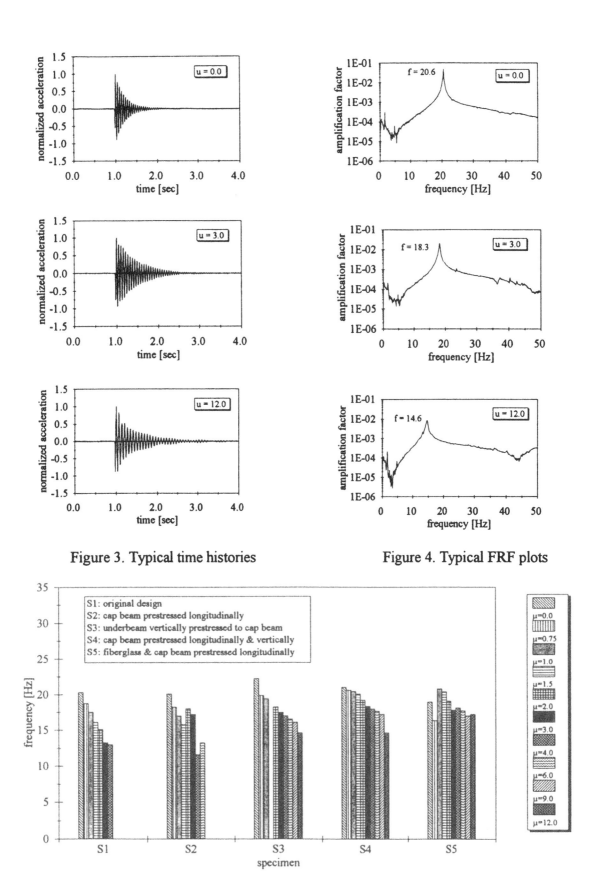

Figure 3. Typical time histories

Figure 4. Typical FRF plots

Figure 5. Frequency sensitivity to structural damage

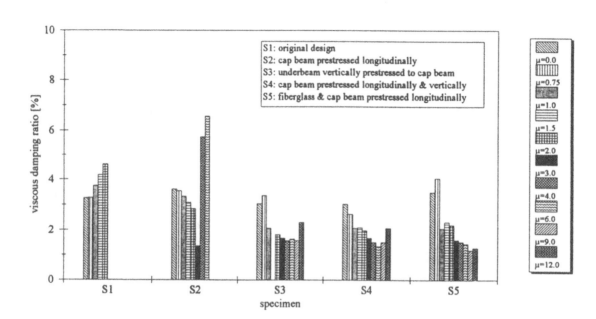

Figure 6. Sensitivity of viscous damping to structural damage

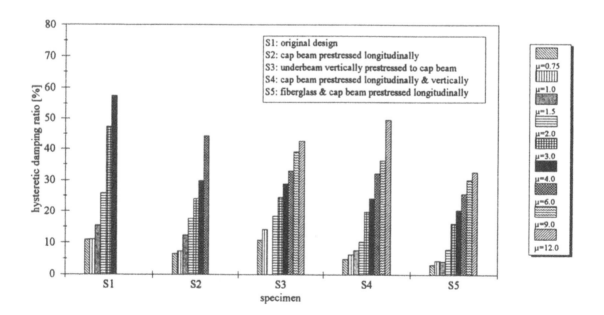

Figure 7. Sensitivity of hysteretic damping to structural damage

evaluated using the logarithm decrement method. The second method assumed structural (or hysteretic) damping. Structural damping calculations were based on the hysteresis loops obtained from the cyclic loading sequences.

Damping ratios evaluated from viscous damping assumption show no particular trend with increasing damage level (see Figure 6). Evaluated damping ratios increased with the increasing ductility level (see Figure 7). Considering that this type of damping is closely associated with internal friction, increasing damping ratios were expected to correlate with increased cracking of the specimen.

CONCLUSIONS

Impact testing was used as a mean of damage assessment in reinforced concrete frames. Experimental data were analyzed to study sensitivity of frequency and damping ratio to structural deterioration. Frequency domain analyses were performed to identify the first longitudinal frequency. Generally, all specimens showed decreasing frequency with increasing ductility levels. Damping ratio sensitivity was analyzed with two types of damping. No particular coherent trend yielded from the viscous damping study. Structural damping evaluated with hysteresis loops provided coherent increasing damping with increasing structural damage.

ACKNOWLEDGEMENTS

The financial support of the Natural Science and Engineering Research Council (N.S.E.R.C.) of Canada is gratefully acknowledged. This includes a Postgraduate Scholarship as well as Research Grants awarded to both Dr. C.E. Ventura and Dr. R.G. Sexsmith. The research was also made possible by financial support of the Ministry of Transportation and Highways of British Columbia.

REFERENCES

Anderson et al. (1995), "Tests of alternate seismic retrofits of Oak Street Bridge", 7th Canadian Conference on Earthquake Engineering.

Horyna, T. (1995), "Dynamic Analysis of Bridges with Laminated Wood Girders", Master Thesis (to be published), University of British Columbia.

Villemure, I. (1995), Master Thesis (to be published), University of British Columbia.

Non-Destructive Evaluation of Fatigue Cracking

E. C. KASLAN

ABSTRACT

California has approximately 1500 steel girder bridges and
400 steel truss bridges on the public highway system. Of
these, 769 have been identified as possessing either fatigue
prone or fracture critical details. Fatigue cracking is
becoming evident on our high ADT steel bridges. Most of this
cracking is associated with out-of-plane bending
predominately occurring at fatigue prone connections that are
subject to high cycle rotations and displacements. Code based
fatigue analysis gives little guidance for these situations.
Repairing or retrofitting a specific location seems to drive
the cracking to other locations. Once cracking occurs,
widespread cracking seems to quickly initiate at similar
details. Retrofits are difficult to effectively design as
structural behavior and driving forces are not well
understood and are difficult to model. In some cases,
reliable retrofits are not clear or evident. Environmental
and health concerns associated with lead based paints are
making repairs of steel bridges costly; a definitive strategy
before work begins should be determined to reduce these
costs.

   CALTRANS would like to present current ideas and
information on inspection programs, effective NDE methods for
early crack detection, case studies, fatigue analysis
strategies, effective retrofit designs, ideas as to the cost
effectiveness of rehabilitation vs. replacement, and other
related topics. The purpose of this presentation is to
exchange ideas and information with engineers, researchers
and transportation officials in the United States to better
understand the best strategy to take in tackling fatigue
cracking problems.

Erol C. Kaslan, Caltrans Division of Structures, Office of Structure Maintenance and Investigations

PROGRAM OVERVIEW

As a quick overview of our fatigue management program, in the 70's, Caltrans identified and grouped all bridges which posses either fatigue prone and/or fracture critical features, which we call Group A bridges.

On these bridges we perform special investigations, that is, those requiring a more detailed effort than the routine biennial inspections.

These investigations are a tactile 24" investigation of suspect details, with appropriate non destructive evaluation of a statistical sample of details; this NDE work, usually dye Penetrant testing or ultrasonic testing, is performed by experienced NDE technicians called our Fracture Critical Inspection team, who are also experienced bridge inspectors.

If any indications are found, then a 100% investigation of all details on a bridge is performed.

Inspection intervals range from 3 to 48 months, and are determined by a maintenance engineer depending on the level of distress.

CASE STUDIES:

While we have several different types of fatigue problems in the State, the majority of fatigue cracking problems in California highway bridges are related to either out of plane bending or high cycle/low amplitude vibrations from traffic loads.

In primary elements, most of our cracks occur at web plates or web to flange welds,

and in secondary elements, we see commonly see fatigue problems in bracing or bracing connections.

I'd like to present a few illustrative case studies of these problems, show you what we've done about it, and perhaps stimulate some discussion afterward.

OUT OF PLANE BENDING:

For these examples, I'd like to show both a box girder and open plate girder bridge; the box girder, not a very typical animal in California, is the I-80 Sacramento River bridge at Bryte Bend, and the open girder bridge is the West Merced Overhead on SR99, one of our highly skewed bridges with staggered cross frames that is very typical of steel highway bridges in this state.

BOX GIRDER-BRYTE BEND:

These twin bridges are 4050' long, carrying I-80 traffic over the Sacramento River just north of the City of Sacramento.

The structure typical section is a two cell trapezoidal box girder with a composite concrete deck. Two details to note here are little "detail V", the cope at the base of the stiffener, and the fact that the upper horizontal element of the cross frame is cast and shear studded into the deck. These details factor into the problem.

There are 16 simple approach spans 146' long with 6 longer continuous spans over the river. Our problems are primarily in the simple spans, where the A36 web plates are only 3/8" thick.

While our problems in the longer continuous spans are minimal at this time, it is noteworthy that the negative moment flanges of these spans contains high strength steel (A517) and that during construction in 1970, while the deck was being poured, a brittle fracture occurred in one of these A517 flanges, cracking clean through the flange and into the web plate.

Our problems here are cracks in the web plates at web stiffener locations where cross frames are connected to the webs. As I mentioned, these are primarily in the simple spans, and were first detected in 1991. Of 448 cross frame locations in these spans, about 75% are cracked. Since initial detection in 1991, we have been inspecting regularly at 6 month intervals, and have seen slow but steady growth in the number and size of cracks.

The cracks occur at the bottom of the weld connecting the web stiffener to the web at the cope hole accommodating the web to flange weld, little detail "V". These cracks appear to initiate in the weldment, then migrate the web plate.

For the uninitiated, this general out of plane bending slide shows the common fatigue mechanism associated with this detail. As the frames and stiffeners are loaded, bending is induced into the web and at the "weak" unstiffened area at the cope, her called the web gap. This bending is very small, and cycles constantly. To compound the problem, this area usually contains a weld or heat affected zone, setting up all the ingredients for fatigue cracking. The loading of the frames and stiffeners usually occur from girder deflections; Significant to Bryte Bend is that we have a compound loading mechanism, both from girder deflections and direct loads to the frames, as you saw from the frame cast into the deck slab.

Well, what have we done and what are we doing: all the through web plate cracks were drilled with 1" diameter holes immediately after detection, which seemed a good temporary solution to buy time. This proved to be not really effective , as shown here; while we were successful in catching crack tips in the arrest holes, a new crack developed between these two holes; besides, we were not always successful in catching the crack tip as you can see here.

We felt that a effective long term rehabilitation was necessary to prevent crack initiation, or we could eventually drill over 300 cracks. Of course, the big concern is that we risk missing a crack and having it migrate into the bottom flange, which could be a real problem.

Since this bridge is large and critical to the transportation system in the area, we decided to approach this rehabilitation as a full blown research and design project instead of applying a "from the hip style" repair. This project would consist of creating a three dimensional finite element model of the bridge and frames, calibrated and combined with strain gauge load tests, to try better

understand the problem and help model possible solutions.
Once modeling was completed, the plan was to design and
fabricate a prototype rehabilitation, install it on a test
location, and then instrument the cracked areas using
Acoustic Emission technology to determine it's effectiveness.
Based on this, we will then move on to a whole bridge
retrofit.

Just a quick word on our experiences using FEM modeling
for fatigue that some of you might find informative.

We learned that FEM is beneficial in that modeling gives
good results of general load distribution in the bridge.

However, a meaningful model should be calibrated with
instrumentation and load tests to smooth out modeling
assumptions.

Our efforts of modeling showed it can be a valuable
design tool in that it allows the analytical examination of
various alternative rehabilitation strategies, but

We found that this capability and the ability to
accurately model the local areas of cracks requires special
software, which unfortunately, we didn't have available.

FEM can be very tricky, and a good model requires an
experienced analyst and a fair amount of time to build.

For these reasons, we feel that we probably won't use
FEM routinely for analyzing fatigue.

We are now in the middle of this project. Using our
models as a guide, we pooled our experience in round table
type strategy meetings to come up with an effective retrofit.
From sheer experience, we knew that the web stiffener and web
was moving out of plane from differential girder deflection,
and from modeling and load tests, we knew that the cross
frames were loaded an order of magnitude further by direct
deck loads. Our rehabilitative strategy was to both restrain
the OOP movement of the stiffener and reduce or eliminate the
direct frame loads.

We came with several alternatives to accomplish this,
with one possible solution being to eliminate the frames
altogether and thus eliminate the problem (you may hear me
say this over and over); this is where the modeling was a
strength, as it showed that in this box girder the frames
were contributing a great amount to the torsional stiffness
of the entire cross section under traffic loads. If the
frames were eliminated, this would cause increased web
distortions that could make the problem appear at all
stiffener locations.

Our design solutions were to restrain OOP movements by
bolting gusset plates from the web stiffeners to the bottom
flange, and to eliminate direct traffic loads from the frames
by uncoupling the upper horizontal frame element from the
deck. Thorough not easy to design, all connections will be
bolted to eliminate fatigue prone field welds. A big cost in
this is the paint removal required to ensure a slip critical
connection between the restraining gussets, stiffeners and
flanges; in the total retrofit, this could be the biggest
cost on the job. We are scheduled to fabricate, install and

evaluate this rehab in four months, and I'll let you know how it works this time next year.

The down side to this rehabilitation is we don't have a clear idea of what this retrofit will do to the bridge, that is, will we be driving our fatigue problem somewhere else that we cannot foresee at this time? We plan to strain gauge likely areas during the prototype evaluation for guidance, but it will still be a question of time before we know if we licked the problem or simply created another one.

OPEN GIRDERS-WEST MERCED OVERHEAD:

This will be a few quick slides to generally illustrate the common out of plane bending problem at web stiffeners/cross frame connections. Shown here is the West Merced Overhead, which has typical problems found at similar bridges. While not all our problem bridges are as heavily skewed as this (which greatly increases differential deflections between the girders), most have the common detail of staggered cross frames as shown here.

The cracking occurs at the bottom of the stiffener weld at the flange cope in frame locations, similar to Bryte Bend, as well as the web to flange welds shown here; this slide gives you an idea of the "tree cracking" that can develop at these areas.

Our standard procedure in these cases is to drill the crack ends to arrest cracking. I'm not certain that drilling these "out of plane" cracks is an appropriate repair or not , as it is difficult to capture the crack tip, and trying can lead to the "Swiss cheese" syndrome.

Our standard retrofit for this type of cracking is to install a clip angle connector to restrain the stiffener to the bottom flange and weld up the web plate cracks, which is shown in its most modern incarnation here; if this looks like the same idea as Bryte Bend, this bridge is where we first tried this and learned from it. We also learned a few other things from this bridge.

When we first applied this on West Merced in 1990, we only installed it on the cracked side of the frame; we found 1 year later that cracks developed in the webs on the opposite side of the frames. Now we are finding cracks on the non retrofitted frames: lesson here is that when applying this retrofit, apply it to all frame locations and on both sides of the frame. Also, we learned that we need at least two bolts on each leg to restrain movement, as our first designs only had one to preserve net section on the tension flange, and these probably weren't as effective as they could have been. We also learned that we should remove the paint from the angle-web-flange contact surfaces, as any minute slip will probably greatly reduce the required restraint from the retrofit. Again, paint removal costs could substantially affect the cost of this retrofit when we apply it to other bridges, which are beginning to do.

Some discussion as to whether to put back to back cross frames across the section is always around the office, and seems a logical answer to try to balance stiffness across the section in addition to distributing forces and reducing the rotations of the stiffener.

HIGH CYCLE/LOW AMPLITUDE VIBRATIONS/DEFORMATIONS/LOADS:
        Now I'll show some examples of high cycle low amplitude vibration induced fatigue problems. The two case studies are the West End Viaduct carrying I-5 traffic through Sacramento and the I-5 Sacramento River bridge at Antlers over Lake Shasta.

CONNECTION GUSSETS-WEST END VIADUCT:
        The West End Viaduct in Sacramento is a plate girder bridge that has cracks in the cross frame connection gussets. This bridge was a real leaning experience for this type of problem.
        The cracking is occurring at these locations , where we have sway bracing one side of the web plate and cross frames on the other side. The significance is that we see cracking only at the areas where we have a large stiffness change across the section on the weak side.
        Here is a few shots of the failed gussets, which you can see are fractured clear through.
        Because of the bridge is over a large rail yard and rail flagging costs are tremendous, we could not easily repair these gussets one at a time and thought that it would be appropriate to come up with a rehabilitation for all the similar details on the bridge to handle under contract.
        So, we set up some strain gauges to help us get a handle of the driving mechanism to guide us to an effective retrofit.
        We learned alot from this instrumentation work: the first thing is that these cross frames get significantly stressed from differential deflection of the girders under traffic loads as shown here, upwards of 6000 psi; we also saw that we could not balance loads across the section, which suggests that this is a dynamic problem and difficult to analyze; we also saw that while loads to the frames are high, the cycles from traffic vibrations in the frames and gussets are extremely high, and were measured to be about a constant 4 cycles per second; "from the hip" analysis shows that we reach the fatigue life of this element in about 6 days at current traffic and load levels.
        We feel that the fatigue mechanism here is that the cracks are initiated from high cycle low amplitude loads and vibrations, and then driven to propagation and element failure by high direct frame loads. We are now trying to design a way to mitigate the problem by damping the vibrations in the frames and connections as well as reducing frame loads. But the solution is not clear, and is a interesting problem: about half of us feel that we can remove the cross frames and eliminate the problem with no ill effects to the bridge, and the other half us feel that we should put additional sway bracing across the entire bridge and not remove a thing.

BRACING ELEMENTS-SACRAMENTO RIVER AT LAKE SHASTA (ANTLERS):
    This is the I-5 Sacramento River Bridge at Antlers at
Lake Shasta in North California. This cantilevered deck
truss was built in 1945 by the COE, and has since been
strengthened and widened. Ever since construction, this
bridge was reported to deflect an unusually large amount
under traffic loads. After widening and strengthening, loads
increased and these deflections were worse. You don't need
strain gauges to see the that all wind bracing and sway
bracing deflects and vibrates excessively back and forth for
several cycles on this bridge, and it's obvious that these
vibrate long after traffic loads have left the bridge.
    The problems occur at the gusset connections of this
bracing. shown typically here. The deflections and
vibrations of the bracing from traffic loads are causing the
gusset plates to crack and fracture, as well as breaking
bracing elements themselves.
    This bridge has been a maintenance headache for years,
as we found that we would repair a single gusset or element
only to find another broken one somewhere else. In some
cases, even our repairs failed.
    We tried damping the deflections and vibrations in the
braces by coupling this bracing together with cables, and
tied the cables to the deck. This concept probably would
have worked better if we had used rigid elements instead of
cables, because cables still allowed movement in at least one
direction, as these recent repairs show. as we had limited
success . What we will do is to again try damping this
bracing by bolting it together with rigid tube or pipe
connections, and tie everything together in a balanced frame.
As with most problems of this type, we not sure if we will be
creating another fatigue problem by this retrofit, it could
even be maintaining the retrofit itself. On a bridge this
loose, there may be nothing to do but constantly repair the
failures.
    So now I'd like to stimulate some discussion, and pose
these statements:
    -repairing a single area seems to drive cracking to
another area, so we should try to come up with a
comprehensive rehabilitation strategy for all fatigue prone
details on a bridge when we first detect fatigue.
    -but there is a general feeling that rehabilitation's
are not really effective, and we may just be buying time
    -paint removal is a new player in this game, and the
costs of this are high; if we have to repeatedly repair
fatigue problems, is the costs worth it
    -maybe we should program our bridges for replacement
concurrent to initial fatigue repairs or rehabilitation.
I've heard that other DOT's with older inventories than in
the West have done this; maybe we should too.

The Use of Still Image Transmissions as an NDT Tool

R. J. SCANCELLA

ABSTRACT

This paper presents the findings of a research program by the New Jersey Department of Transportation , Bureau of Materials, Inspection Unit on the use of still image transmissions as a viable tool for the non-destructive inspection of newly fabricated and existing structures. The discussion is based upon the "*Picasso*" still image phone transmitter manufactured by AT & T.

INTRODUCTION

A large percentage of the structures fabricated for the State of New Jersey are made out of State where travel to and from the plant takes at least one full day. The NJDOT then provides for overnight lodging, meals and travel to these fabricators. Because of reduction in State work forces and the cost associated with sending an inspector or engineer to a fabrication plant, the NJDOT has a minimum number of inspectors at the fabrication site.

Prior to still image transmissions, when serious problems occurred it required expertise from one of the managing engineers and that engineer would travel to the site for a few days and return after the problem was solved. If the problem is not serious enough to warrant a trip, Polaroid photographs can be taken and mailed with an overnight carrier. When the pictures are received they are usually not clear and often require a phone call to the inspector at the site. It is difficult for two people to always see the same item in a photograph.

Usually when a design problem occurs, or a problem warrants inspection on an existing structure, a design engineer can not get to the site for direct observation of the problem. In most cases the problem is documented with pictures, which are then taken to the design unit to for a resolution of the problem.

Robert J. Scancella, PE, New Jersey Department of Transportation, Project Engineer Materials, 1035 Parkway Ave., Trenton, NJ 08625

GOAL

A State agency a limited scope when trying to solve a problem. The problem must be approached using available manpower and minimum cost. The goal of the Inspection Unit for this problem is to arrive at an economical solution where all parties concerned can look and solve a problem at the same time. By reviewing the problem simultaneously, questions can receive answers in the shortest time possible. This procedure insures that the contractor's schedule is not interrupted and maintains cooperation between the State DOT and the contractor.

SOLUTION

A presentation on still image transmission was given to the NJDOT at the Trenton headquarters. The presentation lead the Inspection Unit to request the State and AT &T to use the "*Picasso*" equipment on loan for determination of its use in construction and fabrication of materials. Other alternatives such as computer generated pictures and transmissions were also reviewed, but all positive marks pointed to the "*Picasso*". After discussion with State officials and AT &T representatives, the *Picasso* system was approved for use on a loan/trial basis.

EQUIPMENT

The equipment that was used during the test trials were as follows:

Using two *Picasso* Still Image Phones, manufactured by AT&T, one phone was connected to the standard telephone line in the Trenton headquarters. The other phone was transported to the various fabrication/construction sites and the inspector using this phone used standard telephone service that was provided by the fabricator. It should be noted that fabricators are required to supply the State inspector with a field office that includes telephone service.

One Panasonic Palmcorder, S-Video type camera was also used to record the activity and easily connects to the *Picasso* Unit. This palmcorder is not the only type of recording device that can be used with the *Picasso* unit, however the palmcorder best fit our needs.

Two Gateway 2000 Colorbook laptop computers were required to be compatible with the *Picasso* unit and be able to run the Windows version 3.1. It is also best to have a color screen for better picture resolution. These two laptop computers were used at the fabrication site and at the Trenton headquarters. At the Trenton headquarters it was used to transport picture files to various offices that could not attend the test trial. The field use of the laptops was for documentation at the site.

It was determined after the first test trial that the field inspector should have a monitor to capture the images for transmission. Trying to capture the images with only the laptop created a problem since it was necessary to look through the viewer of the palmcorder while trying on capture the image to the laptop's screen. When using the monitor, the video image is seen on the monitor when performing the capturing function.

At the Trenton headquarters the use of a full size Gateway 2000 personal computer was also used to maintain the picture files and perform the test trial reviews.

TEST TRIALS

The first test trials were set up at nearby fabrication plants of both structural steel and structural concrete. Although there were not any major problems which occurred that required discussion between the materials inspector, materials management, and design personnel at the same time, these trials did afford the inspector and the Bureau's Inspection Unit to resolve any minor problems in using this type of equipment at a plant location.

After the test trials were arranged, we were able to use the equipment in an actual problem solving situation. A steel fabricator in western Pennsylvania reported a problem with proper adherence of structural paint. We were fortunate to have an inspector leaving from Trenton to perform an inspection at a fabricator near the site of the problem. One of the *Picasso* units was sent with him to our inspector. Once he received the unit, a video camera, and small monitor he was given instructions by phone on how to use the equipment. This took approximately one-half hour. By the end of the day we were receiving video photographs of the problem. A meeting was scheduled for the following morning when the Bureau's paint experts would be available. At the designated time our inspector again transmitted the video photos and several questions were asked of the inspector. In addition more specific photos were requested. The inspector procured the requested pictures and transmitted it to our meeting within twenty minutes. Several solutions were then proposed. The fabricator, also involved in the meeting, was able to perform the necessary adjustments on the same day. Another morning meeting indicated that the previous day's solutions were helping to solve the problem. After several refinements to the solution, the fabricator was able to work nearly a full day. A final meeting showed that the problem was solved. By use of the image transmission, the NJDOT was able to keep two paint experts at headquarters and take them away from their daily routine for a total of 4 hours. Since the *Picasso* unit is compatible with IBM type computers, a permanent record of the transmissions was made. We were able to have the fabricator working rather than awaiting the arrival of a problem solving expert at the plant. Overall, the test trial of the *Picasso* unit was a great success.

Following this problem solving exercise, the Inspection Unit decided to expand the use of the *Picasso* Unit to on site inspections and in-depth inspections. Videos were made of the Route 295 over Route 1 bridge damaged by a vehicle, and the Route 37 Mathis Bridge over the Barnegat Bay in-depth paint systems. When the Route 295 bridge damage inspection was performed, *Picasso* transmissions enabled immediate decisions to be made involving repairs. The bridge flaws were able to be located and repaired in one day, saving the general public several days of road inconvenience. For the in-depth inspection of the paint systems on the Route 37 Mathis Bridge, the *Picasso* unit allowed the Department to have a permanent record that can be reviewed using a personal computer. The need for a video player or slide projector for later viewing is eliminated and reproduction of the computer disk is relatively simple and cost effective.

COST ANALYSIS

The cost of the *Picasso* unit is approximately $3000 each. The cost for the laptop computers, the personal computer, and image recording devices at this time is not considered since these items were in existence in the Materials Bureau and are used for more than just the *Picasso* Unit. Costs to send one *Picasso* unit to a particular fabricator's site varies from $0 (if the unit is sent with the inspector when a problem is occurring) to approximately $25 for overnight shipping.

Without the use of the *Picasso* unit the cost involved to the state would be the travel expenses for at least one expert. In most cases two experts are needed. The cost for two experts are approximately $1000 or more. This cost includes travel time in salary, cost of travel, and overnight per diem accommodations.

CONCLUSIONS

By use of the test trials, it was determined that the clarity of the still image transmission ranged from very good to excellent. Little or no problem occurred when viewing the images. This is most likely due to simultaneous viewing of the picture which points out discrepancies so that solutions can be made quicker.

The cost of each unit is slightly more than that of a personal computer. However, this cost would be offset by keeping experts from traveling to the fabricators site. The actual dollar savings would pay for two (2) units after four or five uses. Additional savings are gained by keeping expert inspectors available in headquarters performing normal work tasks. If enough still image phone units are available, an expert inspector could conceivably solve several problems at different locations in one day.

Overall the **still image transmission phones** appear to be a viable solution to save money for State Department of Transportations, ensure quality workmanship, and solve problems while involving all units concerned.

REFERENCES

AT&T, May 1993, Picasso Still Image Phone, "Installation and Use Manual"

SPECIAL THANKS

New Jersey Department of Transportation: Robert Skalla (Principle Engineer), Jay Patel (Senior Engineer), Frederick Lovett (Principal Engineer), John Shaner (Construction and Maintenance Tech 2), Rich Smetanka (Principal Engineer), Michael Schillaci (Assistant Engineer), Wayne Carragino (Senior Engineer), and Godfrey Tisseverasinghe (Senior Engineer)
Diane Smith, AT&T, Picasso Representative; High Steel Inc., Lancaster, PA; Colonial Steel, Canonsburg, PA; Precast Systems, Lakewood, NJ

Objective Bridge Condition Assessment

A. E. AKTAN, V. DALAL, D. N. FARHEY, V. J. HUNT, M. LENETT AND A. LEVI

ABSTRACT

A methodology and its associated tools for the objective condition assessment and reliability evaluation of bridges are discussed. Integrated analytical-experimental research, conducted to explore the serviceability and damageability behavior, is summarized. An effective research and development strategy is described, comprising field and laboratory applications, designed with a structural identification approach.

INTRODUCTION

Many new bridges require extensive rehabilitation after only 10-20 years of service , while officials expect a service life of 75-100 years with routine maintenance. The average age of the nearly 575,000 bridges in the U.S. is over 35 years. Practically, 50% of the bridges are deemed deficient by the FHWA, and $5 Billion is needed every year for their rehabilitation. While the state-of-the-practice has provided a satisfactory level of public safety, inaccurate condition assessment has been identified as the most critical technical barrier to effective management (*Technology*, 1993; *Civil*, 1993). Condition assessment of bridges is generally conducted visually, lacking conceptualization of the complete system. Bridges are then typically rated by idealized models. This approach does not permit bridge reliability evaluation, which incorporates actual state and behavior parameters, relating local damage to global health.

OBJECTIVES AND SCOPE

New tools for condition assessment have been demonstrated on seven actual bridge test specimens. Analytical and experimental research has been integrated to explore the serviceability and damageability behavior. The strategy comprised field laboratory techniques designed with the structural identification concept. Field-calibrated analytical models represent the real structural behavior, enabling quantification of the actual relative contributions of different resistance mechanisms. Simulations by field-calibrated models help us to optimize the safety for all the bridge components, in the context of global performance.

A. E. Aktan, D. N. Farhey, V. J. Hunt, M. Lenett and A. Levi, University of Cincinnati Infrastructure Institute, P.O. Box 210071, Cincinnati, OH 45221-0071
V. Dalal, Ohio Department of Transportation

CONDITION ASSESSMENT METHODOLOGY

Although recent AASHTO Guides and Manuals encourage nondestructive field testing, no specific guidelines or technologies are described. A new manual for diagnostic and proof testing by loaded trucks for rating has been issued by AASHTO (*Manual*, 1994). While such tests are beneficial, the reliability and limitations of such tests are not yet fully understood. In the case of bridges, which have structural systems that cannot be analytically modeled or conceptualized, the reliability and safety of a proof test for condition assessment is highly questionable.

The writers formulated a structural identification methodology, applied to reinforced concrete slab bridges (Aktan et al., 1993b; Shahrooz et al., 1994; Miller et al., 1994) (Fig. 1) and steel-truss bridges (Aktan et al., 1995a) (Fig. 2) tested to damage and failure. Additionally, nondestructive field tests have been applied to three steel-stringer bridges (Aktan et al., 1993a) (Fig. 3). Structural identification of the steel-stringer bridges determined that the actual drift under legal truck loads is generally far smaller than L/800 (*Standard*, 1989), since many mechanisms providing larger than expected stiffness to bridges are ignored (Aktan et al., 1995b). The methodology, schematized in Fig. 4a, utilizes experimental and analytical tools transformed from various disciplines. The methodology leads to bridge management with the additional steps shown in Fig. 4b.

CONDITION ASSESSMENT RESEARCH

The proposed methodology permits bridge-specific and bridge-type-specific applications, offering excellent benefits especially for recurring bridge-types. Applications include statistical sampling and classification of bridge types; first or periodical follow-up applications; long-term on-line; or, intermittent monitoring applications. Electronic instrumentation and technologies have been field-proven for each application. The first six steps shown in Fig. 4a lead to a geometric analytical model which serves for evaluating bridge reliability throughout the serviceability limit states. For the vast majority of highway bridges, which belong to types with large populations, it is possible to develop type-specific condition assessment approaches that would be considerably more practical than the bridge-specific approach.

Bridge-type-specific condition assessment will be suitable for bridges whose state parameters at limit states and behavior can be expressed in terms of a small number of geometric and structural parameters. The recurring operations of the bridge-type-specific condition assessment and management methodology are shown in Fig. 5. The statistical sample would be comprehensively tested and 3-D finite-element models identified, similar to the bridge-specific process described earlier. Based on a careful study of the state parameters at limit states and behavior of the statistical sample, it would be possible to design practical experimentation for the complete population that would reveal an objective quantitative measure of condition and reliability. This data, in conjunction with the knowledge-base generated by comprehensive studies on the statistical sample, should reliably provide a quantitative measure of structural condition.

Destructive tests are the only realistic means of determining the state parameters at advanced limit states and for properly calibrating the tools which will project the results of linearized identification to future performance and remaining bridge capacities. More importantly, destructive tests are the only means, other than costly bridge failures during service, that reveal the influence of defects, deterioration, and damage on the structural capacities and failure modes.

Figure 1.   Figure 2.

Figure 3.

#	ANALYSIS	TEST	OPPORTUNITY
1	Create database & initial FEM	Digital image proc. & rev. CAD	*Conceptualize*
2	Classify modal parameters	Material & prel. modal test	*Conceptualize*
3	Parameter sensitivity studies	Rigorous modal testing	*State parameters*
4	FEM calibration	Instrumentation & monitoring	*Damage evaluation*
5	Completeness & reality check	Diagnostic test, local response	*Damage localization*
6	Optimized grid mod. calibration	Local probes	*Quantify local damage*

Figure 4a.

7	Nonlinear FE analysis	Destructive testing of samples	*Maintenance/renewal design*
8	Reliability evaluation	Intelligent system applications	***Management***

Figure 4b.

#	ANALYSIS	TEST	OPPORTUNITY
1		Digital image processing	Update database
2			
3			
4		Diagnostic testing, or instrumented monitoring by passive sensors	State/performance changes diagnosed
5			
6	Select grid model & customize		
7			Maintenance/renewal design
8	Reliability evaluation		Management

Figure 5.

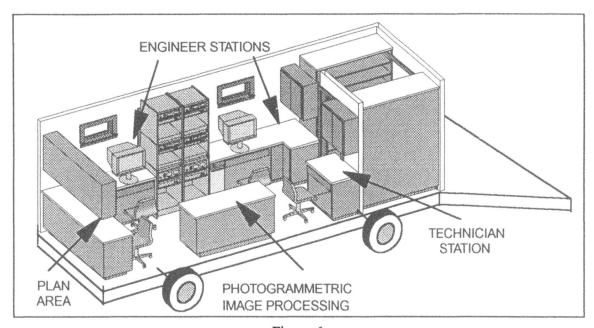

Figure 6.

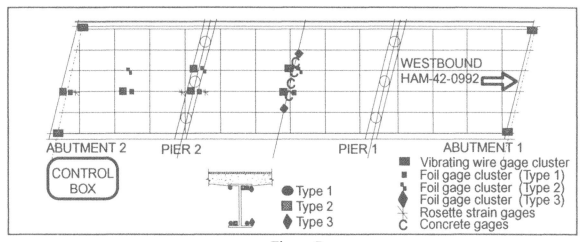

Figure 7.

CONDITION ASSESSMENT APPLICATIONS

A mobile field laboratory has been designed to conduct condition assessment research (Fig. 6). Its tools include modal testing by impact and by multi-input forced-excitation, instrumentation and diagnostic testing or short-duration monitoring under truck loading, digital photogrammetric image-processing, and local sampling for material characterization. Analysis capabilities include experimental modal models, 2-D grid and 3-D FEM, and parameter identification.

Instrumented health monitoring is expected to complement inspection methods, provide an objective measure of the state-of-health, and alert to deterioration or failure (Aktan et al., 1995c). Structural identification is a prerequisite for instrumented health-monitoring, since a calibrated FE model is needed for reliably designing the monitor. The most promising sensors and data acquisition systems have been investigated through a rigorous system-calibration conducted in the laboratory. Then, a 64-sensor monitor system was designed and installed on a 180-ft-long, three-span, reinforced concrete slab-on-steel grid type typical bridge which was identified (Fig. 7).

The pilot system for highway bridge monitoring will be used for long-term monitoring of the traffic and environment effects and the corresponding responses of a typical steel-stringer bridge. Static tests are intermittently performed with known truck loads to confirm monitoring results and to calibrate finite-element and section analysis models of the bridge.

REFERENCES

Aktan, A.E., Chuntavan, C., Toksoy, T., and Lee, K.L. (1993a). "Structural identification of a steel-stringer bridge for nondestructive evaluation." *Transportation Research Record*, 1393, TRB, 175-185.

Aktan, A.E., Zwick, M., Miller, R., and Shahrooz, B. (1993b). "Nondestructive and destructive testing of decommissioned R/C slab highway bridge and associated analytical studies." *Transp. Res. Rec.*, 1371, TRB, 142-153.

Aktan, A.E., Lee, K.L., Naghavi, R., and Hebbar, K. (1995a). "Destructive testing of two 80-year-old truss bridges." *Transportation Research Record*, 1460, TRB, 62-72.

Aktan, A.E., Farhey, D.N., and Dalal, V. (1995b). "Issues in rating steel-stringer bridges." To appear in *Transportation Research Record*, TRB, 1996.

Aktan, A.E., Dalal, V., Farhey, D.N., Hunt, V.J. (1995c). "Bridge reliability evaluation in the 21st century," *Proc., Research Transformed into Practice*, Colville and Amde eds., Arlington, VA, June 14-16, 493-505, ASCE Press.

Civil infrastructure systems research: Strategic issues. (1993). Report of the Civil Infrastructure Systems (CIS) Task Group, National Science Foundation.

Manual for condition evaluation of bridges. (1994). AASHTO, Washington, D.C.

Miller, R.A., Aktan, A.E., and Shahrooz, B.M. (1994). "Destructive testing of decommissioned concrete slab bridge." *J. of Struct. Engrg.*, ASCE, 120(7), 2176-2198.

Shahrooz, B.M., Ho, I.K., Aktan, A.E., de Borst, R., Blaauwendraad, J., van der Veen, C., Iding, R.H., and Miller, R.A. (1994). "Nonlinear finite element analysis of deteriorated rc slab bridge." *J. of Struct. Engrg.*, ASCE, 120(2), 422-440.

Standard specification for highway bridges. (1989). AASHTO, Washington, D.C.

Technology for America's economic growth, a new direction to build economic strength. (1993). President William J. Clinton and Vice President Albert Gore, Jr., February 22.

Bridge Inspection and Assessment—Virginia Department of Transportation

J. E. COLEMAN

ABSTRACT

The infrastructure of the United States is getting older, loads are increasing and funding is insufficient to meet the needs of repair, rehabilitation and replacement. Through practical applications of existing and new technologies, valuable information is acquired which assists personnel involved in bridge inspection, bridge management, bridge design and budgeting in providing for public safety and protecting the public investment.

HISTORY OF BRIDGE INSPECTION

In the early 1950's Virginia's bridge inspection program had been initiated and a standard reporting form had been developed. As early as 1954 Virginia issued a "Manual of Instructions For Movable Bridges". In 1957 a more comprehensive bridge inspection policy was put into place which covered the approximately 9,000 structures on the Primary and Secondary systems of Virginia. The Interstate system was not yet in existence. Structures were inspected, at intervals not to exceed one year, by a Bridge Maintenance Superintendent and findings reported on a standard form. This evolved into inspections being performed by Highway Construction Inspectors during the winter months, when most construction projects had been shut down. Underwater inspections were to be performed by qualified divers on a standard schedule. These divers were most often volunteer personnel who were state employees. Even with no formal training, inspections were being performed by those with some knowledge of structures. In 1968, Virginia placed special emphasis on truss span inspection and a two man team headed by an engineer was required to perform these inspections. Virginia's first formal training was held for those individuals who were to perform these truss span inspections. This training typically took the form of a one day exercise to concentrate on the important points to look for regarding truss span inspection.

To keep track of Virginia structures, a "Bridge Inventory Record" system was developed in 1960. With this system each structure was assigned a number according to its location. The cards on which the information was stored were color coded as to the type and capacity of the structure. Each card contained information on the individual structures geographic location, age,

John Coleman, Virginia Department of Transportation, Structure and Bridge Division, 1401 East Broad Street, Richmond, Virginia 23231 type of structure and physical characteristics.

On December 15, 1967 the 2,235 foot Silver Bridge crossing the Ohio River at Point Pleasant, West Virginia collapsed during rush hour traffic killing 46 people. This catastrophe influenced the United States Congress to revise the "Federal Highway Act of 1968". This revision required the Secretary of Transportation to establish standards for bridge inspection nationwide. In 1971, with the passage of the "Federal Highway Act of 1968" the "National Bridge Inspection Standards" (NBIS) became law. The National Bridge Inspection Standards set national policy concerning many aspects of the states bridge safety inspection program. It did this by mandating that each state's highway department include a bridge inspection organization, that structures should be inspected at intervals not to exceed two years and it set guidelines for the qualifications of those who would perform the inspections and oversee the program.

Virginia incorporated the federal standards into its own, more comprehensive, program. At that time the states were only required to inspect structures on the federal-aid system which met the federal definition of a bridge. In the early 1970's the federal program was expanded to included those structures not on federal-aid routes. Virginia continued to inspect additional structures, not required by the federal mandate, and included in its inventory all bridges and drainage structures regardless of length, functional classification of the road or the highway system on which a structure was located. The only exception was to exclude culverts and multiple pipes with total openings of less than 36 square feet.

In the late 1970's, Virginia created a computer database for the storage of all federally reportable items and those uniquely Virginia items. In the mid 1980's, Virginia renovated its structure inventory system. Data collected in the field enters the system by direct input or by uploading of computerized inspection reports. More change is on the horizon. The adoption of the "Intermodal Surface Transportation Efficiency Act of 1991" (ISTEA) by the United States Congress requires all states to develop six management and monitoring systems. Among these is a bridge management system. Virginia has chosen to subscribe to the PONTIS bridge management software which has been developed by the Federal Highway Administration (FHWA) and will be maintained and further developed by the American Association of State Highway and Transportation Officials (AASHTO). This gives Virginia the perfect opportunity to investigate more up-to-date methods of expediting and optimizing our data collecting procedures. At present a task force is investigating various techniques and systems to augment the data collection and data presentation aspect of bridge safety inspection.

INFRASTRUCTURE

There are 590,232 federally reportable bridges in the United States with 186,733 or 31.6% being defined as substandard (Better Roads, 1994). Substandard is defined as a sufficiency rating of 80.0 or less. The Federal Highway Administration developed the sufficiency rating formula. This formula assigns a sufficiency rating to each structure dependent on the information gathered for that particular structure. The ratings are assigned from 0.0 to 100.0, with 100.0 being good. Virginia has 12,429 federally reportable structures with 3,617 or 29.1% having a sufficiency rating of 80.0 or less (Better Roads, 1994). In addition to those structures federally mandated for inspection, Virginia inspects 7,513 additional structures making for a total amount of structures inspected in Virginia of 19,967.

The infrastructure of the United States continues to age and age is one of the most

TABLE I - BREAKDOWN OF STRUCTURES IN VIRGINIA BY AGE

Built	Years of Service	Number of Structures
1990 - 1995	00 - 05	1098
1980 - 1989	06 - 15	2175
1970 - 1979	16 - 25	3652
1960 - 1969	26 - 35	4596 *
1950 - 1959	36 - 45	1923
1940 - 1949	46 - 55	998 **
1930 - 1939	56 - 65	4647 ***
1920 - 1929	66 - 75	662
1910 - 1919	76 - 85	137
1900 - 1909	86 - 95	66
pre 1900	95 plus	13

*   the interstate system begins
**   World War II
***   due to the depression, government work programs begin (WPA, CCC, etc.)

comprehensive culprits in the weakening of our structures. In Virginia, the breakdown in age is shown in Table I. The life of a structure, prior to the structure requiring replacement or major repairs, is estimated to be 35 to 50 years. In Virginia, 42.3% of the structures are greater than 35 years old. In the next decade that number could grow to over 65% of all structures. Millions of dollars are spent every year in the constructing of new structures and/or the repair and rehabilitation of existing structures. A new structure can cost 75 times that of an equal length of roadway. Repair contracts for a single small simple structure can easily top $500,000 and for the more complicated and larger structures can easily top several million dollars. The use of existing funds in the most efficient manner should certainly be a top priority in any maintenance organization. It then becomes extremely important that the information the inspector gathers is quantified accurately. This information should then be used to determine the priority of the necessary bridge repair or if replacement may be a more economically viable alternative.

FIELD INVESTIGATION

A bridge safety inspection team is composed of 2 or 3 individuals for a typical above water inspection. Size of the team can vary dependent on the size and complexity of a structure. Each team is led by an individual who is either: 1) a registered professional engineer or 2) an individual who has a minimum of 5 years experience in bridge inspection assignments in a responsible capacity and has completed a comprehensive training course based on the "Bridge Inspector's Training Manual" or 3) an individual who has current certification as a Level III or IV Bridge Safety Inspector by the National Institute for Certification in Engineering Technologies (NICET) which is a program sponsored by the National Society of Professional Engineer's. The remaining portion of the team is made up of members with varying levels of experience.

Included in the total number of structures in Virginia, are 500 structures requiring an underwater inspection. The majority of the underwater inspections are performed by our own divers. Some of the underwater inspections are performed by consultant inspectors when the structure is very large or the dive is more complicated. We have one underwater bridge safety inspection team composed of four individuals who perform 100 to 150 underwater inspections per year. This underwater team is lead by an individual who has the same qualifications as the leader of the above water bridge safety inspection teams. In addition, this individual has extensive training in diving.

The field investigation begins with the review of the plans and the past reports to determine the need for specialized equipment. Access must first be gained to each element of the structure. With some structures carrying twelve lanes of traffic over an interstate route or with structures 165 feet above groundline or with water depths which approach 100 feet, easy access to the elements of the structure so a hands-on inspection can be made cannot be taken for granted. Access to some areas of some structures can be very dangerous or strenuous. When the element of danger is present or when the inspector is fatigued, it is very difficult for the inspector to concentrate on the detection of a problem. Special rigging, scaffolds, underbridge inspection vehicles, bucket trucks and traffic control may have to be scheduled for the inspector to be able to perform the job safely and accurately. If entry is required into confined spaces then special air sampling equipment will be required along with the need for individuals trained in access to and egress from these areas. Non-destructive testing equipment may be needed if there are fracture critical or fatigue prone details. The structure may need to be scheduled for inspection at a special time due to vehicular traffic (rush hour) or there may be navigational consideration for our 14 movable spans (swing spans and lift spans). These special access considerations are usually very costly. In addition to being costly they add additional time and safety considerations to the inspection. Any additional time spent exposes our inspectors to not only the elements but more importantly to traffic. Each year many individuals nationwide who are involved in road and bridge construction or inspection are killed by motorists, who, for one reason or another, fail to obey the advanced warning signs. Lane restrictions, which are often necessary, not only expose inspection personnel to traffic but causes delays for the traveling public and can result in traffic accidents. Anything the inspector can due to reduce exposure time to traffic and to reduce the inconvenience to the traveling public is certainly going to be appreciated by all.

The major part of any inspection is visual. What the inspectors see can tell them much about what is occurring on or in the rest of the structure. Different types of material will be inspected visually for problems characteristic to that type of material. There are no specialized tools for the bridge safety inspector. The tools used have been adapted from other industries. A typical list of tools would include whisk brooms, wire brushes, shovels and scrapers for cleaning; pocket knife, ice pick, drill, chipping hammer, increment bore, inspection mirror, binoculars, ladders, brush axe and flashlight for inspection; rulers, calipers, crack gauges, paint

TABLE II - SUPERSTRUCTURE TYPES AND MATERIAL

Type/Material	Bridge	Culvert
Concrete	5142	4782
Steel	7693	2269
Timber	61	0
Masonry	13	10
Wrought Iron or Cast Iron	4	0
Other	0	3
Total	12913	7064

film gauges, thermometer, plumb bob and levels for measuring; pens, pencils, clipboard, camera, chalk, keel and spray paint for documenting. Currently available to each inspector in Virginia are the following more specialized equipment:

Potential meters - these meters determine the possibility that corrosion is or will affect the reinforcing steel in concrete.

Ultrasonic thickness gauges - used for determining the thickness of metal remaining in a corroded area where access to both sides of the member is not possible and therefore a direct measurement cannot take place, such as where the top flange of a steel beam contacts the deck or where there is a built-up member.

Dye penetrant - used to determine the length of an existing crack.

D-meters - for determining the depth of reinforcing steel in concrete.

The structures of the United States are made up of many materials. In Virginia, the breakdown of superstructures by type and material is shown in Table II. Many material types are difficult to inspect properly. In timber and concrete it can be difficult to determine the extent of some problems without the use of destructive testing. The destructiveness of the test can cause a condition worse than the original problem. For instance, the destroying of timber piling to quantify a problem may cause a condition worse than that which originally existed. Breaking the protective coating of the treated timber can cause a problem to progress at an accelerated rate. Another consideration for the team leader is the weather. Clear, cool, sunny days with no fear of bees, wasps and snakes are rare. If traffic control or climbing is required, it should not be scheduled if rain, sleet or snow is in the forecast. Climbing wet steel or up muddy embankments can pose a serious threat to the inspector. Very hot or very cold days increases the possibility of **heat or cold related first-aid emergencies.**

PUBLIC SAFETY AND INVESTMENT

The task of inspecting a bridge is, in most cases, a complex technical assignment requiring specialized skills and experience. Moreover, each bridge safety inspector shares a tremendous dual public responsibility. The inspector is relied upon to provide for public safety and to protect the public investment with respect to bridges. The inspector discharges these responsibilities through inspection and the reporting of bridge conditions and situations which would impair, or potentially impair, either the publics safety or the publics investment. If what the inspectors find is not fixed in an appropriate manner and time period, it could lead to longer closure times for repairs, permanent closures rather than temporary closures, permanent or more prolonged reduction in capacity of structures, more costly repairs and accidents which could lead to injury or death of the traveling public.

From 1967 to 1993 there have been 196 fatalities from structure collapses across the United States. These collapses have occurred from West Virginia to California and from Alabama to Connecticut. Virginia has been fortunate that there have been no fatalities on our system of structures due to structural or mechanical problems. But no state is exempt from this possibility. The "low bid" system along with the desire to build the most economical design has led to structures being built with less redundancy and increases the possibility of catastrophic failures.

In Virginia, our inspectors have been responsible for finding possible life threatening problems. A six foot crack was found in an eight foot girder in one of only three girders of a structure on Interstate 64 in Rockbridge County, on Route 712 in Brunswick County a concrete cap had sheared off causing the deck to settle 1 1/2", and on Route 460 in Giles County half of the beams had their steel pins shear. In all cases, traffic on each structure was immediately restricted and repairs were developed. A well organized inspection program and individual inspectors who were diligent, knowledgeable and experienced diverted certain disaster.

CONCLUSION

It is important that we realize that the job an inspector does affects the lives of hundreds if not thousands of families everyday. Families that the inspector may never see and surely seldom, if ever, will see the inspector. But there is a trust that those who cross a particular structure have in the inspectors knowledge, skills and abilities. We must show stewardship of this trust by doing our job as well as we possibly can. By doing so we will continue to protect the publics safety and the publics investment.

REFERENCES

"Better Roads", Annual Bridge Report, November 1994.

Federal Highway Administration (FHWA), "Bridge Inspector's Training Manual 90", Report No. FHWA-PD-91-015. Washington, D.C.: United States Department of Transportation, 1991.

Use of Impact-Echo Method in Bridge Monitoring

M. S. AGGOUR AND H. M. FOUAD

ABSTRACT

A study performed for the State of Maryland Department of Transportation, SHA, for utilizing proven nondestructive testing techniques to monitor the performance of concrete bridges as a methodology of their routine periodic inspection and evaluation, identified two types of evaluations: global evaluation and local evaluation. The paper presents an analysis of the local evaluation aspect, i.e., the material integrity. The most practical and feasible technique in the determination of material integrity was found to be the Impact Echo method, which is based on emitting a signal at a point in a bridge's element and picking it up at another point. Experimental as well as numerical analysis of the response of concrete elements to such signals for both sound and artificially flawed concrete elements are presented. The methodology of the determination of the size and location of internal flaws in concrete elements is also presented. The method can thus enable the determination of physical properties and physical defects of the main bridge elements and any degradation that occurs during the bridge's life.

INTRODUCTION

Assessing the remaining life and verifying the structural integrity of bridges is of great importance. This is essential for an economical decision regarding their rehabilitation. Many of the nation's nearly half a million bridges may suffer from a variety of serious structural defects and engineering problems. In addition to the problem of inadequacy of inspection, engineers are now implementing higher strength materials when designing, while methods of inspection have not kept up with these developments. Therefore, a quantitative inspection technique with specific guidelines needs to be developed. This new technique must rely on quantitative values to define and evaluate the safety and condition of a structure. Evaluation of the safety of a structure should cover at least two areas, determination of the integrity of the materials used (i.e., local evaluation), and the deformation and rigid-body motion of the whole structure (i.e., global

M.S. Aggour, Professor of Civil Engineering, Civil Engineering Department, University of Maryland, College Park, MD 20742-3021

H.M. Fouad, Assistant Professor of Civil Engineering, Higher Technological Institute, Tenth of Ramadan City, Egypt

evaluation) (Aggour and Fouad, 1991). Since there are devices and techniques readily available with complete data acquisition systems to assess the global condition of a structure, our discussion will be limited to assessing the integrity of the concrete itself (local evaluation). A structure's integrity of material includes the existence of any type of flaws (e.g., cracks, honeycombs, voids, etc.).

This study was initiated to assess the possibility of using the impact echo method as a quantitative nondestructive testing technique for routine inspection and evaluation of concrete elements. By instrumenting a structure with sensors and transducers during construction such an inspection could be undertaken. An experimental as well as numerical analysis of the response of both sound and artificially flawed concrete elements to impact signals was investigated. Results were used to plot relationships between the resonant frequencies, length of the tested concrete element, and sizes and locations of internal flaws. These plots would provide the quantitative relationships required.

EXPERIMENTAL ANALYSIS

To provide the required quantitative evaluation needed for routine inspection the relationship between the resonant frequency and the location, size, and thickness of the internal flaws must be developed. To obtain such a relationship, both sound, good quality concrete cylinders 12 inches long and 6 inches in diamter, and artificially cracked cylinders were tested. The testing was accomplished using an impulse hammer, an accelerometer, and a fast fourier transform (FFT) analyzer. The impulse hammer was used to strike the cylinder on one side at its center. The traveling signal was then received by the accelerometer attached to the opposite side of the cylinder, as shown in Fig. 1. Testing was then repeated with the accelerometer attached to the same cylinder surface as the hammer.

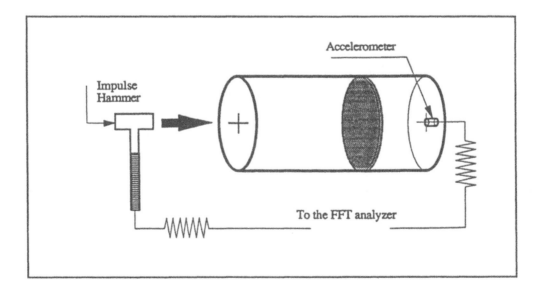

Figure 1. Schematic of the Impact Echo Testing

Relating the resonant frequencies on one hand with crack sizes and locations from the surface on the other hand provided a quantitative relationship that could be used for routine inspection and evaluation of concrete elements. After testing a number of concrete cylinders having artificial cracks with different thicknesses, it was found that a crack width of 0.2 inches was too wide to allow a pulse wave to pass through. The detection of cracks having sizes equal to or larger than that size was fairly simple as the pulse wave was totally reflected back. The emphasis here was placed on small thickness cracks that allowed part of the wave energy to pass through. Four sets of artificially cracked cylinders were prepared with crack thicknesses of 0.04, 0.08, 0.12 and 0.16 inches, respectively. For each set, cylinder lengths ranged between 5 and 24 inches with a 1 inch interval for the location of the artificial crack.

An example of the results is shown in Fig. 2. A number of observations from these experimental results were determined and can be summarized as follows: First the recorded resonant frequency peaks were not affected by the locations of the impulse hammer and the receiver (i.e., whether they were on the same side or on opposite sides). This was expected since the measured frequency was the resonant frequency for a certain concrete length. Second, when the impulse hammer and the accelerometer were on the same side there was another frequency peak. This peak represented waves reflecting back and forth between the concrete surface and the embedded crack. This frequency peak was used to locate the exact position of the embedded crack from knowledge of the concrete's dynamic modulus. Third, placing the transmitter and receiver on opposite sides could not be used to locate cracks efficiently. This was because when testing a section having a crack at a certain distance from the transmitter, then testing another section having a crack at the same distance from the receiver, two identical readings were recorded. The only difference was in the amount of power stored in each signal. The closer the crack was to the transmitter, the lower the power.

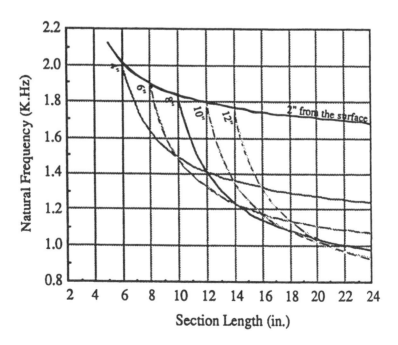

Figure 2. Section Length Versus Resonant Frequency (crack 0.04" thick), Experimental

NUMERICAL ANALYSIS

The experimental study showed that the impact echo method can provide quantitative relationships between resonant frequencies, section lengths, and the thicknesses and locations of embedded cracks, which can be used for routine inspection of concrete elements. However, these relationships were bounded by the maximum, yet practical, length of concrete element that can be tested in the laboratory, i.e., 24 inches. Therefore, in order to be able to implement the proposed technique for a wide variety of concrete structures, these relationships had to be extended to cover concrete sections of greater lengths.

One way of achieving this goal was by using finite element analysis. First the tested concrete cylinders were modeled by the finite element method. A comparison between the results from the finite element solution and those from the experimental study was made. Good agreement between them was noted. The finite element models were then used to expand the experimentally determined relationships to cover a wider range of concrete lengths. In this study a general purpose finite element package (ANSYS), developed by Swanson Analysis Systems, was used, and all the finite element analysis was performed on a VAXstation 3100.

The magnitude of the resonant frequency determined from the modal analysis for sound concrete with no artificial cracks was found to level off at a concrete length of 60 inches or more, see Fig. 3. Therefore it was decided to use this technique for concrete sections less than or equal to 60 inches. Relationships between resonant frequencies, crack thicknesses, and crack locations from the surface using the finite element models were then determined, an example of which is shown in Fig. 4. A quick glance shows the similarity between this figure and Fig. 2 from the experimental study.

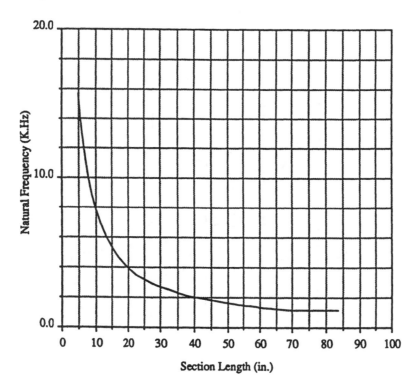

Figure 3. Resonant Frequency Versus Section Length for Sound Concrete

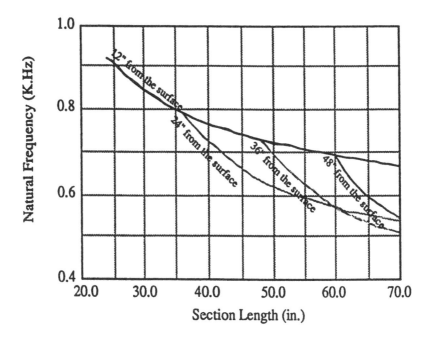

Figure 4. Section Length Versus Resonant Frequency (crack 0.04" thick), Analytical

DEVELOPED INSPECTION TECHNIQUE

The preceding information helped in developing a quantitative inspection technique with specific guidelines for the evaluation of the performance of concrete structures under service loads. The procedure is presented in detail in Fouad, 1994. It can be summarized as: (1) Determine the distance between the emitting signal and the receiving point. (2) Determine the frequency response. (3) Check the value with Fig. 3 or one similar to it. If the value agrees with the figure, the concrete is sound. (4) If the value does not agree with Fig. 3 refer to one like Fig. 4. From this figure, the crack's thickness as well as its location from either of the two concrete surfaces can be determined. (5) By exchanging the places of the transmitter and receiver, and comparing the power spectrums of the two setups, the signal that has the lower power spectrum indicates that the crack is closer to the transmitter for that particular setup. (6) If no signal was received this means that the crack is wide enough to prevent the energy of the pulse wave from passing through. To determine the location of such a crack, testing has to be repeated while placing the transmitter and the receiver on the same concrete side.

This technique will enable engineers to determine the integrity of the material at the time of construction of a structure and at different times during its life. The integrity of the material includes the determination of the uniformity and strength of the concrete and the existence of any type of flaws (e.g., cracks, honeycombs, voids, etc.).

CONCLUSIONS

This research study investigated the use of a new quantitative nondestructive testing approach in the routine inspection and evaluation of concrete structures throughout their life. This work, coupled with earlier work and ongoing research, can be used as the basis for a reliable, versatile, and practical method for the conditional assessment of concrete structures. It can be used for detecting a variety of cracks with different thicknesses, locations, and cross-sectional areas. It can be used to detect and locate discontinuities such as lack of consolidation, segregation, voiding, and material nonuniformity, as well as to detect crack propagation. It can also be used to assess damage due to such unusual occurrences as fire, flooding, and collisions.

ACKNOWLEDGMENTS

The research reported herein was sponsored by the Maryland Department of Transportation State Highway Administration. The authors are grateful for their support. Sincere appreciation is due to Mr. Earle S. Freedman, Chief Engineer, Bridge Development, State Highway Administration.

REFERENCES

Aggour, M.S. and Fouad, H.M., 1991. "Use of Nondestructive Testing Techniques as a Routine Inspection Procedure", Final Report No. MD-92104, Maryland Department of Transportation. SHA, June, 95 pages.

Fouad, H.M., 1994. "Nondestructive Evaluation of Concrete Foundations and Structures", dissertation presented to the University of Maryland in partial fulfillment of the requirements for the Ph.D. degree.

Development of a Wireless Global Bridge Evaluation and Monitoring System

K. MASER, R. EGRI, A. LICHTENSTEIN AND S. CHASE

ABSTRACT

A system for remotely monitoring the condition and performance of bridges is being developed and tested. The system consists of small, self-contained battery powered transducers that measure and record displacements, strains, rotations, and accelerations of key bridge elements. The recorded measurements are transmitted via a radio transponder to a local controller near the bridge, which in turn transmits the data via satellite, cellular link, or phone line to the owner agency. The system is intended to provided real time data collection and processing for bridge load testing, and to permit long term monitoring of overloads, consumed fatigue life, and movement of piers and abutments. The long term monitoring configuration can also record and report critical events, such as impact by a barge, which may require immediate action. The data will be integrated into current bridge management systems to provide accurate and timely condition information which will translate into more effective maintenance and rehabilitation programs.

BACKGROUND

Accurate monitoring of bridge conditions is an essential part of implementing a cost-effective safety conscious bridge management program. Bridge condition monitoring, however, is a complex task. Bridges consist of many components, the critical components differ from bridge to bridge, and there are many bridges to monitor. Biennial inspections provide qualitative ratings of the conditions of each of these components based on visual, hands-on observations. More in-depth information about bridge behavior can be obtained with a bridge testing program requiring the installation of gauges throughout the bridge and the wiring of these gages to a central data collection system (Moses, et. al., 1994; El Shahawy and Garcia, 1992; Bakht and Jaeger, 1990). The cost and time involved limits the number of structures which can be economically evaluated by bridge testing.

Kenneth Maser, President, INFRASENSE, Inc., 14 Kensington Rd., Arlington, MA 02174
Robert Egri, Research Fellow, M/A-COM, Inc., 100 Chelmsford St., Lowell, MA 01853
Abba Lichtenstein, Bridge Consultant, 26 Trafalgar Rd., Tenafly, NJ 07670
Steven Chase, Structures Division, FHWA, Turner-Fairbanks Highway Reasearch Station, 6300 Georgetown Pike, McLean, VA 22101

Bridge Management Systems (BMS's) have now been implemented which can evaluate large quantities of bridge data for the purpose of projecting and allocating maintenance and rehabilitation resources (PONTIS, 1994; NCHRP, 1994). Information on bridge conditions comes from regular biennial inspection programs, which may not necessarily reveal critical conditions that may lead to precipitous failure. Small fractures can propagate rapidly over a short period of time, and scour can quickly reduce the bearing capacity of pier foundations. It is important that these rapidly developing conditions be monitored and reported in a cost-effective manner so that major failures can be prevented and repairs can be carried out in a timely fashion.

The availability of low-cost, highly accurate sensors, together with data acquisition and data transmission systems offers a new potential for monitoring of bridge conditions. With these systems a higher level of information can be obtained on a larger number of structures, and the data can be transmitted at any desired rate to a central management facility. This information will enhance the day to day implementation of a bridge management program, improve current bridge load rating procedures, and provide a safety net for timely identification of critical conditions.

SYSTEM DESCRIPTION

A generic data collection system has been developed that is applicable to a variety of measurements and can be implemented on a large number of bridges at relatively low cost. The system concept is shown in Figure 1. The wireless telemetry system eliminates the cost of communications wiring and provides the user with deployment and relocation flexibility. The proposed *Wireless Global Bridge Evaluation and Monitoring System* (WGBEMS) is based on the use of small , self contained, battery operated transducers which are easily attached to key structural elements. Each transducer contains (a) a sensor (which measures acceleration, strain, or some other parameter); (b) a small radio transponder, which transmits the measurement to a nearby receiving station; and (c) a battery.

To conserve battery power the transponders will be designed to report their status back to the local controller at preset time intervals, say once every hour or day, depending on the application. One unique reporting time slot is assigned to each transponder by the local controller. The data gathering units will contain a low power clock that will periodically activate the instrument. Also, a sufficient time gap will be left between slots to prevent collisions caused by clock drift and timing errors. By carefully budgeting power requirements and information requirements, a transducers can remain operational for up to five years with a single lithium battery pack.

As shown in Figure 2, the system will consist of a local controller placed off the bridge and several transducers distributed throughout the bridge. Each transducer consists of a sensor, a processor, and a radio that will relay the signal measured by the sensor to the local controller . The data collection at the transducer involves signal conditioning, filtering, sampling, quantization and digital signal processing. A robust communication link ensures the accurate radio transmission of the collected data. The proposed radio link uses a wide band direct sequence spread spectrum waveform with a frequency agile carrier in the 902 to 928 MHz ISM band. The effective information rate in the transducer-to-controller link will be 20 kbits/sec. The reverse link, which is used exclusively for short control messages, will operate at 200 bit/sec.

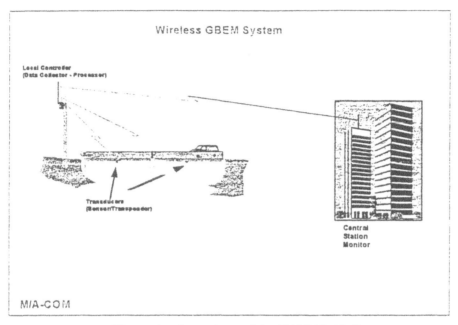

Figure 1 - Overview of the WGBEMS Concept

Any number of these transducers can be attached to a bridge depending on the nature of the structure and the information to be measured. No external wires are involved, since the data from each transducer is transmitted by radio to a local controller. The local controller, located near the bridge will receive data from all the transducers on a given bridge, and will transmit that data via cellular or satellite transmission, or via regular telephone line to a central station for downloading, processing, and eventual entry into the bridge management system (BMS). The local controller will also transmit control signals to the transducers to initiate and terminate data collection. To minimize the total quantity of transmitted data, some preprocessing will be carried out at the local controller, and some can be carried out within the transducers themselves.

The WGBEMS is based on the implementation of existing technology for the measurement of strain, rotation, and displacement, and is designed to accommodate any one of a number of available sensors. The number, location, and type of sensor, and the interpretation of the measurement, will be dictated by the type of structure and the information needs of the bridge management system. In production quantities, the cost of the proposed WGBEMS including bridge sensors, processors, radio transponders and power supplies is estimated not to exceed $1000 per transducer plus $2000 per local controller.

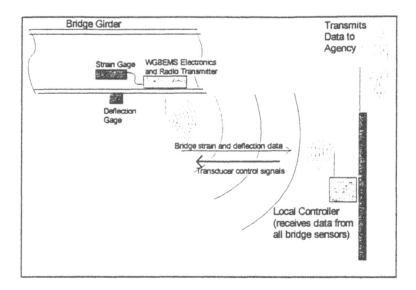

Figure 2 - Bridge Transducers and Local Controller

IMPLEMENTATION

Five implementation concepts have initially been proposed for the Wireless Global Bridge Evaluation and Monitoring System (WGBEMS). Two of the concepts involve bridge testing, in which conventional wired instrumentation systems are replaced by wireless data acquisition for the duration of a test. The three remaining concepts involve implementation of wireless data collection for long term monitoring of overloads, fatigue life, and substructure movement.

LOAD RATING USING DIAGNOSTIC TESTS

Key members of the bridge are instrumented, and their response to a known truck loading is recorded and compared to theoretically computed values used in standard bridge rating procedures. The computed bridge rating is then adjusted using the results of the load test (Lichtenstein, 1993).

Using the WGBEMS, wireless strain sensors attached to bridge members transmit bridge response information via radio link through the local controller, to the test truck, where it is recorded on a PC along with longitudinal truck position. The loading truck is outfitted with a fifth wheel to measure longitudinal travel distance, a PC/data acquisition system, and an antenna.. The local controller is set up to the side of the bridge with line-of-sight to the WGBEMS bridge sensors and to the truck's antenna. The operator can observe the bridge response directly on the computer screen, and quickly check the performance of all of the sensors. One or more slow crossings of the truck and/or static tests over the bridge can be made with the test truck in a reasonably short period of time. Once crossings are completed, the test is done and the measured results can be transferred to the DOT or owner agency for further processing.

LOAD RATING USING PROOF TESTS

In this application, the response of bridge elements is measured using a sequence of known truck loadings, each of progressively increasing magnitude, up to the calculated target load. This target load is computed specifically for the bridge being tested. The linearity of the measured responses are observed. The test is discontinued when nonlinearities are observed, at which time the load on the bridge is recorded.

The measurement is implemented in the same way as application (1) above. In this case, the test trucks cross the bridge or are placed statically at various locations, and more than one truck may be utilized in multiple lanes. During each loading the data are plotted and displayed so that the operator can ensure that all sensors are functioning properly. The PC in the truck will plot the response of each instrumented element vs. test truck load, so that linearity can be quickly evaluated. In some circumstances, a combination of proof test and diagnostic test may be beneficially implemented.

MONITORING OVERLOAD OF POSTED BRIDGES

In this application, sensors are mounted to key structural elements in order to monitor the total vehicle load applied to the bridge (like a bridge Weigh-in-Motion system). The relationship between load and sensor response will be calibrated upon initial installation (in fact, the load tests of applications (1) and (2) can be the calibration step for this application). The sensors will then count the number of overloaded trucks crossing the bridge and the strain level associated with each overload.

In this application, the local controller transmits data to the DOT or owner agency, rather than to a test truck. The sensors are operational for representative monitoring periods; e.g., one or two weeks at different times of year. The period will be determined based on local traffic data. Overload detection requires an algorithm which reports only those events where a threshold strain or displacement has been exceeded. At the end of each day of operation, the recorded events will be transmitted from the transducers to the local controller. At a transmit rate of 20 kbits/sec., 100 overload events occurring at one transducer in one day of operation produces 4000 bits of data which will require 0.2 seconds of transmission time. For the rest of the time, the radio can be "off", thereby conserving battery capacity.

MONITORING STRAIN CYCLES ON DETAILS OF FATIGUE-SENSITIVE BRIDGES

In this application, fatigue-sensitive bridge details are instrumented with WGBEMS transducers containing strain gauges. The strain gages provide data that can be used to carry out fatigue life calculations. The WGBEMS strain gauge transducer will be designed to operate continuously for sample time periods throughout the year (e.g., two one-week periods). A strain histogram is recorded and transmitted to DOT or owner agency after each sample period via the local controller, where it is processed and interpreted.

DETECTION OF ROTATION OF PIERS

The rotation of piers caused by loss of foundation support or by impact from water vessels can be monitored using a WGBEMS sensor containing an inclinometer or tiltmeter. The tilt measurement can be made at specified intervals (e.g., 4 times per year). The inclinometer can be attached to each pier, with a direct line of sight to the local controller. The system will be timed to make readings at regular time intervals. Each "reading" may have to be averaged over a 24 hour period to factor out the influence of daily temperature cycles. The tilt data will be transmitted to the DOT or owner agency where it can be processed to check for undesirable trends.

INTEGRATION WITH BRIDGE MANAGEMENT SYSTEMS

The value of a the WGBEMS can be realized by integrating the collected data into bridge management systems (BMS's). The following concepts have been proposed for this integration:
1) Use the Load Ratings generated from WGBEMS data to adjust values in those BMS's which incorporate predicted changes in load rating over time (e.g., BRIDGIT)
2) Use WGBEMS measurements of fatigue cycles to adjust condition ratings of specific details and bridge components.
3) Use WGBEMS measurements of the frequency and magnitude of overloads as a basis for possible "emergency response" inspections or remedies.
4) Use WGBEMS measurements of both fatigue cycles and overloads to recommend changes in regular inspection intervals for particular bridges.

Items 1 and 2 can be integrated into current BMS structures, while items 3 and 4 require the implementation of additional capabilities.

FUTURE DIRECTIONS

An experimental WGBEMS prototype is being fabricated, and field testing of this prototype on an in-service bridge is planned for early in 1996. Upon successful completion of these tests, future work will include the conduct of a pilot study on a number of bridges using a second generation prototype system.

ACKNOWLEDGMENTS

The authors would like to acknowledge the Federal Highway Administration for their support of this work under contract DTFH61-94-C-00217.

REFERENCES

Bakht, B. and Jaeger, L. (1990) "Bridge Testing- A Surprise Every Time." *Journal of Structural Engineering,* Vol 116., No. 5.

El Shahawy, M. and A. Garcia, A. (1992) "Structural Research & Testing in FLA." *Transportation Research Record* No. 1275. Transportation Research Board, National Research Council, Washington, D.C.

Lichtenstein, A.(1993) Bridge Rating Through Nondestructive Load Testing. NCHRP 12-28(13)A. Manual For Bridge Rating through Load Testing. Final Draft.

Moses, F., Lebet, J-P, and Bez, R. (1994) "Applications of Field Testing to Bridge Evaluation." *Journal of Structural Engineering,* Vol. 120. No. 6.

NCHRP (1994) "BRIDGIT Technical Manual," Draft Technical Report, National EngineeringTechnology Corporation, Pre-Release Copy.

PONTIS (1994) "PONTIS Version 2.0 Technical Manual," Report FHWA-SA-94 031, Federal Highway Administration, Washington, D. C.

Bridge Non-Destructive Testing Using an Optical Fiber Sensor System

R. L. IDRISS and M. B. KODINDOUMA

ABSTRACT

An optical fiber sensor system is being developed for bridge monitoring. The research study is an on-going multi year program, to be implemented in several phases, (a) large scale component testing, (b) full scale bridge testing in the laboratory and finally © full scale implementation in the field on existing bridge structures. In the current phase of the project, embedment of the sensors in structural elements is investigated. Large scale concrete beams are constructed in the lab, and damages in the form of delaminations are modeled in the beams during construction. The sensors are used to monitor the beams.

INTRODUCTION

A tough task that faces the bridge engineer when inspecting a deficient structure is damage assessment. The objective of this research project is to investigate the feasibility and effectiveness of an integrated optical fiber sensor system for bridge monitoring. The passive sensor system would be used to:
- Assess the loading history
- Check the performance compared with the design assumptions
- Detect abnormal conditions during routine surveys
- Assess the effectiveness of repairs and maintenance programs.

Use of embedded optical fibers can result in the non-destructive measurement of internal stress and strain. On the other hand, for monitoring existing structures, fibers can be bonded to the surface of the structure.

To date, efforts to integrate optical fibers in structures have mainly focused on small scale component testing (Fuhr et al.,1992) , and only a very limited number of large scale testing has been performed (Maher and Nawy, 1993; Measures and Alavie, 1994). To adapt this technology to infrastructure problems further development and engineering is needed.

R. L. Idriss and M. B. Kodindouma, New Mexico State University, Department of Civil Engineering, Box 3CE, Las Cruces, NM 88003-8001

FIBER OPTIC SENSING SYSTEM

Optical fibers are geometrically versatile and can be configured to arbitrary shapes. Besides flexibility, and extremely small size (250 micron in diameter), optical fibers are immune to electric and electromagnetic interference, which can seriously affect the performance of electric and piezoelectric sensors. Probably one of the most attractive features of fiber optic sensors is their inherent ability to serve as both the sensing element and the signal transmission medium, therefore greatly simplifying the instrumentation of large structures.

The sensor selected for use in this study is the fiber optic bragg grating sensor. It is an intrinsic sensor, i.e. sensing occurs along the fiber itself, while the light remains guided through the fiber. The fiber optic intracore bragg grating is a segment of an optical fiber that has been internally modified by exposure to UV light so that it reflects light at one wavelength.

The bragg grating sensor is used for point sensing. The sensor's wavelength depends on the strain and temperature imposed on the optical fiber at the location of the bragg grating. Therefore, as the strain changes the wavelength at which the light is reflected shifts. The Bragg grating sensors have a linear response up to rupture, and were proof tested at 5,000 micro strains. The gage factor accuracy is ±0.5% to 0.8%. A gage length of 0.04 in. (1 mm) was used.

A multi channel fiber laser demodulation system, the FLS3000 developed by Electrophotonic corporation is used to interrogate the fiber optic sensors. The instrument was calibrated to cover a measurement range of ±4,000 micro strains.

To check the accuracy of the fiber optic sensors, strain readings using the Bragg grating sensor were compared to strain measured using the resistive gage. An intracore Bragg grating sensor was mounted next to a foil gage at the center of an aluminum plate. The specimen was 10 in.(25.4 cm) long, 1 in.(2.54 cm) wide and 1/8 in.(0.318 cm) thick. A uniaxial direct tension test was performed on the specimen. Results correlated well (Fig.1).

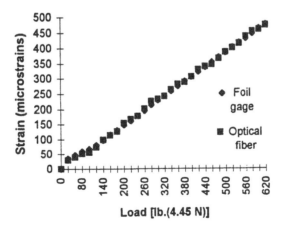

**Fig 1. Correlation between Optical fiber and
Foil gage**

LABORATORY COMPONENT TESTING

The objective of this phase was to investigate the feasibility of optical fibers as an embedded sensor system for monitoring of bridge components. Optical fiber sensors are embedded in reinforced concrete beams. Intact as well as damaged beams are monitored in a test to failure. Damages in the form of delaminations resulting in loss of cover and bond over a certain length of the beam are modeled in large scale laboratory beams.

The test beams are 10 ft (3.05 m) long, with a 6 in x 10 in (15.2 cm x 25.4 cm) cross-section. The tension reinforcement consists of 2 No. 6 rebars at an effective depth of 8 in (20.3 cm). Stirrups are placed at 4 in (10.2 cm) center to center on the outer third of the beam to avoid premature shear failure. For a loss of both cover and flexural bond in the laboratory beams, concrete is blocked out in the casting process, in order to simulate deterioration in such a way as to leave about one inch (2.54 cm) between the tension bars and the cast concrete. A control beam, and three damaged beams with center delaminations of 2 ft, 4 ft, and 6 ft (0.61 m, 1.22 m, and 1.83 m) respectively were studied.

To measure the strain in the steel, sensors were bonded to the tension steel. Two optical fiber gages were used. The first was embedded at midspan of the control beam and the second was embedded at the delamination edge in the 4 ft (0.22 m) center delamination beam. Optical fiber sensors and strain gages were placed side by side to compare results. Additional resistive gages were placed on the tension steel at 1 ft (0.305 m) intervals.

BEAM TESTING

A concentrated load was applied at midspan of each beam using a structural frame, and a hydraulic jack. The load was increased, by 200 lbs increments, up to failure.

In the elastic range, the strain measured in the tension steel over the delamination was much larger then the strain measured in the intact beam (Fig.2). At the delamination edge, the tensile stress is redirected into the steel, causing stress concentrations. Over the delamination, the strain in the steel is constant since there is no bond to the concrete.

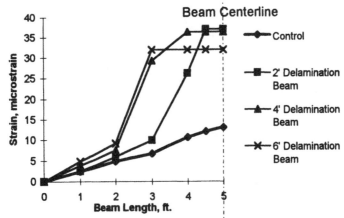

Fig 2. Strain in Steel due to 200# @ midspan

The strain-load curves in the tests to failure show a distinctive pattern. A large discontinuity can be seen at the onset of failure (Fig. 3 & 4). When a large crack suddenly opens and spreads, there is sudden added surface stresses at the interface between the steel and the concrete. The sudden surge of frictional forces at the interface is picked up by the optical fiber gage as an added compressive stress, causing the sudden discontinuity in the curve.

In the control beam, the optical fiber gage shows a large discontinuity at 9,400 lbs (41.8 KN). The ultimate load for the beam was 15,800 lbs (70.3 KN). In the 4 ft (1.22 m) center delamination beam, the optical fiber gage shows a large discontinuity at about 9,000 lbs (40.1 KN). The ultimate load was 13,800 lbs (61.4 KN). The foil gage at the vicinity of the optical fiber gage shows the same pattern, but at a slightly later time, and to a lesser extent. This could be due to the fact that although close, the gages were not at exactly the same location.

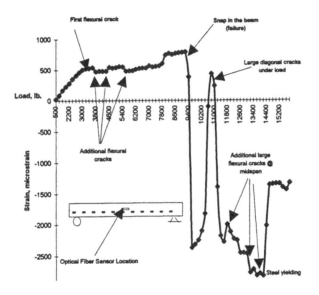

Fig 3. Optical Fiber Strain in Control Beam

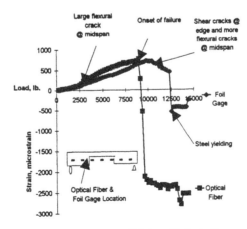

Fig 4. Optical Fiber and Foil Gage Strains in 4'
Delamination Beam

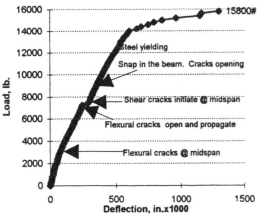

Fig 5. Control Beam Deflection @ Midspan

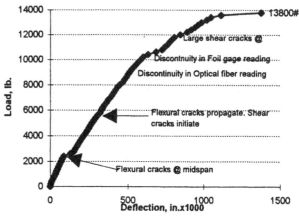

Fig 6. Deflection @ Midspan of 4' center delamination beam

The failure of the beams was ductile as can be seen in the load deflection curves (Fig.5 & 6). The control beam developed typical flexural cracks at midspan, and a final diagonal shear crack at ultimate load. The damaged beams showed typical shear cracks emanating from the delamination edges as well as flexure cracks at midspan (Fig.7 & 8).

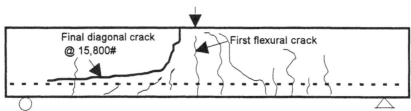

Fig 7. Control Beam Crack Pattern

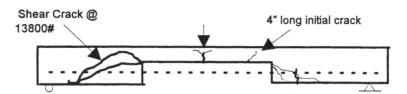

Fig 8. Crack Pattern in 4' Delamination Beam

Non-Destructive Testing Technical Internet System

M. K. CHANEY

ABSTRACT

We need to develop a computer linked internet of technical data involving non destructive testing techniques to assist in problem solving and developing testing procedures. This would include existing and advanced, state of the art testing equipment and technology.

Presently there is the National Technical Information Service. The services are for dissemination to the private sector, state and local governments.

However the one case scenarios of non destructive testing sometimes requires more complex and in depth information, such as technique procedures not commonly used for routine testing, which are not standard specifications and codes. Even the utilization of just completed research and development and advanced technology in non destructive testing applications could render the problem of the untold, within reach. Commonly known limitations of certain non destructive testing methods can then be either replaced or enhanced to give more comprehensive data to the testing industry.

With the application and joint sharing of technical data in a internet system, past trouble shooting and technique development in testing would shorten the acquisition time and delays in testing set up and operations. This pool of information would be very valuable to the structural testing industry.

Research and development, either in process or finalized could be tapped for information solving inquiries by other professionals via this computer link up system. By providing a non destructive testing internet that provides solutions for testing that can be implemented prior to expensive delays, shutdowns or even catastrophic failures on our highways. With this data available to all involved parties, quick repair or retrofit of our infrastructure and continued service of it, with minimal disruptions to traffic usage and public safety. This system will provide the technical edge for the future.

Michael K. Chaney, C.W.I.,ASNT Level-II.,4701 Lawntara St. Hbg, Pa. 17111

BRIEF CASE HISTORIES

West End Bridge, Allegheny County

This is a sixty year old bridge with the main span is supported by suspended cables. In which the structure is solely supported. The critical socket and cable assemblies were to be tested for structural soundness. There was concern that the cable connections into the sockets could be corroded or cracked. If all cable strands had to be removed, the cost of repairs to the bridge would be an additional two to three million dollars. This would also delay the scheduled opening of the bridge. Quite possibly by one year.

Current portable radiographic equipment could not penetrate the thick socket castings. The critical sockets were filled with lead spelter cores. The cables were attached to them by this process. A special Minic-6 , high energy x-ray unit was utilized for radiographic testing. One foot of cable above the sockets and the sockets themselves were tested. Test results were conclusive as the forty six cable sockets were found to be acceptable. Very costly delays were averted on this project. Digitally-enhanced images of the radiographs were provided by the consultant testing company. Physical testing at Lehigh University's laboratory, by proof load testing and sectional cutting of a specimen, confirmed results that the cables and sockets were structurally sound.

Claysville Bridge - Washington County

A twenty year old bridge, located on a heavy traffic area, was found to have out of plane bending cracks located on the cross-support girder beams. These beams were determined to be Fracture Critical members on the structure. The bridge inspection crew's only means of non destructive testing was visual and dye-penetrant exams. Initial inspection and dye tests were inconclusive. The bridge inspection unit contracted the department of transportation's non destructive testing division to ask for assistance in locating and sizing the crack flaws, to determine their evaluation of the structure. Ultrasonic testing and magnaflux testing was conducted on the beams and defects were precisely located. The fatigue cracks were mainly located below the vertical stiffners and at the terminations of the fillet welds. The Pennsylvania Department of Transportation engineering department evaluated the flaws and determined that stop-drilling the crack ends, and retrofitting web-plate stiffner connection plates, to counter the out of plane bending. By properly determining the extent of defects, a possible bridge shut-down, and possible replacement with costly repairs were prevented.

Interstate 90 - Erie County

Sixteen year old high mast luminary light poles, fabricated with 588 steel was visually found to have section loss by corrosion on the bottom inner diameter, sleeve section , inside the poles , next to the base-plate area . The poles are of a two part design as an inner sleeve is pressed into the outer pole , to strengthen the structure, as a man-hole is present at three feet up the pole. Initial non destructive testing exams by an ultrasonic digital meter were futile as the inner diameter surfaces had extensive rust packing , and the surfaces were too rough and pitted to take accurate readings . An independent testing lab was contacted to perform radiographic testing , to determine section loss on the corroded locations . This can be accomplished by means of a technique called edge radiography of the structures . Density readings would accurately gauge thicknesses to assess the amount of corrosion on the poles . This technique will allow proper evaluation of the structures , to determine repair or replacement by the department .Test results are pending.

South Bridge - Dauphin County

A pin hanger link which is a fracture critical member was found to have active internal defects by acoustic emission testing. It was performed by an independent commercial agency. The agency stated the hanger had internal cracks which were growing inside the hanger due to cycle fatigue loading . The Pennsylvania Department of Transportation , non destructive testing division was contacted to verify the results . Ultrasonic testing on the internal linear flaws were located and sized at 1/64" to 1/16" in widths , and 1/8" to 1/4" in lengths. As tension and torsional loads worked the hanger link , the flaws were found to have grown and expanded .The defects showed to have doubled in size widths. The defective link was removed and replaced prior to possible catastrophic failure of the structure. The scheduled retrofit of a catch beam system was installed on the structure. A possible disaster, and loss of life and property was prevented due to quick decisions , and non destructive testing systems at work .

CONCLUSIONS

All these non destructive testing case histories were conducted by testing professionals who have a common working knowledge of their technical abilities . They can only perform testing within the scope of their experience and the limitations of their equipment . Testing technique problem solving is by association with other technicians and experts , or by lengthy assistance searching by written or verbal means , through contacts .

RECOMMENDATIONS

The standardizing of research and development and case history procedures for non destructive testing of structures should be pursued and developed. Acquisition of this data and category filing within a computer internet system , which is linked to all involved professionals. The internet system would need to focus on reliable, fast basic data. Also it should be designed for the structural testing industry, and adaptable to user suggestions and ideas.

Therefore user comments and suggestions concerning testing technologies are not only welcomed, but necessary in contributing to the systems success.

An example of how a user might benefit from using the ndt-net, is a potential problem in testing would be introduced. Involved parties could view the situation via a bullentin board. Recommended technology or testing requirements could be shown,and or other practicies that would be benificial in testing scenarios. Users would then be alerted to subsequent changes in technology or techniques, that can resolve concerns besetting present problems.Case history files, along with reference files could be generated thus supplementing the data flow for the base program. A software exchange could enhance the documentation of the system too.

As involved parties come to rely more on the internet, all must count on their interactive imput to identify exactly what information is the most valuable for each need. Eventually e-mail, file transfers and usenet group meetings can become introduced into it.

The shared technology and ideas would assist and improve the industry to better cope and deal with testing problems which are encountered in the future .

A brief case history is occurring today as we continue to strive in research and development , along with the technology exchange at this conference .

The collection of ideas and the convergence of minds is a powerful source to work with and use.

ACKNOWLEDGMENTS

The author would like to sincerely that all professionals involved in the structural testing industry for their dedication and spirit.

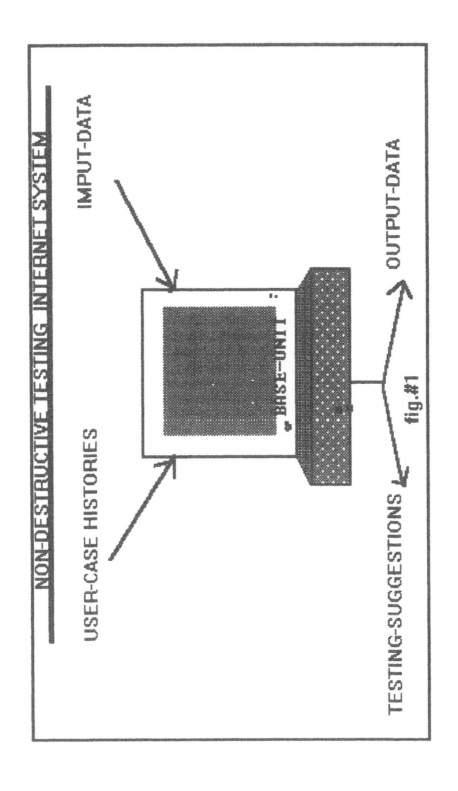

NON-DESTRUCTIVE TESTING INTERNET SYSTEM

IMPUT-DATA

USER-CASE HISTORIES

BASE-UNIT

OUTPUT-DATA

fig.#1

TESTING-SUGGESTIONS

Monitoring of Truck Loads

A. S. NOWAK AND S. KIM

ABSTRACT

Knowledge of the expected extreme loads is an important element of bridge evaluation. In general, analytical methods without the actual data cannot provide reliable results. The paper deals with a procedure for field measurement of truck load and truck load effect in bridge components. The major truck parameters include gross vehicle weight (GVW), axle weights, and axle configuration (spacing). These parameters are measured using weigh-in-motion technique, invisible to the drivers (to avoid bias). Stress spectra are important for prediction of the remaining life for bridge girders and details. These measurements are performed over longer periods of time and they are carried out using strain transducers.

INTRODUCTION

Evaluation of bridge performance involves several elements including: (a) estimation of load history, (b) prediction of future loads, (c) estimation of the actual strength, and (d) estimation of future rate of deterioration (corrosion, fatigue strength). In particular, knowledge of the actual loads is essential for evaluation of bridges with regard to fatigue and ultimate strength limit states. Design loads and design load frequencies are typically used as inputs to the bridge evaluation process. However, the bridge location determines the actual loads, load frequencies, and truck configurations that an existing bridge will experience. Data on truck load show considerable variation with respect to the functional road classification and locations on which they are collected. More realistic evaluations of bridges may be possible by considering site-specific truck load.

The objective of this study is to determine the actual truck loads and their effects. The approach is demonstrated on selected bridges in Michigan. The results are based on the papers by Kim, Sokolik and Nowak (1995) and Laman and Nowak (1995). The measurements were taken using a weigh-in-motion (WIM) system. The equipment is calibrated using a truck with known gross vehicle weight (GVW) and axle weights. The accuracy of GVW measurements is estimated at 5 percent for most types of trucks. The accuracy of axle weights is estimated at 20 percent. For each measured truck, the record includes vehicle speed, axle spacings and axle loads. Selected bridges were instrumented and measurements were taken for two or three consecutive days for truck loads. Fatigue load measurements were carried out for up to three week periods at each location.

Andrzej S. Nowak, Department of Civil and Environmental Engineering, University of Michigan, Ann Arbor, MI 48109-2125
Sangjin Kim, Department of Civil and Environmental Engineering, University of Michigan, Ann Arbor, MI 48109-2125

WEIGH-IN-MOTION EQUIPMENT

Truck axle weights, gross vehicle weights, and axle spacing were obtained with a WIM system from Bridge Weighing Systems, Inc. A general view of a bridge cross section with WIM system components is shown in Figure 1. This system collects strains in each of the girders lower flange and decomposes the strain time history using influence lines to determine vehicle axle weights. The vehicle speed, time of arrival, and lane of travel is obtained by the system through lane sensors on the roadway placed before the instrumented span of the bridge.

The WIM system consists of three basic components. The analog front end (AFE) acts as a signal conditioner and amplifier with a capacity of 8 input channels. Each channel can condition and amplify signals from strain transducers. During data acquisition, the AFE maintains the strain signals at zero. The auto-balancing of the strain transducers is activated when the first axle of the vehicle crosses the first axle detector. As the truck crosses the axle detectors the speed and axle spacing are determined. When the vehicle reaches the bridge, the strain sampling is activated. As the last axle of the vehicle has exited the instrumented bridge span, the strain sampling is turned off. Data received from strain transducers are digitized and sent to the computer where axle weights are determined by an influence line algorithm. These data do not include dynamic loads. This process takes from 1.5 to 3.0 seconds, depending on the instrumented span length, vehicle length, number of axles, and speed. The data is then saved to memory.

The WIM equipment was calibrated using state weigh scale calibration trucks. The readings are verified and calibration constants are determined by running a truck with known axle loads over the bridge several times in each lane. The comparison of the results indicates that the accuracy of measurements is within 13 percent for 11-axle trucks. GVW accuracy for 5-axle trucks is within 5 percent, however, the accuracy is $\pm$ 20 percent for axle loads.

WIM TRUCK MEASUREMENTS

The results of measurements include thousands of data points. Statistical data can be presented in a form of histograms and cumulative distribution functions (CDFs). It is convenient to use CDFs to present and compare the critical extreme values of the data. They are plotted on normal probability paper (Benjamin and Cornell, 1970).

The horizontal scale is in terms of the considered truck parameter (e.g. gross vehicle weight, axle weight, lane moment or shear force). The vertical scale represents the probability of being exceeded, p. It is more convenient to plot CDF's using computer. Then, the probability of being exceeded (vertical scale) is replaced with the inverse standard normal distribution function, $\Phi^{-1}(p)$. For example, $\Phi^{-1}(p) = 0$ (vertical scale), corresponds to the probability of being exceeded, $p = 0.5$; $\Phi^{-1}(p) = 1$, corresponds to $p = 0.841$; and $\Phi^{-1}(p) = -1$ corresponds to $p = 0.159$; and so on.

In general, truck load on bridges is strongly site-specific. There is a considerable variation in traffic volume and weight of trucks. The estimated average daily truck traffic (ADTT) varies from 500 to over 1,500 in one direction. The maximum observed GVW's vary from 80 kips (360 kN) to 250 kips (1,125 kN). The maximum observed axle weights vary from 20 kips (90 kN) to almost 50 kips (225 kN). The cumulative distribution functions of the GVW's and axle weights are shown on normal probability paper, for a convenient statistical interpretation.

The percentage of trucks exceeding the legal limits varies depending on the road. The heaviest GVW and axle weights were observed on interstate highways. On surface roads with lower volume of traffic, the weight of trucks is mostly within the legal limits. The percentage of overloaded trucks varies from 0 to 40 percent, depending on the number of axles. The largest percentage of overloaded trucks was observed for 11 axle vehicles.

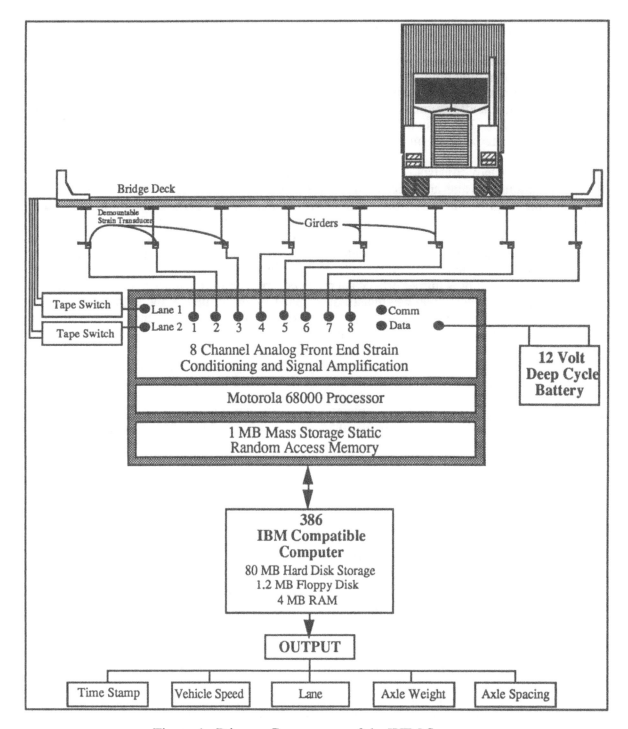

Figure 1. Primary Components of the WIM System.

Results of GVW WIM measurements for six bridges considered in the study are shown in Figure 2 in a form of cummulative distribution functions on the normal probability paper. The distribution of truck type by number of axles will typically bear a direct relationship to the GVW distribution; the larger the population of multiple axle vehicles (greater than five axles) the greater the GVW load spectra. Past research has indicated that 92% to 98% of the trucks are four and five axle vehicles. The data obtained in this study indicates that between 40% and 80% of the truck population are five axle vehicles, depending considerably on the location of the bridge. Three and four axle vehicles are often configured similarly to five axle vehicles, and when included with five axle vehicles account for between 55% and 95% of the truck population. Between zero% and 7.4% of the trucks are eleven axle vehicles in Michigan.

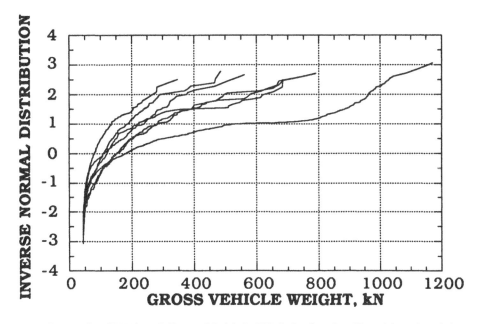

Figure 2. CDF's of Gross Vehicle Weight for the Considered Bridges.

Most states allow a maximum gross vehicle weight (GVW) of 80 kips (355 kN) where up to five axles per vehicle are permitted. The State of Michigan legal limit allows for an eleven axle truck of up to 165 kips (730 kN), depending on axle configuration. There were a number of illegally loaded trucks measured during data collection at several of the sites. Maximum WIM truck weights (265 kips, 1192 kN) exceeded legal limits by as much as 63%.

Potentially more important for bridge fatigue and pavement design are the axle weights and spacing for the trucks passing over the bridge. Figure 3 presents the distributions of the axle weights of the measured vehicles. All distributions include axles with weights greater than 5 kips (22 kN).

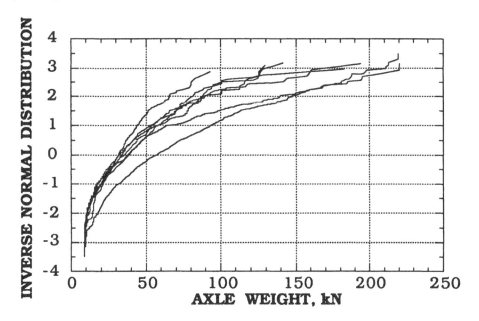

Figure 3. CDF's of Axle Weight for the Considered Bridges.

EQUIPMENT FOR FATIGUE LOAD SPECTRA

The data was collected and recorded using Stress Measuring System (SMS) with the main unit manufactured by the SoMat Corp. The SMS compiled stress histograms for the girders and other components. Strain transducers were attached to all girders at the lower, mid span flanges. The configuration of the system is shown in Figure 4.

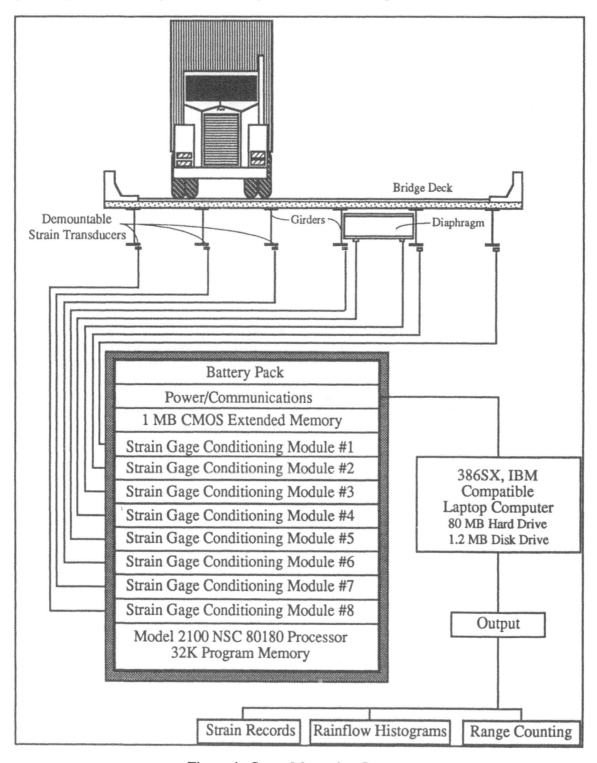

Figure 4. Stress Measuring System.

The SMS collects the strain history under normal traffic and assembles the stress cycle histogram by the rainflow method of cycle counting, and other counting methods. The data is then stored to memory and down loaded at the conclusion of the test period. The rainflow method counts the number, n, of cycles in each predetermined stress range, S_i, for a given stress history. The SMS is capable of recording up to 4 billion cycles per channel for extended periods in an unattended mode. The batteries used for this study enabled testing to continue uninterrupted for as long as three weeks.

STRAIN DATA

Strain histories were collected continuously for at least one week periods and reduced using the rainflow algorithm. Data was collected for each girder in the bridge and presented in the form of CDFs.

For each bridge, the girders are numbered from 1 (exterior, on the right-hand side looking in the direction of the traffic). The number of girders in each bridge vary from 6 to 10. The average observed strain was less than 50×10^{-6} for all girders and all bridges, however, the largest strains were observed in girders supporting the right traffic lane (girder numbers 3, 4 and 5) and nearest the left wheel of traffic in the right lane. As expected, the exterior girders of each bridge experience the lowest strain extremes in the spectrum.

CONCULSIONS

On the basis of WIM measurements, site-specific load spectra are presented for several bridge sites. Strains are measured to determine component-specific load spectra.

The truck load spectra for bridges are strongly site-specific. Bridges located on major routes between large industrial metropolitan areas will experience the highest extreme loads. Routes where vehicles are able to circumvent stationary weigh stations will have very high extreme loads. Bridges not on a major route, that are very near a weigh station, or that are within a metropolitan area experience much lower extreme loads.

Live load stress spectra are strongly component-specific. Each component experiences a very different distribution of strain cycle ranges. The girder that is nearest the left wheel track of vehicles traveling in the right lane experiences the highest stresses in the stress spectra and decreases as a function of the distance from this location. This information can be useful to target bridge inspection efforts to the critical members.

ACKNOWLEDGMENTS

The research presented in this paper has been partially sponsored by the Michigan Department of Transportation, the Great Lakes Center for Truck Transportation Research, and the University of Michigan, which is gratefully acknowledged.

REFERENCES

Benjamin, J.R. and Cornell, C.A., (1970), Probability, *Statistics and Decision for Civil Engineers*, McGraw-Hill, New York.

Kim, S-J., Sokolik, A.F., and Nowak, A.S., (1995), *"Measurement of Truck Load on Bridges in the Detroit Area"*, Transportation Research Record, submitted

Laman, J.A., and Nowak, A.S., (1995), *"Fatigue Load Models for Girder Bridges"*, ASCE Journal of Structural Engineering, accepted.

Nondestructive Search for Existing PC Cable in Shinkansen Elevated Track

K. AZUMA

ABSTRACT

A new elevated track has been built next to the existing Shinkansen track, and the two tracks were joined by adding prestress strength. We had to pass the PC cables through the existing Shinkansen track. We have examined the methods of non-destructive search for the PC cables in the thick concrete to avoid the existing PC cables, and have used these methods. As a result, the whole process was successfully completed.

1. Introduction

A project with advanced technology is being undertaken by East Japan Railway Company. This project will inaugurate through operation of Shinkansen trains(standard-gauge) into the Morioka-Akita area, using the Tazawako line, which has been narrow-gauge (Figure-1).

A viaduct is being constructed at Morioka station as an approach to connect the Tohoku Shinkansen with the Tazawako line(Morioka approach, Figure-2). The substructs have been widened so that a new switch and crossing could be added, and a new elevated track has been built next to the existing Shinkansen track. Then, the Shinkansen track was joined to the newly-built track by adding prestress strength to the cross beams.

Photo-1 Akita Shinkansen train

Kotaro Azuma, East Japan Railway Company, Tohoku Construction Office, 1-1-1 Itsutsubashi, Aoba-ku, Sendai-shi, 980, JAPAN
Keiichi Saito, Hideaki Tada

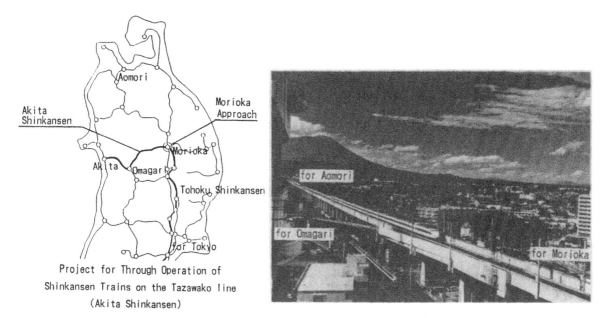

Figure-1 Location map

Project for Through Operation of
Shinkansen Trains on the Tazawako line
(Akita Shinkansen)

Photo-2 Morioka approach

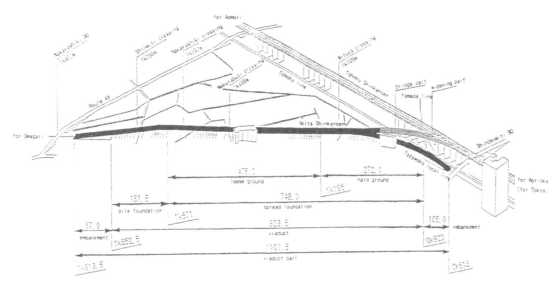

Figure-2 Morioka Approach

In this process, it was necessary to confirm the exact position of PC cable in the thick concrete in order to join the two tracks. We examined how to confirm these positions, and the whole process was successfully completed.

2. Location of the search

It was necessary to make some holes in the webbing near the cross beams of the existing track in order to insert the prestressed cables, and the newly-built track was joined to existing track. The webbing part is 900mm thick. The prestressing steel bars will be arranged every 1.6m in the overhanging slab of the existing track. The overhanging slab part is 400mm thick. The holes are 90mm in diameter in the webbing part and 70mm in diameter in the slab part. The arrangements of the PC cables are shown in Figure -3(webbing part) and Figure-4(slab part). Each hole's position is fixed from the plans of the existing track of the Shinkansen so as not to cut off the PC cables, the stirrups, and the precaution reinforcements of the existing track in the webbing and the slab part. In some of the webbing parts, we must make holes whose allowable maximum is 70mm in the 900mm thickness. However, this type of construction is unprecedented. We expected that this operation would be difficult.

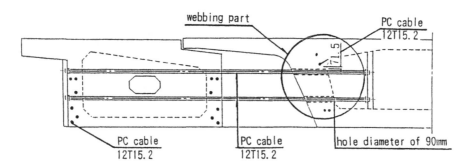

Figure-3 Arrangement of the PC cables in the webbing part

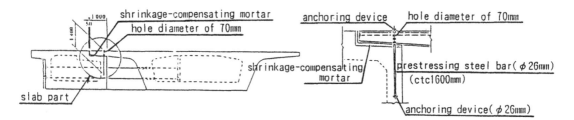

Figure-4 Arrangement of the prestressing steel bars in the slab part

3. Examination of the methods

1) Comparison of the methods

The methods of non-destructive search for the PC cables laid in the webbing and the slab of the concrete bridge are showen in Table-1.

Table-1 Non-destructive search

	radar	supersonic waves	electromagnetic waves	X-ray
searching ability	60cm	20cm	20cm	45cm
characteristic	If there are obstacles on the near side of the object of search, may not be able to search	Good for detecting cracks, voids, and the thickness of the concrete	If there are obstacles on the near side of the object of search, may not be able to search	necessary to take counter-measures for safety
economy	◎	○	○	△
application	○(webbing part)	×	×	○(slab part)

◎:Best ○:Good △:High costs for deep search ×:Not satisfactory

We used radar for search in the webbing part, and X-rays for search in the slab part, because the webbing has a maximum thickness of 900mm, and the slab has a thickness of 400mm. Moreover, radar for search was used on both sides of the webbing.

2) Specific techniques used in the research

If there are obstacles(reinforcing bars) on this side of the search object(PC cable) to confirm the exact position of the mark is very difficult because these interfere with the radar. We made a test piece, and tested the searching ability. As a result, if the interval between the reinforcing bars is 20cm(ϕ13mm, the same as the interval on the existing track), and the depth of the PC cable(ϕ75mm) is as great as 30cm, we learned that the position of the PC cables is confirmed as the exact position, but the searching ability deteriorates when the water content of the concrete is high. But to confirm a deep position is difficult, and so we considered that we could small holes using a mistdrill for the PC cables and locate them through a fiberscope. The mistdrill spouts water from the tip and makes holes in the concrete. We learned that we could tell the existence of the PC cables in the concrete through the differences in vibration when making holes. We tested X-rays for search in an actual slab. As a result, we learned that X-rays used for search missed the position of the PC cables by 40mm out of 400mm thickness of the slab. We used radar for search and making holes by the mistdrill in the webbing part, and X-rays for search and radar for search in the slab part.

3) The process of searching for the PC cables

The process of searching for the PC cables in the actual track is shown in Figure-5. The webbing part is divided between what we call the neighboring part and the common part. The neighboring part is the part where the positions of making holes are separated by not less than 71.5mm nor more than 111.5 mm from the PC cables in existing track. There are 9 neighboring parts. In the common part, the positions of making holes are separated by more than 188.5mm from the PC cables. There are 19 common parts. Eight of the 9 neighboring parts are near to the anchoring device of the PC cables in the existing track. There are many reinforcing bars that strengthen the anchoring device. So we used only radar for search in the common parts, and radar for search and making holes by the mistdrill in the neighboring parts, for confirming

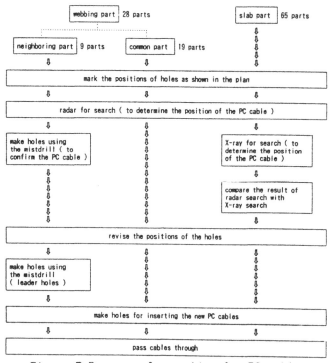

Figure-5 Process of searching for PC cable

their exact position to avoid the existing PC cables in the existing track. We used 11 mistdrills that are different in length and size. The mistdrills were used in sequence from short to long in order to go straight. And we developed a new trestle to make holes straight. After the mistdrill reached the PC cables, we confirmed them again through a fiberscope(Photo-3). Furthermore, as a safety measure, before making holes for inserting the new PC cables using a ϕ90mm diameter drill in the webbing, we made up and down holes where this was planned, using a ϕ9mm mistdrill(Figure-6, a leader hole). When making holes using a ϕ90mm diameter drill, we make it our concern to avoid damage to the existing PC cables.

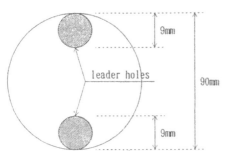

Figure-6 A leader hole

Photo-3 PC cable through a fiberscope

4. Search in the actual track

1) The webbing part

The result of radar for search is shown in Figure-7 and Photo-4. The reinforcing bars are near the surface, and the PC cable is behind them. In this Figure, we can judge the position. In some other places, however, it was impossible to find the exact position of the existing cables. In those places, we made holes to reach the PC cable, and confirmed the position. An example of the result is shown in Figure-8. To complete the search, we made holes.

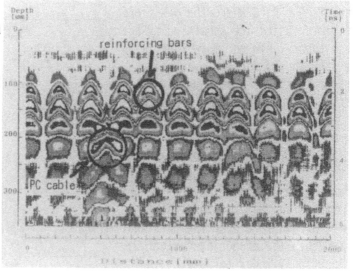

Figure-7 Result of radar for search

Photo-4 Radar for search

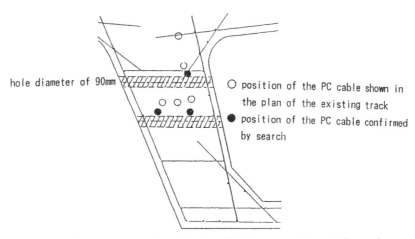

hole diameter of 90mm

○ position of the PC cable shown in the plan of the existing track
● position of the PC cable confirmed by search

Figure-8 Cross-section of the Shinkansen elevated track

2) The slab part

The result of X-rays for search is shown in Figure-9. Eventually, we were able to confirm the exact position of the PC cables. We made holes and were able to insert the new PC cables into the holes, exert prestress, and successfully join the two tracks.

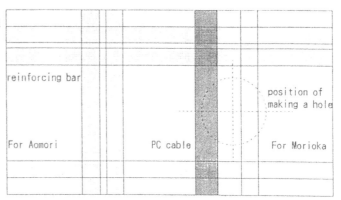

reinforcing bar

For Aomori PC cable

position of making a hole

For Morioka

Figure-9 Result of X-rays for search

5. Conclusion

We will be happy if this report will be of service to you when you must search for PC cables in thick concrete.

Fiberscope Inspection of Concrete Box Girder Bridge Sections Following the Northridge Earthquake

P. E. HARTBOWER

ABSTRACT The Department of Transportation is responsible for the safety and integrity of every bridge in the states highway system. The task of evaluating damaged bridges becomes critical after a disaster or failure of a structure. The use of simple tools can make an enormous difference in the amount of time need to assess the damage and open structures that are safe for the public. The fiberscope is a relatively simple tool to use and has allowed us to investigate detail on many bridges that were impossible to inspect due to access. This paper discusses the application of fiberoptic technology for bridge inspection and the time and manpower it saved.

SACRAMENTO, CA.--Almost immediately following the devastating earthquake that centered just north of Los Angeles on January 17, 1994 a team of technical experts from CALTRANS Division of Structures Maintenance was dispatched to determine the safety of every bridge and overpass in the complex highway system that spans the area. The Northridge earthquake had a magnitude of 6.8. Approximately nine bridges collapsed. For the Office of Structure Maintenance and Investigations (OSM&I), the collapse of one bridge impeaches the structural integrity of all bridges within a 40 km (25 miles) radius. The Inspection began in the areas where Bridges had collapsed and the focus was on the expansion joints. Critical to the safety of each bridge section was the condition of the heavy steel blocks with earthquake-restraining cables that prevent the self-supporting highway sections from separating.(See Photo 3)

The enormity of the problem can be appreciated when you consider the many hundreds of overpasses, exit and entrance ramps, as well as bridges and raised sections of highway. Each section of prestressed concrete box girder seats on the next section to allow for temperature expansion. To help withstand future earthquake shocks, adjacent sections were retrofitted some years ago with cable restrainer units. However, no structure could withstand dramatic shifts in the earth where the movement exceeds the limits of reasonable design.(See Photo 1) So the inspection teams from CALTRANS had to determine the condition of each restainer, as well as the other structural condition at each overpass.

When I arrived at the first bridge the Engineers were accessing the inside of the box girders by opening a .6 meter square hole in the deck at each location.(See Photo 2) Inspecting each restrainer would have meant using jackhammers to open up 4 to 6 holes large enough to climb inside. Where the hinge and restraining cables were intact, extensive repairs to the reinforced concrete bridge deck would have been necessary. I Suggested using a fiberoptic instrument that would enabled them to look through a 16 mm .hole drilled through the bridge deck, near the restrainer

Paul E. Hartbower, Department of Transportation, Engineering Service Center, Office of Materials Engineering and Testing Services, 5900 Folsom Boulevard, Sacramento, CA   95819

I first started using the Olympus fiberscope about eight years ago to look at pins that supported some of the bridges in California's highway system. Knowing the instrument was available led to additional uses, including applications for other agencies of state government. It was even used to investigate structural components in the Governor's Mansion, an historic building in the Capital City.

The Fracture Critical Bridge Inspection Team from the Welding and Metal Technology Branch of the Office of Structural Materials--a special bridge inspection unit from the Division of New Technology, Materials and Research was sent to Los Angeles. Under normal circumstances, the team members perform highly specialized bridge inspections for the OSM&I. Using Nondestructive Testing (NDT) techniques such as Ultrasonics, Dye Penetrant, Acoustic Emission, and Radiography, the team finds hairline fatigue cracks on fracture critical steel bridges and performs the necessary welding repairs thoughout the state. After completing the inspection of the steel structures, the team began assisting in the inspection of the steel cable restrainer systems installed inside the concrete box girder bridges.

The procedure involves drilling a 5/8 inch diameter hole into the concrete deck, as close as possible to the location of the restainer assembly. (See Photo 5) An 11mm flexible fiber optic borescope was lowered into the concrete box girder and maneuvered into position to allow the operator to view the inside of the chamber. Caltrans engineers and Fracture Critical Team inspectors examined the cable restrainer assemblies from different angles and the inside of the bridge for cracks or damage. (See Photo 6) The holes were patched quickly after the inspection was completed. About 45 minutes to an hour was required for the entire inspection procedure.

The original fiberscope was 6 mm diameter and only about .9 meters long. After the initial investigation the Department of transportation purchased two new fiberscopes that are 11 mm diameter and 1.7 meters long at a cost of $40,000. The visibility was excellent and I also recorded images on videotape, and by simply using a 35-mm camera.

When possible the hinges are inspected by opening an existing soffit access door at the bottom of the concrete box girder bridge. An inspector would then crawl inside and visually inspect the cables, seats, anchors, and diaphragms. However, in many cases when the soffit access did not exist or was over live traffic, inspectors would have to cut a 2 foot by 2 foot opening through the concrete deck to reach these restrainers from the top of the bridge. This process takes about eight hours to complete and the road must remain closed to traffic to allow the concrete to cure.

The inspection proceeded quickly, starting in the area near the epicenter (Northridge) and fanning out to the outlying areas. Hundreds of damaged cables were found (See Photo 4), and any unsafe section of highway were closed or demolished to protect the public. To keep traffic moving, sections of highway were detoured, using nearby roads to move the traffic. As a result, the famous Interstate 5 was opened to traffic in less than three weeks.

As the inspection teams moved out from the epicenter, the number of damaged expansion joints diminished. However, the enormity of the total problem can be appreciated from the fact that more than 75 concrete bridges with over 200 hinges were inspected during the two weeks following the quake. The borescope was used on 60 percent of the hinges; only 15 of the total bridges inspected were found to have severe hinge damage and 38 percent of the restraining joints inspected need to be repaired. Without the use of the fiberscope, this job would have taken many more weeks, and the cost of repairs even to those sections with good joints would have been enormous. On the other hand, the patching of the 16 mm diameter inspection holes was minimal. The inspection team from the Welding and Metals Technology Branch made a significant contribution to the quick reopening of these bridges once their safety was assured.

Photo 1 Overpass section destroyed by earthquake
(Note severed ends of restrainer cables)

Photo 2 Deck opening for inspection
(.6 meter square hole)

Photo 3   Intact earthquake restrainer unit
(steel block and seven cable ends)

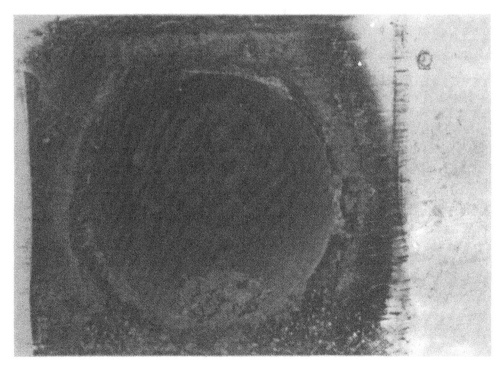

Photo 4 Failed restrainer unit
(Broken cable ends and grouted
steel pipe exposed)

**Photo 5 Drilling holes in deck to access
            interior of box.**

**Photo 6 Inspection of earthquake restrainer using
            fiberscope.**

Developing a Portable Crack Evaluation Tool Based on Shearography

A. K. MAJI AND D. SATPATHI

ABSTRACT

Nondestructive Inspection techniques based on laser interferometry are highly sensitive, noninvasive and can inspect a large area simultaneously. This makes them attractive for remote monitoring and inspection of structures. Electronic Shearography (ES), which is an implementation of laser interferometry using an image shearing device and digital image processing has shown extensive promise for the inspection of aging aircraft structures. The major advantage of this method lies with the fact that it yields strains directly, eliminates the necessity for vibration isolation and minimizes the specimen size limitations created by the coherence length of the laser beam. These advantages coupled with electronic image acquisition and processing, make electronic shearography a strong contender for meeting the non-destructive evaluation needs for civil infrastructure.

Steel highway bridges have been known to be susceptible to fatigue cracking at welds and other connections. Our ability to retrofit these members depends on our ability to detect the crack-tip and to determine the severity of stresses at the crack-tip. The effectiveness of retrofit measures such as hole-drilling can also be evaluated if the crack-tip strain field can be rapidly and accurately evaluated. We are developing a portable tool based on electronic shearography that can address such needs.

In electronic shearography, one image is subtracted from another to obtain fringes that yield information on surface strains. The success of this technique depends on the correlation of the two interacting images, and on the size distribution of the 'speckles' produced by the object surface. The various system attributes that need to be studied in the process of developing a portable system will be discussed here. The proposed instrument can then be used for evaluating cracks (determining strain and stress field) in steel bridges.

A cracked (retrofitted) plate-girder bridge in New Mexico was analyzed using a combination of a commercial finite element package, ANSYS, and a fracture mechanics code. The analyses provided data on the crack-tip stresses, crack opening, bridge deflection, etc.. These numbers helped us evaluate the applicability of Electronic Shearography to inspect bridge structures.

Arup K. Maji, Department of Civil Engineering, University of New Mexico, Albuquerque, NM 87131. (ph : 505-277-1757).
Debashis Satpathi, Department of Civil Engineering, University of New Mexico, Albuquerque, NM 87131. (ph : 505-277-7349).

BACKGROUND OF ES

The setup for shearography (Hung, 1982) involves an object lighted by an expanded laser beam, and an imaging system (Figure 1). Images are stored with a CCD (charge coupled device) camera, and processed by a personal computer and an image analysis system (Figure 1). The electronic processing of laser interferometry data is relatively recent (Jones and Wykes, 1983). The integration of electronic technology with optics has increased the field applicability of shearography (Hung, 1989, Newman, 1991).

An image of the object is first acquired by the image analysis system and stored in computer memory. The video camera takes only 1/30 seconds to acquire an image, therefore, vibration isolation and ambient light are not problems encountered with the ES technique. The image consists of numerous dark spots called 'speckles', caused by the interference of the laser beam with the object's surface roughness; therefore no specimen preparation is necessary. The specimen is thereafter deformed by loading while the ccd camera continues to acquire images. The image processing system continuously subtracts the initial image stored in memory, and displays the image resulting from the subtraction on the video screen. This produces a real time fringe pattern on the video screen.

In shearography, the image at a point on the image plane is caused by the interference of the speckle patterns arising from two points on the object plane, and vice-versa (Figure 1). This is done by an image shearing device (hence the name shearography), a birefringent crystal (Laser Tech Inc., 1995) that separates two orthogonally polarized components of the laser light going to the image plane (Figure 1). The technique therefore detects displacement derivatives or strains. Also, since any rigid body motion effects the two optical paths equally, the setup is not as sensitive to rigid body motions.

The ES system used involved a 35 mw He-Ne laser, with associated optics. A Shearography camera (model SC4000) and control system (model CCU4000) (Laser Tech. Inc., 1995) were used. Digital image processing was performed on a personal computer containing a frame grabber and the 'Imaster' image processing software. For out-of-plane illumination and observation, the sensitivity of the system per fringe is given by :

$$\delta w/\delta x = (\lambda/2) /\delta x \qquad (\lambda = 0.63 \text{ mm for He-Ne laser}) \qquad (1)$$

where δx is the amount of image shear on the object plane.

SPECKLE DECORRELATION

An investigation of the speckle images revealed that the intensity of the pixels in the subtracted image varied only from 0 to 50 (using less than 6 bits of resolution). This meant that the speckle contrast was low. Secondly, rigid-body motion of the I-beam inspected (≈ 1 cm) was decorrelating the two images that were being subtracted, obliterating the fringe pattern. A systematic study of the speckle size and decorrelation was therefore necessary.

Theoretically the fringe visibility reduces to zero as the rigid-body motion X approaches d_R, which is the resolution diameter of the optical system (Butters et al., 1978). For an ideal lens, $d_R = \lambda u/a = (\text{f-stop}) \times \lambda / m$, $m = v/u$, f-stop = f/a (2)
Another limitation on the allowable rigid-body-motion comes from the requirement that the speckle size in the image plane should be significantly larger than the speckle movement (in the image plane) in order for interferometry to be practical, i.e., $Xv/u << d_{sp}$ (3)

where d_{sp} is the mean speckle diameter (in the image plane), $d_{sp} = 2.4 \lambda v / a$ (4)

Expressions 2 and 4 both indicate that the object movement should be significantly less than $2.4\lambda u/a$. Hence, greater rigid-body-movement can be tolerated with smaller aperture and larger object distance. An added requirement in electronic image acquisition is that this speckle size should be should be significantly larger (by a factor of 3 or more) than the dimensions of the pixels on the ccd camera : pixel size $<< d_{sp}$ (5)

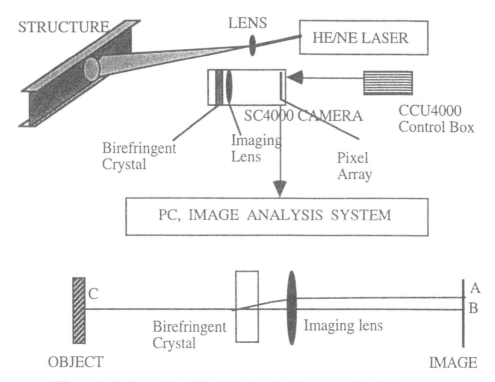

Figure 1. Schematic of the Electronic Shearography Set-up

SPECKLE SIZE STUDY

The objective of this study was to determine the speckle size as a function of the parameters in Equation 4. Various surfaces were illuminated and photographed. The specimens were mounted on a specially fabricated translation stage. Three micrometers on the translation stage could move the specimen independently along the three axis. Three different surfaces were studied; a steel plate sprayed with Magnaflux developer, a steel plate with no surface treatment and a concrete surface sprayed with Magnaflux.

The object distance and the f-stop were varied. The Imaging device consisted of a Nikon SLR camera with a 35 mm 1: 2 lens, similar to the lens on the ccd camera used in our shearography set-up. The recording medium was high resolution (2500 lines/mm), 35 mm 8E-75 holographic film (Agfa Corp.). The developed film was observed under an optical microscope in transmitted light. The observed speckle pattern was digitized by a ccd camera. Subsequently the image was thresholded and the speckle size statistically determined. Figure 2 compares the measured speckle size with predictions of equation 4.

DECORRELATION STUDY

These studies were done using the translation stage and our electronic shearography system. A differential out-of-plane displacement was first applied so that a fringe pattern was visible using the shearography set-up. A rigid-body translation δ_x or δ_y (object plane) was then applied until the fringe pattern was no longer visible due to decorrelation of the speckle pattern. Representative data corresponding to different values of f-stop and object distance (u) have been shown in Table 1, for the sprayed steel specimen. The d_R and d_{sp} shown here were obtained from Equations 2 and 4, and d_C is the physical size of pixels on the photosensitive element of the imaging camera.

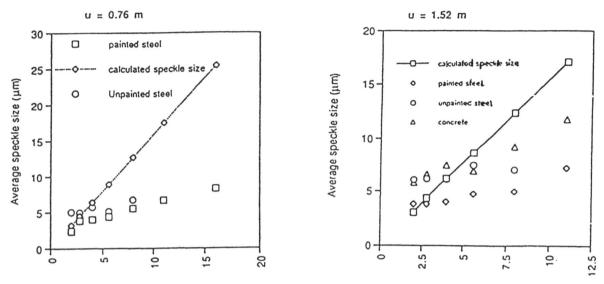

Figure 2. Variation of Speckle-size with f-stop (u = 76cm & 152cm)

CONCLUSIONS OF SPECKLE STUDY

It may be seen from Figure 2 that the speckle size does increase with numerical aperture as predicted by equation 4 , although not as much as predicted. For smaller values of f-stop (< 5.0) the experimental speckle size matches or exceeds theoretical value. Table 1 shows that the ccd camera does not have the ability to properly resolve the speckles due to the large pixel size d_C. The speckle size can be increased by reducing the aperture, hence our 35 mw laser could only inspect areas less than 20cm square. The movements producing complete speckle decorrelation are converted to image plane movements in Table 1. This movement is comparable or slightly smaller than the theoretical speckle size. Although the tolerable movement is low, it is better than prediction of equations 2 and 3.

STRUCTURAL ANALYSES

It was necessary to estimate the magnitude of displacements, and strains in a real structure. A retrofitted composite (steel girder and concrete deck) highway bridge (interchange N.M.P. I-040-2STA. 675+18.0) in the vicinity of Albuquerque, on Interstate 40 was analyzed with design obtained from the state transportation department (NMSHTD). In order to estimate the deflection and crack-tip strain field, a combination of two Finite Element (FEM) codes were used (Zawaydeh, 1994). Using the FEM software ANSYS (revision 5.0), a HS20-44 load was positioned to produce the maximum moment. A convergence study was conducted to check accuracy. In order to study the effect of cracking a two dimensional fracture mechanics FEM code 'CoMet' was used. Linear Elastic Fracture Mechanics (LEFM) with 8-noded quadratic elements were used along with 6-noded singular triangular elements at the crack-tip. The boundary conditions (applied moments) on this mesh were obtained from the results of the ANSYS program. A crack of 1/4 the beam height was introduced to simulate actual cracks observed in such bridges.

The values of stress intensity factor (SIF), was close to 100 MPa√m, which is the limiting value of SIF for A36 steel. This implied that the crack was propagating when it was retrofitted. The in-plane COD was about 1.34 mm., and the vertical deflection from the ANSYS analysis was about 1.0 cm. Theoretically, the maximum movement under dynamic loading could be twice these values. The strain field immediately ahead of the crack-tip was subsequently obtained from the SIF.

TABLE 1. DECORRELATION STUDY
($\lambda = 0.632$ μm, f = 35 mm, $d_C = 11$ μm)

F-stop	u(mm)	$\Delta x(\mu m)$ object plane	$\Delta x(\mu m)$ image plane	$\Delta y(\mu m)$ object plane	$\Delta y(\mu m)$ image plane	$d_R(\mu m)$	$d_{sp}(\mu m)$
3.2	546	51	3.5	76.2	5.2	31.4	4.8
3.2	890	89	3.6	55.8	2.3	51.3	4.8
1.8	546	63.5	4.3	63.5	4.3	17.7	2.7

RELEVANCE TO SHEAROGRAPHY

The dynamic rigid-body motion was 2cm, crack opening displacement was 1.34 mm, and vibration frequency was about 14.3 HZ. This implies than between two frames of the ccd image (1/30 sec), the object could move about 1.0 cm and the relative opening will be 640 mm. Since the 1.0 cm movement is large, the ES system should be attached to the structure to eliminate the rigid-body effect as much as possible. Also, the 1.34 mm COD is large enough to result in numerous fringes, and crack will be visible as a white patch over the width of image shearing. Therefore, due to the high sensitivity of the ES system, a low weight vehicle is sufficient load for the crack to be visible. It will be preferable to shut down traffic and use a smaller known load during the application of ES. Sufficient number of fringes can be generated at the crack-tip to make a quantitative assessment of SIFs possible during inspection of a crack. The sensitivity can be adjusted by changing the angle of shear of the birefringent crystal (or other form of image shearing device).

ACKNOWLEDGMENT

This research was supported by the National Science Foundation Grant no. MSS 9212733, administered by Dr. K. P. Chong. Dr. Maji is currently supported as senior research scientist by the Air Force Phillips Laboratory, Albuquerque, NM.

REFERENCES

Butters J. N., Jones R. and Wykes C., "Electronic Speckle Pattern Interferometry", in 'Speckle Metrology', ed : R. Erf, 1978, Academic Press, pp. 111-158.

Jones R. and Wykes C. (1983) "Holographic and Speckle Interferometry", Cambridge University Press, pp.165-197.

Hung Y. Y. (1982) "Shearography: A New Optical Method for Strain Measurement and Nondestructive Testing", Optical Engineering, V 21, No. 3, pp. 391-395.

Hung Y. Y. (1989) "Shearography : A Novel and Practical   Approach for Nondestructive Inspection", J. of Nondestructive Evaluation, V 8, No. 2, pp. 55-67.

Newman J., "Shearographic Inspection of Aircraft Structure", Materials Evaluation, V49, No.9, 1991, pp. 1106-1109.7.

Laser Technology Inc., 'Users Manual for SC4000', Norristown, PA, 1995.

Zawaydeh S., "Evaluation of a Mode I Crack in Highway Steel Bridges", MS thesis, University of New Mexico, Summer, 1994, 106 pages.

Finite Element Modeling of a Pin-Hanger Connection

A. EL-KHOURY, G. WASHER AND T. A. WEIGEL

ABSTRACT

Since the collapse of the Mianus River Bridge in Connecticut in June 1983, defects in pin-hanger connections have been detected in many bridges in the United States. Pin-hanger connections are used in suspended span configurations in girder bridges and truss systems. The objective of this study is to develop a better understanding of the pin-hanger connection. Using the software package ANSYS 5.0A, a finite element model of a typical connection is generated to examine possible failure modes under several loading conditions.

INTRODUCTION

Pin-hanger connections were used in some multi span steel bridges built before the 1980's. The connection permits expansion movement and rotation of a suspended simple span. A typical pin-hanger connection, as illustrated in Figure 1, has an upper pin, a lower pin and two hangers. These structural components connect the web of a suspended beam to the web of a cantilever beam. The pins are loosely fit in the hanger to permit free rotation of the connection system. When the structural components of the steel bridge expand or contract due to thermal cycles, the hangers rotate around the upper and lower pins without inducing any torsional stresses in either the pins or the hangers. However, these connections are located under expansion dams through which damaging materials such as rainwater, deicing chemicals, and other debris may leak. Constant exposure to the atmosphere, moisture, and chemicals causes excessive corrosion to occur at the pin-hanger connection. Expanding corrosion, termed pack rust, may cause the joint to lock. Additionally, it may cause large tensile stresses in the pin, and may even push the hanger off the pin, as occurred in the catastrophic collapse of the Mianus River Bridge in Connecticut in June 1983 (Demers and Fisher, 1990). Several other failure modes

Armando El-Khoury, Graduate Research Fellow, FHWA, 6300 Georgetown Pike, McLean, VA, 22101.
Glenn Washer, P.E., Research Structural Engineer, FHWA, 6300 Georgetown Pike, McLean, VA, 22101.
Dr. T.A. Weigel, Associate Prof., CE Department, University of Louisville, Louisville, KY, 40292.

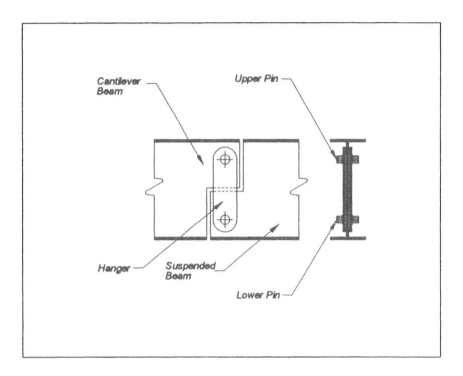

Figure 1. Typical pin-hanger connection.

have been documented, including tensile failure of the pin, cracking of pins due to combine torsion and shear stresses, and cracking in hanger plates.

OBJECTIVE

The objective of this study is to develop a better understanding·of the pin-hanger connection. Using the software package ANSYS 5.0A, a finite element model of a typical connection is generated to study the effect of temperature change and locking conditions on the torsional moment, and stress distribution on the pin and the hanger.

FORCES APPLIED TO THE CONNECTION

In a pin-hanger connection, the hanger acts, theoretically, as a two force member. Therefore, the primary design force applied to the hanger is axial. A torsional force, which will overcome friction and initiate sliding between the pin and the hanger, is also present. When corrosion is present, the contact area between the mating surfaces becomes rough, increasing the coefficient of friction. Pack rust may also increase the pressure between the pin and the hanger. A higher coefficient of friction along with a higher pressure naturally increases the friction force, and larger forces are required to rotate the hanger around the pin. Eventually, these forces interfere with the normal operation of the connection and locking occurs. This state is sometimes termed a "fixity" condition. This change in the structural behavior of the deteriorated connection may be a source of cracking in both the pin and the hanger. Under this locked condition, live-load and thermal cycles produce unanticipated torsional and bending stresses in the pins and the hangers.

The pin-hanger connection is considered free when the hangers can rotate freely about the pins. Locking occurs when the rotation of the hangers is restricted. In this study, locking occurs when (1) the hangers cannot rotate around the lower pin but can rotate around the upper pin, (2) the hangers cannot rotate around the upper pin but can rotate around the lower pin, or (3) the hangers are restrained from rotating around the upper and lower pins. Furthermore, in a free connection the beam webs are assumed to rotate freely around the pins, and in a locked connection the beam webs are assumed to be restricted from rotating around the pins.

When the connection is not locked, the pins undergo shear stresses and bending stresses. The bending stresses are due to the offset distance of the line of action of the load transferred from the hangers to the pins and from pin to the web. The hangers, which normally act as two force members, experience axial tension. When locking occurs, the pins experience not only bending and shear stresses, but torsional stresses, as well. The hangers experience bending and tension.

Expansion joints at the abutments are also susceptible to becoming locked due to corrosion and build up of debris. When these joints fail to perform as expected, they reduce the structure's ability to accommodate thermal changes, and effect the forces in the pin-hanger connection. Therefore, the state of these joints must be considered. The expansion joints at the abutments are considered free when they expand and contract under thermal cycles. These joints are considered locked or restrained when they fail to expand and contract under thermal cycles.

FINITE ELEMENT MODEL

A typical pin-hanger connection of an existing bridge was analyzed using the finite element method. The bridge is along Route 29 over Robinson river in Madison County, Virginia. The fifteenth edition of *Standard Specifications for Highway Bridges* (AASHTO) was used to determine the loads to be applied to the finite element model of the bridge.

BRIEF DESCRIPTION OF STRUCTURE

The bridge over Robinson river consists of two approach spans, two anchor spans, and one suspended span. It carries two traffic lanes. Each lane is to provide support for HS20-44 loading. Span 2, which is the suspended span, is connected to Span 1 using the pin-hanger connection and to Span 3 using a pinned connection. The main girders are W26x85 steel sections. Girder webs are reinforced using two 3/8" steel plates at the pin-hanger assembly.

PREPROCESSING

Three finite element models were created. A global bridge model is used to determine the forces on the pin and the hanger. These forces then are applied to a local pin and hanger models. The global bridge model consists of three-dimensional tapered unsymmetrical beams and shell elements. The approach and anchor spans are modeled using beam elements. A section of the cantilever beam and the suspended beam are modeled using elastic shell elements. Rigid beams are used to ensure continuity between the shell and beam elements. The pins and hangers were modeled as beams in the global model, and as three-dimensional solid and elastic shell elements in the local pin and hanger models, respectively.

LOADING

The loads used in the global bridge model are dead-load, live load, and temperature load. Live-load is the weight of a HS20-44 truck placed at a location on the suspended beam that provided maximum rotation in the pin-hanger connection. Temperature loading was applied by a temperature range of 150°F, providing an upper bound solution. The shear forces obtained from the global model were applied to the local pin and hanger models as contact pressure using Equation 1 (Young, 1989). An elliptical pressure distribution was assumed.

$$P_0 = 0.418 \sqrt{ P_1 E (\frac{1}{R_1} - \frac{1}{R_2}) } \qquad \textbf{(1)}$$

Where:

P_0 = maximum compressive stress (Ksi)
P_1 = load per axial inch on cylinders (Kips)
E = Young's modulus of elasticity (Ksi)
R_1 = Radius of pin (in)
R_2 = Radius of hanger (in)

RESULTS

The finite element models were solved for a variety of locking conditions associated with the pin-hanger connection. Figure 2 indicates the maximum principle stresses that occur in the model under these conditions. In this figure, *Stress Ratio* is used to define the ratio of the maximum principle stress resulting from a certain condition, to the maximum principle stress that occur in a model containing no restraints. Condition A, which is characterized by having both pins and both abutments free, has a stress ratio of 1. The stress ratio of condition B, in which the abutments are both locked and both pins are free to rotate, is similarly 1, assuming sufficient clearance is allowed at

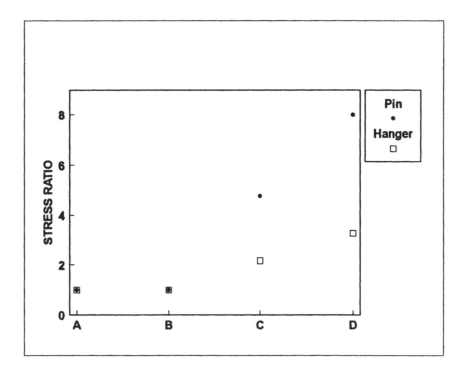

Figure 2. Effect of locking on stress ratio.

the pin and hanger for thermal expansion. Condition C, in which both pins are locked and both abutments are performing normally, a significant increase in the stress ratio is seen. Condition D, in which the abutments and the pins are locked, results in the greatest stress ratio.

Figure 3 indicates the principle stress in the hanger (condition D). These stresses are plotted along a path perpendicular to the hanger's longitudinal axis, indicated as path A. This area would normally be expected to have the highest stresses, and therefore be a likely area for crack initiation. However, due to the high rotational forces created by locking, path B, which is a path 45° from path A, has a higher stress in the area at the pin hole.

CONCLUSIONS

A locked connection as well as locked abutments effect the stresses in the pin and the hanger. In this study, locking of the pin-hanger connection increased the stress in the pin and the hanger by a factor of about 4.75 and 2.15, respectively. Locking of abutment increased these factors to about 8 for the pin and 3.25 for the hanger. It is evident that the condition of abutment expansion joints is an important factor in the potential for failure of pin-hanger connections.

It is also important to note that the maximum stresses in the hanger in a locked condition occur an angle 45° from perpendicular to axial stress. Additionally, the

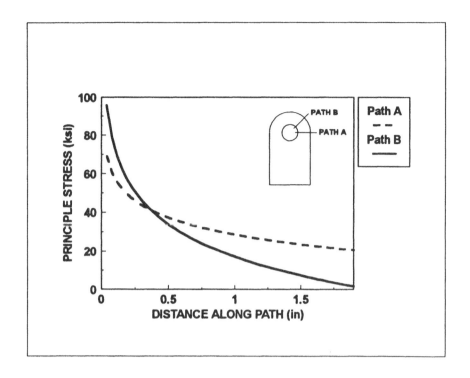

Figure 3. Principal stress along path (in).

stresses in the hanger utilized for this particular model were greater than the stresses in the pin for all conditions. This result is consistent with the fact that the Robinson Bridge is experiencing fatigue cracking in the hanger (Lozev, 1994).

ACKNOWLEDGMENTS

This project was supported by the Eisenhower Grants For Research Fellowships sponsored by The Federal Highway Administration (FHWA).

REFERENCES

Demers, Cornelia E. and Fisher, John W. 1990. *Fatigue Cracking of Steel Bridge Structures Volume I: A Survey of Localized Cracking in Steel Bridges- 1981 to 1988.* McLean, Va: Turner-Fairbank Highway Research Center. FHWA-RD-89-166.

Lozev, M. G. 1994. Private communication.

Young, C., Warren, 1989. *Roark's Formulas for Stress & Strain.* New York: McGraw-Hill.

Session 4

C4—TOMOGRAPHY AND THERMOGRAPHY

S4—ACOUSTIC EMISSION

P4—PAVEMENT

P4A—PAVEMENT

Nondestructive Evaluation of Bridge Foundations

M. F. AOUAD AND L. D. OLSON

ABSTRACT

This paper focuses on the applications of new developments in stress wave techniques for the identification and quantification of defects in bridge foundations. The two methods discussed in this paper are the Crosshole Sonic Logging (CSL) and Crosshole Tomography (CT) methods.

The Crosshole Sonic Logging method measures the travel time of stress waves as they travel between tubes or coreholes in concrete or stone structures. A single CSL log requires 5 minutes or less to test the material between a pair of coreholes or access tubes with lengths to 100 feet and greater. CSL testing identifies soil intrusions, honeycomb, voids, poor quality concrete and other defects between tubes. The most common application for CSL is quality assurance of underwater (or slurry) concrete placement in drilled shafts, slurry walls and seal footings where access tubes are cast-in-place. CSL is also used to locate defects between coreholes in foundations and other large concrete structures. Presently, Crosshole Sonic Logging has been specified for drilled shafts on Department of Transportation construction in 10 states in the USA, and many other states are following.

While CSL technology is very effective in detecting anomalies and defects, it also provides a good indication of defect size, position, and severity within a foundation. In the cases of a critical foundation with a defect, a more precise determination of the defect may be needed. Crosshole tomography is a technique that produces a 'slice' or a plot of the concrete condition between the tubes. This plot clearly shows the defect size, severity, and position between the tubes. Crosshole tomography uses multiple CSL logs taken around a defect and a powerful computer program to calculate the position and severity of the defect. Although this is a new technology, Crosshole Tomography has been successfully applied on several projects.

Discussed in this paper are two case studies on two drilled shaft foundations in California, where Crosshole Sonic Logging and Crosshole Tomography tests were performed.

INTRODUCTION

Quality assurance of foundations, particularly drilled shaft foundations, is becoming an important part of the foundation installation to ensure a good foundation that can transfer the

Marwan F. Aouad, Ph.D., Senior Engineer, and Larry D. Olson, P.E., President, Olson Engineering, Inc., 14818 W. 6th Ave., #5A, Golden, CO 80401

applied loads to the surrounding soil or rock. Until the mid 1980's, quality assurance of driven piles and drilled shafts in the USA was performed at selected shafts and used the Sonic Echo and Impulse Response (SE/IR) test methods to identify anomalies or defects (Davis and Dunn, 1974). The SE/IR method relies on reflection events from a change in impedance. Although the SE/IR method can be applied to identify defects, the method suffers from the following limitations: 1) the strength of the echoes depends on the surrounding soil, 2) echoes depend on the length to diameter ratio of the shaft, 3) the size and location of the defect cannot be determined, and 4) defects located below a major defect cannot be identified.

The drawbacks associated with the SE/IR method have led to the search for other alternatives and the development of the Crosshole Sonic Logging and Gamma-Gamma methods (Olson et. al, 1994) to identify defects in drilled shaft foundations. One of the advantages of the CSL method over the Gamma-Gamma method is that a complete coverage of the shaft conditions can be determined with the CSL method, while the Gamma-Gamma method determines the shaft conditions around the installed tubes in a drilled shaft.

In this paper, the Crosshole Sonic Logging (CSL) and Crosshole Tomography (CT) methods are briefly discussed along with case studies from tests on two drilled shaft foundations in California. It should be mentioned that the CT method is not routinely applied to drilled shaft foundations, and its application is limited to critical structures to produce a better image of defects identified in CSL tests.

CROSSHOLE SONIC LOGGING METHOD

The CSL method was developed in the mid 1980's for quality assurance of drilled shaft foundations, slurry walls and seal footings. The CSL method relies on direct transmission of sonic/ultrasonic waves between two access tubes placed in a drilled shaft prior to concrete placement. Figure 1 shows an illustration for a CSL test setup.

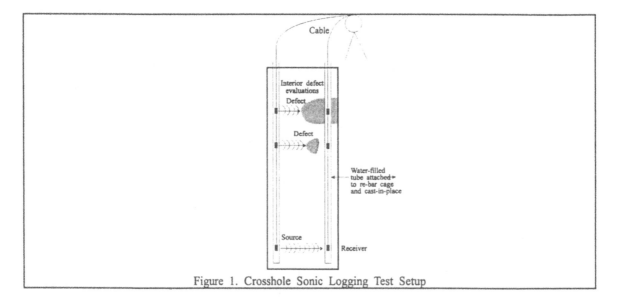

Figure 1. Crosshole Sonic Logging Test Setup

The number of access tubes per drilled shaft is dependent on the diameter of the shaft, typically 1 tube per 1 ft of diameter, and the tubes are installed around the perimeter of the shaft and tied to the inside (or outside) cage of the shaft. To perform a CSL test, two probes (hydrophone) are lowered to the bottom of two access tubes, and are retrieved to the top of the shaft while CSL measurements are taken approximately every 2 inches. The ultrasonic wave pulser is controlled by a distance wheel to control the transmission of waves at preselected distances. Automatic scanning of the collected records produces two plots, time and energy, versus depth. Anomalies and defects between tested tubes are manifested by time delays and energy drops in the scanned CSL plot. Concrete velocities are calculated by simply dividing the distance between the two tubes by the time required for the wave to travel from the source hydrophone to the receiver hydrophone.

CSL tests are typically performed between all perimeter tubes to evaluate the concrete conditions of the outer part of the shaft and between major diagonal tubes to evaluate the concrete conditions of the inner part of the shaft.

CROSSHOLE TOMOGRAPHY METHOD

The Crosshole Tomography method uses the same equipment as the CSL method with more tests being collected (many source and receiver locations). Once a defect is identified in CSL tests, CT tests can be performed to produce an image of the defect between the test tubes. The CT tests are typically performed at depths extending few feet below and above the defect zone as shown in Figure 2.

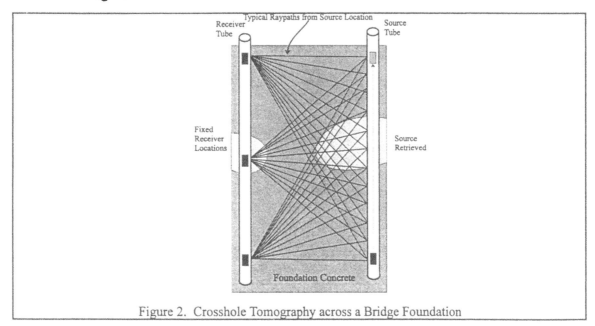

Figure 2. Crosshole Tomography across a Bridge Foundation

The CT data is used to obtain an image of the defect. The test region is discretized into many cells with assumed slowness values (inverse of velocity) and the time arrivals along the test paths are calculated. The calculated times are compared to the measured travel times and the errors are redistributed along the individual cells using mathematical models such as ART and SIRT (Herman, 1980). This process is continued until the measured travel times match the assumed travel times with an assumed tolerance.

CASE STUDIES

Discussed below are results from CSL and CT tests performed by Olson Engineering on two drilled shafts in California. The CSL results between tube pair 1-4 of the first shaft are presented in Figure 3. A significant delay in arrival times of compression waves and a significant drop in energy were observed in this CSL log at depths ranging from 8.3 to 11.2 ft below the top of the shaft. Although the anomaly is well identified in Figure 3, the exact location of the defect between

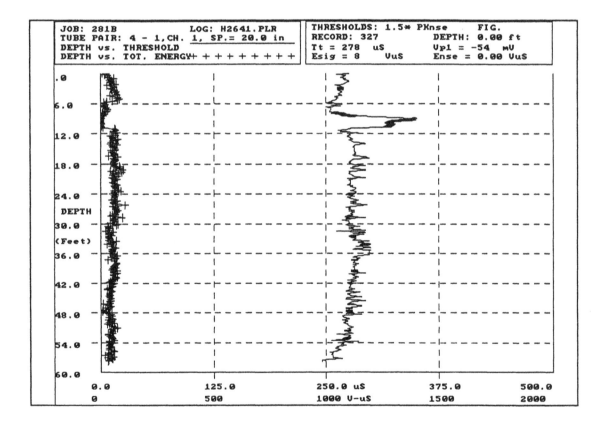

Figure 3. CSL Results Between Tubes 4-1 in a Drilled Shaft in California.

tubes 1 and 4 cannot be determined. For a better characterization of the anomaly indicated in Figure 3, a tomograhic dataset was obtained by Olson Engineering. For this dataset, the source was pulled from a depth of 18 ft below the shaft top and ending at the top of the tubes with the receiver moved at fixed interval locations of 2.25 inches. A velocity tomogram between tube pair 1-4 is presented in Figure 4. The anomalous zone in Figure 4 is represented as the low-velocity area (light area) which extends from a depth of 9.3 ft to a depth of 11.6 ft below the top of the shaft. The apparent low-velocity regions in the middle at the top and bottom of Figure 4 are artifacts resulting from a low ray density in these areas (see Figure 4 for the distribution of the ray densities). Figure 4 clearly shows that the defect occupies the entire distance between tubes 1 and 4, which cannot be inferred from the CSL results.

CSL results performed by Olson Engineering on a drilled shaft foundation at the Sargent Bridge on Highway 101 in Holister, California identified a defect at depths ranging from 15 to 17 ft. The anomaly was more severe between tube pair 1-3 than between other tube pairs (Figure is similar to Figure 3 and not shown here). This anomaly was further confirmed by Gamma-Gamma testing and destructive coring. Tomograhic data was obtained between tube pair 1-3 with the source pulled from 19 ft below the shaft top to the shaft top and the receiver fixed at 49 locations of 2.25-inch separation. Figure 5 shows the velocity tomogram obtained from this tomographic dataset along with the corresponding ray density plot. The anomalous zone in Figure 5 is represented as the low-velocity area which extends from a depth of 15 ft to depth of 17 ft below the top of the shaft. Figure 5 shows that the defect is centered around the tubes with good quality concrete in the interior between the two tubes as opposed to the defect shown in Figure 4 which extends through the entire distance between the two tubes.

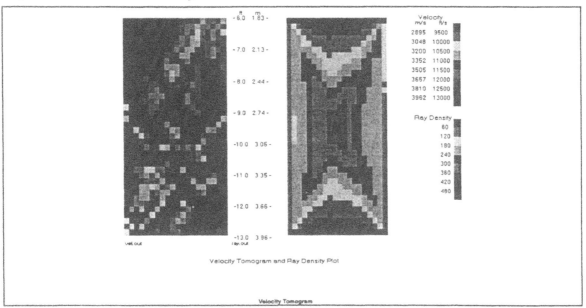

Figure 4. Velocity Tomography Results from the First Drilled Shaft in California.

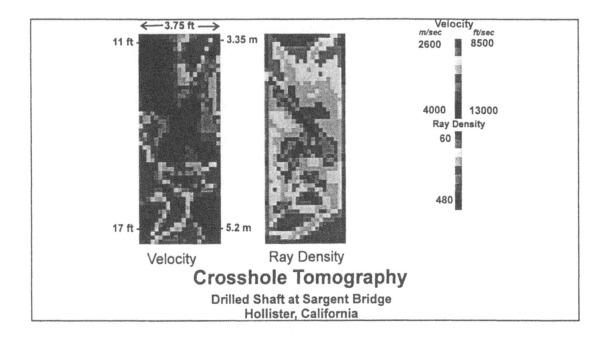

Figure 5. Velocity Tomography Results from the Second Drilled Shaft, Hollister, California.

CONCLUSIONS

The CSL method is an excellent nondestructive method for identifications of anomalous zones in drilled shaft foundations. Many State DOT's are moving towards specifications for CSL tests on new construction of drilled shaft foundations. The method is effective at locating defects between tube pairs, defect depths and extent, but not exact locations of defects between tube pairs. The CT method can be used as a complimentary method to the CSL method to determine a better characterization of the defect. Because of the much greater time required to perform tomographic analysis, the method may not gain popularity and its application will be limited to the more critical structures.

REFERENCES

Davis, A.G. and C.S. Dunn. 1974. "From Theory to Field Experience with the Nondestructive Vibration Testing of Piles," Proceedings of the Institution of Civil Engineers Part 2, 57: 571-593.
Herman, G.T. 1980. "Image Reconstruction from Projections, the Fundamentals of Computerized Tomography," Academic Press, Inc.
Olson, L.D., M. Lew, G.C. Phelps, K.N. Murthy, and B.M. Ghadiali. 1994. "Quality assurance of Drilled Shaft Foundations with Nondestructive Testing," Proceedings of the FHWA Conference on Deep Foundation, Orlando, Florida.

Electrical Resistance Tomography for Imaging Concrete Structures

M. BUETTNER, A. RAMIREZ AND W. DAILY

ABSTRACT

Electrical Resistance Tomography (ERT) has been used to non-destructively examine the interior of reinforced concrete pillars in the laboratory during a water infiltration experiment. ERT is a technique for determining the electrical resistivity distribution within a volume from measurement of injected currents and the resulting electrical potential distribution on the surface. The transfer resistance (ratio of potential to injected current) data are inverted using an algorithm based on a finite element forward solution which is iteratively adjusted in a least squares sense until the measured and calculated transfer resistances agree to within some predetermined value.

Laboratory specimens of concrete pillars, 61.0 cm (24 in) in length and 20.3 cm (8 in) on a side, were prepared with various combinations of steel reinforcing bars and voids (1.27 cm diameter) which ran along the length of the pillars. An array of electrodes was placed around the pillar to allow for injecting current and measuring the resulting potentials.

After the baseline resistivity distribution was determined, water was added to a void near one corner of the pillar. ERT was used to determine the resistivity distribution of the pillar at regular time intervals as water was added. The experiment lasted for about 29 hours and approximately 1.4 L of water was added in total.

The ERT images show very clearly that the water was gradually imbibed into the concrete pillar during the course of the experiment. The resistivity decreased by nearly an order of magnitude near the point of water addition in the first hour, and by nearly two orders of magnitude by the end of the experiment.

ERT has been shown to be an effective means of imaging water infiltration in concrete. Other applications for this technology include monitoring of curing in concrete structures, detecting cracks in concrete structures, detecting rebar location and corrosion state, monitoring slope stability and the stability of footings, detecting and monitoring leaks from storage tanks, monitoring thermal processes during environmental remediation, and for detecting and monitoring contaminants in soil and groundwater.

CONCEPT AND THEORY OF ERT

ERT is a method for determining the electrical resistivity distribution in a volume based on discrete measurements of current and voltage on the boundary. Resistivity data can be

Michael Buettner, Lawrence Livermore National Laboratory, 7000 East Avenue, Livermore, CA 94550
Abe Ramirez, 7000 East Avenue, Livermore, CA 94550
William Daily, 7000 East Avenue, Livermore, CA 94550

taken in a variety of configurations, including borehole-to-borehole, borehole-to-surface, or pure surface. Detailed technical concepts and theory used for ERT can be found elsewhere (Daily and Owen, 1991; Daily et al., 1992).

In order to obtain an image of a body from surface measurements, a number of electrodes are placed on the surface of the body and in electrical contact with it. A pair of adjacent electrodes are driven by a known current, and the resulting voltage difference is measured between other pairs of electrodes. Then, the known current is applied to another pair of electrodes and the voltage is again measured between other pairs. This procedure is repeated until current has been applied to all pairs of electrodes. The ratio of a voltage at one pair of terminals to the current causing it is a transfer resistance. For n electrodes there are n(n-3)/2 independent transfer resistances.

The next step is to calculate the distribution of resistivity in the volume given the measured transfer resistances and to construct an image. However, the calculation for the distribution of resistivity is highly nonlinear because the currents flow along the paths of least resistance and are dependent on the resistivity distribution. Finite element algorithms and least square methods are used to invert the transfer resistances.

Image construction can be on the same finite element mesh used to calculate the measurements or on a different array. An absolute image shows a resistivity structure whereas a comparison image shows changes in resistivity. Comparison images can be used to study dynamic processes in structures, such as imbibition, by comparing data taken at different times. This paper presents comparison images only.

Forward Solution

The forward solution to Poisson's equation uses the finite element method (FEM) to compute the electrical potential response of a two dimensional earth due to a three-dimensional source. To avoid the difficulty of numerically solving a three dimensional problem, Poisson's equation is formulated in the wave number domain via Fourier transformation in the strike direction. The governing equation (Hohmann, 1988) is,

$$\frac{\partial}{\partial x}\left(\sigma \frac{\partial V}{\partial x}\right) + \frac{\partial}{\partial z}\left(\sigma \frac{\partial V}{\partial z}\right) - \lambda^2 \sigma V = -I\delta(x)\delta(z) \tag{1}$$

where V is potential in the Fourier transform domain, σ is the electrical conductivity, λ is the Fourier transform variable, I is the source current, and $\delta(x)$ is the delta function. The two dimensional FEM algorithm is based on the theory described by Huebner and Thorton (1982) and the implementation follows that described by Wannamaker et al. (1987) for modeling two dimensional magnetotelluric data. Using the FEM method, potentials are calculated for a discrete number of transform variables at the nodes of a mesh of quadrilateral elements. The potentials are then inverse transformed back into Cartesian domain using the method described by LaBrecque (1989).

Numerical Inversion

The inversion method uses the modified Marquardt algorithm (Bard, 1974) to jointly solve the non-linear equation,

$$\underline{\mathbf{W}}\mathbf{D} = \underline{\mathbf{W}}*\mathbf{F}(\mathbf{P}) \tag{2}$$

and the equation,

$$\mathbf{P}^{\mathbf{T}}\underline{\mathbf{R}}\mathbf{P} = 0 \tag{3}$$

where **D** is the vector of known data values, $\underline{\mathbf{W}}$ is weighting matrix, **P** is the vector of unknown parameters, **F(P)** is the forward solution and $\underline{\mathbf{R}}$ is the roughness matrix which is a numerical approximation to the Laplacian operator (Sasaki, 1990). To solve these equations jointly, the algorithm minimizes

$$\chi^2 + cL(\mathbf{P}) = 0 \tag{4}$$

where c is a constant and is the Chi-squared statistic which is given by

$$(\mathbf{D} * \mathbf{F(P)})^T \underline{\mathbf{W}}^T \underline{\mathbf{W}}(\mathbf{D} * \mathbf{F(P)}) = \chi^2 \tag{5}$$

Ideally the inverse algorithm would find the maximum value of c for which χ^2 is equal to some known, a priori value. The constant, c, was determined by trial and error by calculating an inverse model with an a priori value of c then adjusting if necessary to achieve the correct value of χ^2.

INFILTRATION EXPERIMENT

Several laboratory concrete pillars were fabricated for testing with ERT. All had square cross sections 20.3 cm (8 in) by 20.3 cm (8 in) and were 61.0 cm (24 in) in length. They had various combinations of standard 1.27 cm (0.5 in) steel reinforcing bars (rebar) and voids (with the same diameter as rebar) running along the length of the pillar. They were made from a standard commercial mortar mix with about 11.4 L (3 g) of water per 27.3 kg (60 lb) bag of mortar. The pillar was allowed to cure for 1 year indoors at room temperature before this experiment was performed. The pillar used for this experiment had 5 pieces of rebar and 3 voids as shown in the sectional view of Fig. 1. Five ERT electrodes were placed on each side of the pillar as shown.

The purpose of this simple experiment was to show how ERT could be used to image moisture infiltration into concrete. Water was introduced into the void at the corner of the pillar as shown in Fig. 1. The bottom of the void was plugged to prevent the water from running out of the bottom.

ERT data were collected prior to the addition of water. These data were used to establish a baseline against which later data sets for the "wet condition" could be compared. Comparison images, as described in the previous section, were constructed based upon ratios of electrical resistivity. That is, the transfer resistances were determined at various times after adding water to the void, and the displayed images show the ratio of resistivity in the wet condition to that in the dry condition. Thus the images displayed show changes in resistivity from the baseline. When moisture enters the concrete, the resistivity is expected to decrease because the water with dissolved minerals from the concrete is much more conductive than the dry concrete. It is then expected that the resistivity ratio displayed will be less than 1 where water has infiltrated.

After collection of the baseline data, water was added to fill the void. ERT data sets were then collected at regular intervals over a period of about 29 hours. Water was added to the void from time to time as it was imbibed into the concrete. The total amount of water added was about 1.4 L.

Comparison grayscale images are shown in Figs. 2 and 3. The elapsed time for the image of Fig. 2 is 1 hour, and the amount of water added up to that time was about 0.75 L. The grayscale can be interpreted by noting that the light end of the scale corresponds to decreases in the resistivity ratio. Thus the light gray area in the upper left-hand corner of the pillar near the void shows the beginnings of water infiltration. Fig. 3 shows the comparison image

29 h and 16 min after water was first added. By this time the total amount of water, 1.4 L, had been added. Note that the region of decreased resistivity, shown as white, is now much larger. Water has apparently been imbibed into a much larger volume of the pillar.

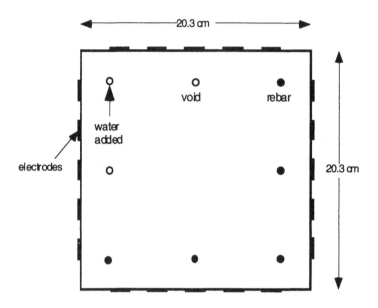

Figure 1. A sectional view of the pillar used for the infiltration experiment.

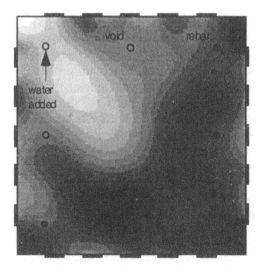

Figure 2. Image of resistivity ratio after 1 hour and the addition of about 0.75 L of water.

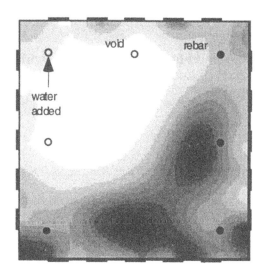

Figure 3. Image of resistivity ratio after 29 hours and the addition of about 1.4 L of water.

CONCLUSIONS AND DISCUSSION

ERT was successfully used to image water infiltration into a concrete pillar. When moisture enters the concrete, the resistivity decreases because the water with dissolved minerals from the concrete is much more conductive than the dry concrete. ERT images show this decrease in resistivity as water is imbibed into the pillar over time. The magnitude of this effect can be rather large. For example, in the area near the water addition point in Fig. 2, the resistivity ratio has a minimum value of 0.123. Thus resistivity has decreased by a factor of $(0.123)^{-1}$ or about 8 in one hour. In Fig. 3, after about 29 hours, the ratio is 0.011, and the resistivity has decreased by a factor of about 90. Resistivity decreases one to two orders of magnitude as water is imbibed, and ERT is a good way to monitor it.

APPLICATIONS

ERT imaging has potential applications relating to structures, soils, pavements, and environmental remediation and monitoring. In the structural field, ERT could be used to monitor the curing of concrete where moisture content is correlated to the degree of curing. ERT should have advantages over other techniques such as seismic and radar. In the seismic case, this is true because water infiltration results in much larger electrical resistivity contrasts (100 to 1) than seismic velocity contrasts. In the case of radar, penetration depth is limited due to highly-conductive, wet concrete, while for ERT this is not a problem. Thus it will be possible to image the interior of large structures using ERT (a concrete dam for example).

ERT should also be applicable for detecting cracking in structures through the changes in resistivity that occur when water infiltrates cracks. In addition, ERT should be capable of detecting rebar location and possibly the state of corrosion. Applications of ERT for soils include monitoring slope stability and the stability of footings. In pavements, ERT has already been used to monitor moisture content and movement (see companion paper). Finally, ERT has been used to detect and monitor leaks from storage tanks, for monitoring thermal processes during environmental remediation, and for detecting and monitoring contaminants in soil and groundwater.

ACKNOWLEDGMENT

John Carbino provided technical support for this project. This work was performed under the auspices of the U.S. Department of Energy by Lawrence Livermore National Laboratory under contract no. W-7405-Eng-48.

REFERENCES

Bard, Y. 1974. "Nonlinear Parameter Estimation," *Academic*, San Diego, CA, pp. 111-113.

Daily, W. D., and E. Owen. 1991. "Cross-borehole Resistivity Tomography," *GEOPHYSICS*, 56(8): 1228-1235.

Daily, W. D., A. Ramirez, D. LaBrecque and J. Nitao. 1992. "Electrical Resistivity Tomography of Vadose Water Movement," *Water Resources Research*, 28(5):1429-1442.

Hohmann, G. W. 1988. "Numerical Modeling for Electromagnetic Methods of Geophysics," *Electromagnetic Methods in Geophysics*, Part 1, Soc. Expl. Geophys, Invest. Geophys., M. N. Nabighian, editor, 3: 313-363.

Huebner, K. H., and E. A. Thorton. 1982. *The Finite Element Method for Engineers*. John Wiley and Sons, Inc., New York, 1982.

LaBrecque, D. J. 1989. *Cross-borehole Resistivity Modeling and Model Fitting*. Ph.D. thesis, Univ. of Utah, Salt Lake City.

Sasaki, Y. 1990. "Model Studies of Resistivity Tomography Using Boreholes," Society of Exploration Geophysicists International Symposium on Borehole Geophysics: Petroleum, Hydrogeology, Mining and Engineering Applications, Soc., of Explor. Geophys., Tucson, AZ.

Wannamaker, P. E., J. A. Stodt, and L. Rijo. 1987. *PW2D Finite Element Program for Solution of Magnetotelluric Response of Two-Dimensional Earth Resistivity Structure.*, Rep. ESL-158, Earth Sci. Lab., Univ. of Utah Res. Inst., Salt Lake City.

Study on Limitations of Thermographic Survey Applied to the Detection of Delamination in Concrete Structures

T. KOMIYAMA, Y. TANIGAWA AND Y. NAKANO

ABSTRACT

As a method of detecting delamination of finishing materials on external walls, the thermographic survey using thermal imager is widely used because of the advantages of easiness, rate of data sampling and safety. However, since this method is based on the lack of heat conductivity of the delaminated area in comparison with that of the sound area of the building wall, which causes temperature differences between their surfaces, the limitations of this method depend largely upon natural weathering.

This paper discusses the applicability and limitations of the method by carrying out an experiment and a heat balance simulation using Finite Element Method. As a result, the applicability and limitations of this method were clarified. In addition, it was proved that the thermographic survey shows a good performance for the detection of delaminated render or tile by choosing fine weather in proper seasons.

INTRODUCTION

The deterioration of concrete structures due to drastic changes in environment or due to poor workmanship has become very serious lately in Japan. In particular, since buildings and several other kinds of concrete structures in this country are finished with render or tile on their facades in order to improve durability and appearance in many cases, the number of accidents resulting in injury or death due to the delamination of such finishing materials continues increasing.

The thermographic survey using thermal imager is helpful to detect delaminated finishing materials. However, the applicability and limitations of the method is not clear.

Tatsuhito Komiyama, Graduate Student, and Yasuo Tanigawa, Professor, School of Architecture, Faculty of Engineering, Nagoya Univ., Furou-cho, Chikusa-ku, Nagoya-City, Japan
Yonezou Nakano, Managing Director, Constec Co., 3-1-46, Nishi-ku, Osaka-City, Japan

In order to confirm the applicability and limitations of the method, an experiment was carried out. In the experiment, in stead of exposing specimens containing artificial delaminations to the sun, temperature differences equivalent to the heat which should be given to the specimens by solar radiation were made.

After the experiment, a heat balance simulation using Finite Element Method was carried out to compare with the results of the experiment. This paper presents the results of the experiment and simulation.

PRINCIPLE OF THERMOGRAPHIC SURVEY

A thermographic survey uses thermal images to locate areas where debonding or delamination has occurred. The method is based on the lack of heat conductivity of delaminated render in comparison with that of the render well bonded onto the parent material. This causes temperature differences on the building surface, which are identified in the thermal images. However, this method requires an sufficient solar radiation bringing about the temperature differences between the delaminated area and the sound area. Also, a large change of air temperature is preferable. In Tokyo, the change of air temperature in a day is about $10\,°C$ under a fine weather condition. However, the quantity of solar radiation varies largely by season and the direction of the wall. An example of the solar radiation curves of the external walls of a building in Tokyo under an ideal weather condition is shown in Figure 1, which were applied to the following experiment and heat balance simulation.

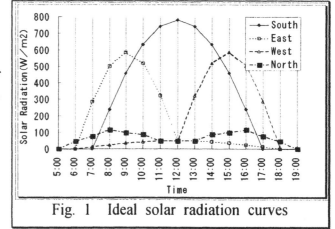

Fig. 1   Ideal solar radiation curves

OUTLINE OF EXPERIMENT

The situation of the experiment is shown in Figure 2. In the experiment, in stead of exposing the specimens to the sun, the change of air temperature equivalent to the heat which should be given to the specimens by solar radiation was made. A solar radiation can be converted into an air temperature by means of Eq.(1).

$$\theta_{solair} = \theta_a + R_{s0}(\varepsilon_{sun} J_{sun} + \varepsilon J_{nit}) \quad (1)$$

where,

θ_{solair} : solair temperature (℃) ／ θ_a : air temperature (℃)

R_{s0} : surface resistance (m²/W) ／ ε_{sun} : solar absorptance (%)

J_{sun} : quantity of solar radiation (W/m²) ／ ε : emissivity (%)

J_{nit} : night radiation (W/m²)

The air temperature calculated by Eq.(1) is called solair temperature. In order to make the change of air temperature obtained by Eq.(1), two curing rooms were used. The larger one was used to make the outdoor solair temperature and the other one was used to make the indoor air temperature. Figure 3 shows the dimensions of the specimens used in the experiment. The thickness of render and the width of the delamination of each specimens are shown in Table 1. D70W1.0, D70W0.5 and D100W1.0 specimens were submitted to the experiments based on the solar radiation curves of east side wall, west side wall and south side wall respectively. The experiment based on the solar radiation curves of north side wall was carried out with D30W1.0 specimen.

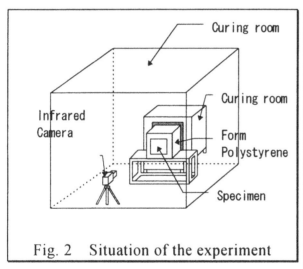

Fig. 2   Situation of the experiment

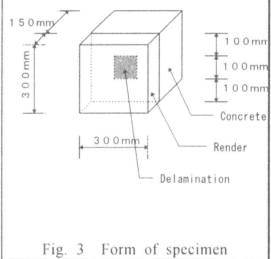

Fig. 3   Form of specimen

Table 1   Render thickness and width of delamination

Name of specimen	Render thickness (mm)	Width of delamination (mm)
D30W1.0	30	1
D70W1.0	70	1
D70W0.5	70	0.5
D100W1.0	100	1

OUTLINE OF HEAT BALANCE SIMULATION

Figure 4 shows the analytical model applied to the heat balance simulation. The ideal solar radiation curves shown in Figure1 were applied to the simulation. The change of air temperature in a day was assumed to be 10℃. In the following discussion, the temperature of Point ① represents the temperature of the delaminated area and the temperature of Point ② represents the temperature of the sound area. The physical constants applied to the simulation are shown in Table 2.

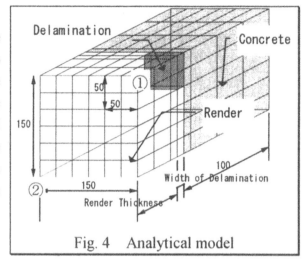

Fig. 4   Analytical model

Table 2   Physical constants

Materials	Heat conductivity (W/m·K)	Specific heat (J/kg·K)	Density (kg/m³)	Heat transfer coefficient (W/m²·K)	Solar absorptance (%)
Concrete	1.59	882	2300	9.3	0.75
Render	1.30	798	2000	9.3	0.75
Air	0.022	1010	1.3	—	—

RESULTS AND DISCUSSION

SURFACE TEMPERATURE DIFFERENCES

The surface temperature differences between the central point of the delaminated area and the edges of the specimens are shown in Figures 5-8 respectively. The surface temperature differences obtained through the experiment were smaller than those of the simulation by about 0.2℃. Those errors were caused by the differences of the physical constants between those applied to the simulation and those of the specimens.

Judging from the results of the experiment, the surface temperature differences occurred in D70W1.0 specimen facing east, west and south were enough to be detected by

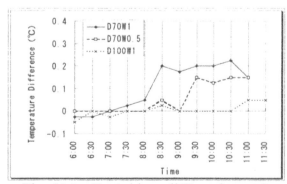

Fig. 5(a)   East side wall(Experimental)

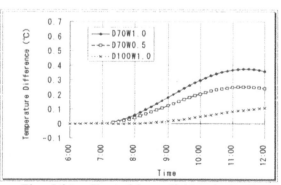

Fig. 5(b)   East side wall (Analytical)

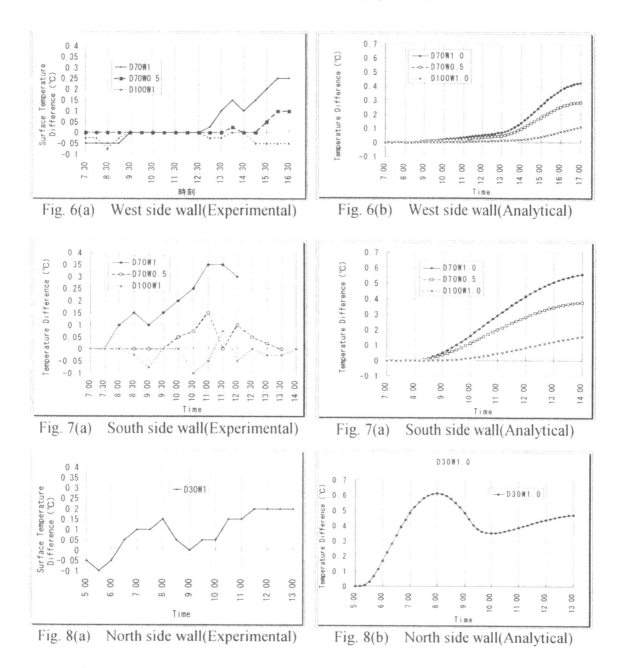

Fig. 6(a)   West side wall(Experimental)

Fig. 6(b)   West side wall(Analytical)

Fig. 7(a)   South side wall(Experimental)

Fig. 7(a)   South side wall(Analytical)

Fig. 8(a)   North side wall(Experimental)

Fig. 8(b)   North side wall(Analytical)

thermal imager. However, those of D70W0.5 specimen were unstable and insufficient. The same thing can be said to D100W1.0 specimen. Finally, in spite of the little solar radiation, the surface temperature difference of 0.2℃ occurred in D30W1.0 specimen facing north.

THERMAL IMAGES

The representative examples of the thermal images of the specimens obtained through the experiment are shown in Figures 9-12. The delaminated area of D70W1.0 specimen can be identified from Fig. 9. However, that of D70W0.5 specimen is not clear. The delaminated area of D100W1.0 specimen is invisible. In spite of the low surface temperature difference, the delaminated area of D30W1.0 specimen facing north is clear.

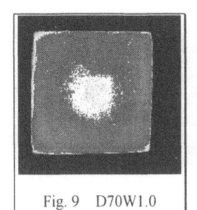

Fig. 9   D70W1.0
facing east (10:00 AM)

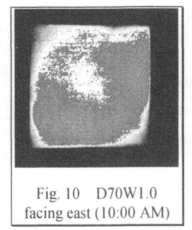

Fig. 10   D70W1.0
facing east (10:00 AM)

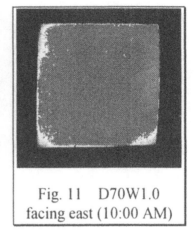

Fig. 11   D70W1.0
facing east (10:00 AM)

CONCLUSIONS

1. In Tokyo, the delamination existing within 70 mm from the surface of external walls facing east, west and south can be detected by thermographic survey if its width is larger than 1 mm and their area is larger than 100 mm $\times$ 100 mm under fine weather conditions.

Fig. 12   D70W1.0
facing north(12:00 AM)

2. The thermographic survey can be applied to the inspection of building walls facing north, which can be given a little solar radiation compared with those of walls facing other directions if the thickness of the finishing material is smaller than 30 mm.

3. There exist cases delaminated areas cannot be recognized in spite of adequate temperature differences. It takes the delaminated areas about an hour to appear in recognizable shape.

4. A heat balance simulation using Finite Element Method is helpful to estimate the applicability of the thermographic survey. However the numerous results obtained by the heat balance simulation can not always represent the easiness of the identification of dclamination.

REFERENCES

Saitou, H. 1993. "Climates Regarding Architecture," Kyoritsu Shuppan, 1-20 (in Japanese)

AIJ, 1978. "Data Book for Architecture Design," Maruzen, 105 (in Japanese)

Yagawa, M. 1983. "Primer of Flow and Heat Conduction," Baihukan, 223-225 (in Japanese)

Discriminatory Analysis of the Ultrasonic Method for the Quality of Pored Cast-in-situ Piles

L. RUIJIE, X. CAIHONG AND M. SHENGDONG

ABSTRACT

This paper presents an improved method for the conventional method of the ultasonics using a discriminatary analysis method of statistics. A discriminatory function and an evaluation criterion were obtained by use of the data of the ultrasonic testing of an imitation pored pile. The contributation coefficients of the two testing indexes(velocity and amplitude) to an evaluation result were also obtained. The function and the criterion can be taken as means of quantitative analysis in quality evaluation on the pored cast-in- situ piles. They can be constantly modified by the test data of piles and the actural flaw information of the piles obtained from the excavation of foundation. Thus, the evaluation reliability can be improved by the increasing of tested pile number. The procedures and the validity of the discriminatory analysis method are illustrated by an example with test data from a pored cast-in-situ concrete pile.

INTRODUCTION

The ultrasonics method is the most widely used nondestructive method for assessing concrete quality(Popovics,et al,1992). Because it is simple, fieldworthy, capable of penetrating deep into concrete, and lack of other better nondestructive method , the method is popular in evaluating the quality of pored cast- in-situ concrete piles. It synthesizes the parameters of the velocity and the amplitude of the first received signal to evaluate the quality of the tested concrete. However, the evaluation is greatly dependent on the experience of a testor. Thus the errors are often made in the evaluation. In a construction field, in fact, the foundation conditions, the raw materials and mix proportions of concrete, the technological processes of construction are similar among piles. Thus, the types, and sizes of the flaws of one pile are similar to those of another pile in the same field. According to above discussion, we can use the test data of piles whose flaw information are known to obtain a discriminatory function by use of the criterion of statistics. The quality of a new pile in the same field can be evaluated according to the function. Its validity and the effect of each parameter of sonics can be evaluated by the criterions of statistics. The function can be constantly modified by use of the new reliable information of testing of piles. Thus, the evaluation errors

Liu Ruijie,Xiu Caihong,Mi Shengdong, Tianjin Port Engineering Institute, Dagunan Road 1474, Hexi District, Tianjin, 300222, P.R.China

due to the artificial factor of testors can be reduced and the reliability of the ultrasonic method may be improved.

In the paper, The discriminatory analysis method was described. An imitation pored cast-in-situ pile was made to establish the discriminatory function. An engineering pored pile was evaluated according to the function.

DISCRIMINATORY ANALYSIS(Han,Y.G.,1989;Liu,R.J.,1994)

Discriminatory Function

A linear discriminatory function with p discriminatory variables(such as pulse velocity, amplitude, etc.) can be written as follows:

$$Y = \sum_{k=1}^{P} C_k X_k \tag{1}$$

where Y is discriminatory function, and X_k is the K-th discriminatory variable, and C_k is the K-th constant coefficient to be determined. The Fisher's criterion is often used to obtain the coefficients. Its main idea is described as follows(Han,Y.G.,1989). A new discriminatory index can be obtained by use of the linear combination of the p variables of the two group elements(one goup is normal —group A and the other is poor —group B). To give the best discrimination on the two groups, the criterion is to make the difference of the new index mean values of the two groups maximum and in the mean time make the interclass quadratic sum of deviation of each group minimum. If group A has n_1 elements and group B has n_2 elements, then discriminatory values$Y_j(A)$ and$Y_j(B)$ can be obtained as follows

$$Y_j(A) = \sum_{k=1}^{P} C_k X_{jk}(A); \quad Y_j(B) = \sum_{k=1}^{P} C_k X_{jk}(B) \quad (j = 1, 2, \ldots, n) \tag{2}$$

where $X_{jk}(A), X_{jk}(B)$ are the k-th test index values of the j-th element of group A and group B respectively. If $\overline{Y}(A)$ is the mean value of Y_j and $\overline{Y}(B)$ is the mean value of $Y_j(B)$,they can be given as follow

$$\overline{Y}(A) = \frac{1}{n_1} \sum_{j=1}^{n_1} \sum_{k=1}^{P} C_k X_{jk}(A); \quad \overline{Y}(B) = \frac{1}{n_2} \sum_{j=1}^{n_2} \sum_{k=1}^{P} C_k X_{jk}(B) \tag{3}$$

The criterion requires the following index I is up to the maximum value.

$$I = [\overline{Y}(A) - \overline{Y}(B)]^2 / \left(\sum_{j=1}^{n_1} [Y_j(A) - \overline{Y}(A)]^2 + \sum_{j=1}^{n_2} [Y_j(B) - \overline{Y}(B)]^2 \right) \tag{4}$$

Sign d_k is: $d_k = \overline{X}_k(A) - \overline{X}_k(B) \quad (k = 1, 2, \ldots, p)$ \hfill (5)

where $\overline{X}_k(A)$ is the mean value of the k-th test index of group A with n_1 elements, and $\overline{X}_k(B)$ is the mean value of the k-th test index of group B with n_2 elements. They can be given as follow:

$$\overline{X}_k(A) = \frac{1}{n_1} \sum_{j=1}^{n_1} X_{jk}(A); \quad \overline{X}_k(B) = \frac{1}{n_2} \sum_{j=1}^{n_2} X_{jk}(B) \quad (k = 1, 2, \ldots, p) \tag{6}$$

Sign S_{kl} is:

$$S_{kl} = \sum_{j=1}^{n_1}[X_{jk}(A) - \overline{X}_k(A)][X_{jl}(A) - \overline{X}(A)]+$$

$$\sum_{j=1}^{n_2}[X_{jk}(B) - \overline{X}_k(B)][X_{jl} - \overline{X}_l(B)] \quad (k,l = 1,2,\ldots,p) \tag{7}$$

The coefficients in Eq.(1) to make I value in Eq.(4) maximum can be obtained from the following equations:

$$\sum_{l=1}^{p} C_l S_{kl} = (n_1 + n_2 - 2)d_k \qquad (k = 1,2,\ldots,p) \tag{8}$$

Discriminatory Index and Criterion

When the coefficients in Eq.(1) are obtained, $\overline{Y}(A)$ and $\overline{Y}(B)$ can be calculated from Eq.(3). A weighted average of them can be regarded as the discriminatory index given as follows:

$$Y_c = [n_1\overline{Y}(A) + n_2\overline{Y}(B)]/(n_1 + n_2) \tag{9}$$

If $\overline{Y}(A) \geq Y_c > \overline{Y}(B)$, then when $Y \geq Y_c$, we can conclude that the element is normal. When $Y < Y_c$, the element is discriminated to be poor. If $\overline{Y}(A) < Y_c \leq \overline{Y}(B)$, when $Y \geq Y_c$, the element is poor. When $Y < Y_c$, it is normal.

Inspection on Discriminatory Function and Contribution Coefficients

Mahalanobis's D^2 statistical value can be used to carry out the inspection on the validity of the discriminatory function erected.

$$D^2 = |c_1 d_1| + |c_2 d_2| + \ldots + |c_p d_p| \tag{10}$$

Statistical index F can be given by:

$$F_{p,n_1+n_2-p-1} = \left[\frac{n_1 n_2}{(n_1 + n_2)(n_1 + n_2 - 2)}\right]\left[\frac{n_1 + n_2 - p - 1}{p}\right] \cdot D^2 \tag{11}$$

It is F distribution with p and $n_1 + n_2 - p - 1$ freedom degrees. Therefore by use of the table of critical value of F distribution in statistics, the valididy of the discriminatory function can be inspected. Contribution coefficients of the test indexes can be defined by:

$$D_k = \frac{|c_k d_k|}{D^2} \times 100\% \tag{12}$$

where D_k is contribution coefficient of the k-th test index. The higher the value is, the greater the contribution of the test index to the results of discrimination is.

ANALYSIS AND DISCRIMINATION ON IN-SITU DATA
Discriminatory Function of an Imitation Pile

An imitation pored pile was made in a field. The flaws of the pile was artificially made in the similar way to that of the actural engineering piles. It was tested by use of the two little steel pipes in the concrete as shown in Fig.1(a).

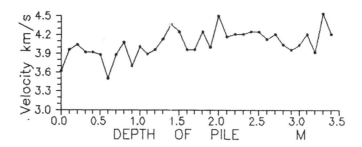

(c) Amplitude Values of Testing Points

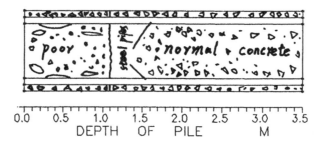

(b) Velocity Values of Testing Points

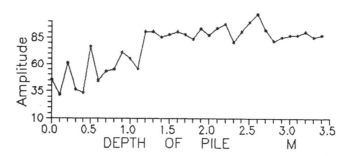

(a) Diagram of the Imitation Pile

Figure 1: Testing Data of an Imitation Pile

The ultrasonic parameters of velocity and amplitude of the pile were shown in Fig.1(b) and Fig.1(c) respectively. The discriminatory function can be obtained from above equations using the data in Fig.1 as follows

$$Y = 42.39X_1 + 24.13X_2 \tag{13}$$

where X_1, X_2 are the dimensionless values of velocity and amplitude respectively. The dimensionless values can be obtained from the values of velocity and amplitude divided by their mean values.
$\overline{Y}(A), \overline{Y}(B), Y_c$ can be calculated as follows

$$\overline{Y}(A) = 67.78; \quad \overline{Y}(B) = 51.26; \quad Y_c = 62.11 \tag{14}$$

Then

$$\overline{Y}(A) > Y_c > \overline{Y}(B) \tag{15}$$

Therefore when $Y > Y_c$, the tested point will be discriminated to be normal. When $Y < Y_c$, the point will be discriminated to be poor.

Inspection and Discussion about the Function

D^2 is :
$$D^2 = 16.52 \tag{16}$$

Fvalueis :
$$F = 63.16 \tag{17}$$

The critical value of F distribution is: $F_\alpha(2, 32) = 19.5, \alpha = 5\%$ $\tag{18}$

As F=63.16> $F_{0.05}(2, 32) = 19.5$, so the difference of mean value of discrimination function between normal concrete and poor concrete is marked highly. Therefore the discrimination function(Eq.(13)) is valid. Contribution coefficients of pulse velocity(D_1) , amplitude(D_2) are:
$$D_1 = 23\%; \quad D_2 = 77\% \tag{19}$$

So the significance of pulse velocity parameter is much lower than that of amplitude parameter. This indicates that even though the velocity of test point is high, the quality of concrete may be poor when the its amplitude is low. So the parameter of amplitude is the main factor to determine the discriminatory result of the concrete quality in the discriminatory analysis.

Discriminatory Analysis of an Engineering Pile

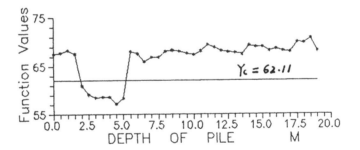

(c) Discriminatory Function Values

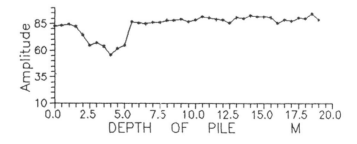

(b) Amplitude Values of Testing Points

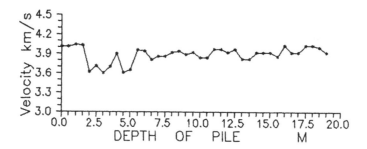

(a) Velocity Values of Testing Points

Figure 2: Test Data and Discriminatory
Result of an Engineering Pile

An engineering pile in the same field as the imitation pile was tested and evaluated according to the function of Eq.(13). The test data and discriminatory result are shown in Fig.2. As Fig.2(c) shows, There would be poor concrete at the depth between 2.0m and 5.0m. The upper part of the engineering pile was excavated and the concrete was detected. We found out that there was a serious necking at the depth between 1.8m and 5.3m.

CONCLUSIONS

It is possible to establish a discriminatory function according to statistics theory ,and to make the assessing for the quality of pored pile quantitative and thus to reduce the artificial effect on the assessing result. The original discriminatory function can be constantly modified by use of the new reliable test data of pored piles. Thus the reliability of evaluation may be improved at the present technological level of the ultrasonic method.

REFERENCES

Han,Y.G.,1989, " Applying Mathematical Statistics , " Peking Aviation and Space University Press, 1st. edn..(in Chinese)

Liu,R.J.,1994, " Discrimination Analysis of the Quality of Concrete Strength of Structural Elements, " Proceedings of the Third International Conference on Inspection, Appraisal Repairs and Maintenance of Buildings & Structures 1-2 June 1994:Bangkok, Thailand, 145-150.

Popovics,S.;Popovics,J.S.,1992, " A Critique of the Ultrasonic Pulse Velocity Method for Testing Concrete, " Nondestructive Testing of Concrete Elements and Structures, Edited by Farhad Ansari and Stein Sture, 94-103.

Study of Fatigue Cracks in Steel Bridge Components Using Acoustic Emissions

H. L. (ROGER) CHEN AND R. D. FULTINEER, JR.

ABSTRACT

Preliminary results of an acoustic emission (AE) study of fatigue cracks in A-36 steel specimens are presented. The fatigue propagation of cracks in three steel beams were monitored with the use of a six channel acoustic emissions system. Each specimen was subjected to three-point bending with cyclic loading. Based on these experimental results, plots were established of crack growth rate and acoustic counts versus change in stress intensity factor. Preliminary field monitorings of fatigue cracks in an actual bridge girder have been conducted. The field monitorings results have shown that the location method being used is effective.

INTRODUCTION

Steel bridge members are constantly subjected to cyclic loadings of variable magnitudes. As a result, material fatigue may be experienced. To prevent sudden failures of bridge members and structures, acoustic emission has been proposed as a passive warning system for detecting fatigue cracking. Several previous studies have revealed possible relationships between AE signals and the fracture parameters (Dunegan, 1968, 1973; Vannoy, 1991). However, only limited data have been reported in the literature for applying AE monitoring to bridge structures (Miller and McIntire, 1987; Gong, Nyborg and Oommen, 1992). This study is being conducted to obtain and to characterize the acoustic emission signals emitted from cracks in steel bridge components. This characterization is intended to enhance the existing damage evaluation of highway bridges. The result of this study is targeted at developing AE as a feasible bridge monitoring technique.

H.L. (Roger) Chen, Associate Professor, Constructed Facilities Center, Dept. of Civil and Environmental Engineering, West Virginia University, Morgantown, WV 26506-6103; (304)293-3031, X631; EMail HLCHEN@WVNVM.WVNET.EDU.

Roy D. Fultineer Jr., Research Assistant, Constructed Facilities Center, Dept. of Civil and Environmental Engineering, West Virginia University, Morgantown, WV 26506-6103: (304)293-3031, X644.

LABORATORY TESTS

INSTRUMENTATION

A LOCAN-AT system has been used for the AE signal acquisition during testing. The complete AE monitoring setup, as shown in Figure 1, contains six piezoelectric transducers (with a sensitive frequency range of 100 kHz to 1 MHz), six preamplifiers corresponding to each sensors and a 6-channel data acquisition system. The system includes three independent channel controllers with 100 kHz - 1.2 MHz shielded band-pass filter.

The laboratory specimens tested were monitored using a threshold setting of 35 dB, which is above the machine and background noises of the loading frame. The internal gains within the system were set at 30 dB. Preamplifiers were set to a 40 dB gain. The system is also capable of locating AE sources using linear, 2-D, and 3-D source location techniques. Standard pencil lead break tests were used to calibrate the AE system. The sensitivity and the effectiveness of the location techniques are dependent on the sensors' sensitivities, the attenuation of the material, and the placement of the sensors. The fact that steel is low in attenuation allows AE technique to monitor a larger area on a subject.

Figure 2 shows the sensor locations on a steel specimen: sensors #1, #2 and #3 were placed in the front panel and sensors #4, #5, and #6 were at the back panel. The sensors were placed in a triangular pattern and were 3 in. away from the midspan. The sensors were attached to the specimens with hot melt glue. The 3-D source location technique was applied.

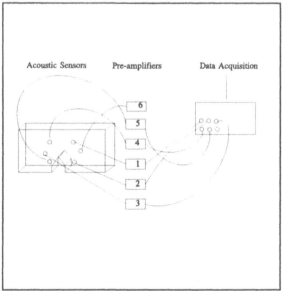

Figure 1 AE Monitoring Setup

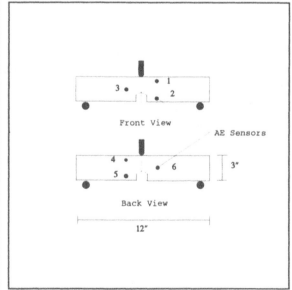

Figure 2 Loading Diagram and AE Sensor Placement

The test specimens used in the experiments were cut from an A-36 steel plate. The steel plate has a yield strength of 50 ksi and a ultimate tensile strength of 67.5 ksi. The plate was cut into 16 in. long sections with a width of 3 in. and a depth of 1.5 in. (see Figure 1). A notch, 1.25 in. in depth was then machined into each specimen. Fatigue precracks of crack lengths of between 45% and 55 % of total depth have been induced into the specimens before testing.

Three point bending tests were conducted on the specimens. Six notched specimens were tested to date, and the results from specimens #3, #4, #5, and #6 are reported in this paper. Specimen #4 was monotonically loaded to failure and specimens #3, #5, and #6 have been successfully fatigued to failure.

The three-point bending tests were conducted on a MTS close-loop loading machine. A MTS microprofiler, which can be used to set the upper and lower load limits along with the loading frequency, was used to control the fatigue load. The notched steel specimens were placed between a loading head and two supporting rollers. The two support rollers were placed with a span of 12 in. distance between them. The loading head was constructed to apply a line load at the midspan of the specimen.

MONOTONIC TEST

Specimen #4 has an initial crack length of 1.44 in. (1.25 in. of notch depth and 0.19 in. prefatigue crack length). The prefatigue cracking was conducted with a sinusoidal load with peaks between 6,000 and 9,000 pounds. After the precrack had been established, specimen #4 was monotonically loaded until it failed. The final load of 17,000 pounds was obtained and used to calculate the stress intensity factor. The stress intensity factor, K_{max}, for specimen #4 is about 92.6 ksi-in$^{1/2}$. The results of the monotonic test was used to determine the fatigue stress level for the fatigue testing.

FATIGUE TEST

For Specimens #3, #5, and #6, the prefatigue and fatigue tests were conducted at a loading range of 700 lbs. to 7000 lbs. at 1.5 Hz., which corresponds to a R ratio, K_{min}/K_{max}, of 0.1. Crack gages were used to obtain crack lengths during the experiments. Two different crack gages were used. The crack gages (Model TK-09-CPA01-005/DP, Micromeasurements) used on specimen #3 have 20 grid lines per gage and the lines were spaced 0.01 inches apart. The crack gages (Model TK-09-CPA02-005/DP, Micromeasurements) used on specimen #5 and specimen #6 have 20 grid lines per gage and lines were spaced 0.02 inches apart. Recording both crack length and time with the data acquisition system during the experiment made it possible to calculate crack growth rate. The measured fatigue crack growth for specimen #3 is shown in Figure 3. Each voltage jump of the crack gage output represents a crack growth of 0.01 inch.

RESULTS AND CONCLUSIONS

Three cracking stages were observed during the fatigue loading of the steel specimens:

the first stage was the crack initiation stage, the second stage was a stable crack propagation stage, and the last stage was a stage where sudden and brittle fracture was accompanied by a large plastic zone. On specimens #3, #5, and #6, cracks were initiated and prefatigued to lengths of 1.56, 1.60, and 1.57 in., respectively. The prefatigue cracking usually takes about 85,000 to 100,000 cycles depending on the length of crack achieved. During initial fatigue testing, stable cracks were observed up to a time that the cracks would reach a total length of 2.10, 2.09, and 2.06 in., respectively. Unstable and accelerated crack growths have been observed afterwards, during which the cracks would grow to total lengths of 2.5, 2.47, and 2.375 in., respectively. During unstable crack growth, necking of the specimen has been gradually observed on the surface of the specimen, indicating the existence of a large plastic zone. Large amount of AE signals were generated during this stage. The specimen finally failed in a brittle manner. The different cracking stages can be identified by studying the textures on the cross sections of the cracked specimens. The portion showing stable fatigue crack growth is usually identified as smooth surfaces, while unstable cracks are shown to have rougher surfaces. The final failure stage is usually indicated by an irregular tearing surface. The K_{max} for the specimens #3, #5, and #6 are: 89.2, 87.7, and 83.5 ksi-in$^{1/2}$, respectively.

Figure 4 shows the relationship between the crack growth rate (da/dn) and the change in stress intensity factor (ΔK) for all three specimens (#3, #5, and #6). The specimens are shown to fail at a rate of approximately 5 x 10^{-5} inches/cycle and the corresponding critical stress intensity factors ΔK are between 62 and 68 ksi-in$^{1/2}$. Figure 5 shows the relationship between acoustic counts and ΔK for the current tests, which indicates that ΔK increased in an exponential fashion, and that only limited acoustic emissions were recorded in the earlier stages of the fatigue crack propagation. Major jumps in acoustic emissions occur after the critical values of ΔK have been reached. Similar trends have been observed for the AE behaviors of other brittle metals, which has been related to the fatigue behaviors as a function of the stress intensity factor (Dunegan and Harris, 1973).

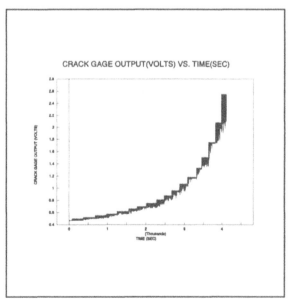

Figure 3 Crack Gage Data for Specimen #3

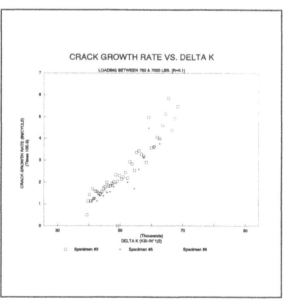

Figure 4 Crack Growth Rate versus Delta K

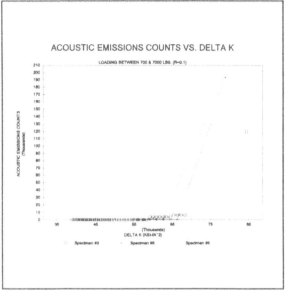

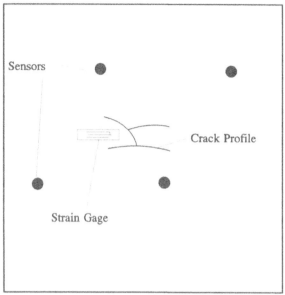

Figure 5 Acoustic Counts Versus Delta K

Figure 6 Sensor and Strain Gage Layout for Bridge Test

FIELD TEST

Preliminary study of AE monitoring of fatigue cracks in a highway bridge with concrete deck and steel girders has been conducted on the Uffington Bridge (I-79) of West Virginia. Located near Morgantown area, the bridge was inspected biannually by the West Virginia Department of Highway, and has been found to contain fatigue cracks in the bridge girders (WVDOH, 1991, 1993). The bridge girders are constructed of built-up sections of three welded plates with stiffeners, and are made of A36 steel. The web of the bridge is 0.25 in. thick and approximately 4 ft. in depth. Several fatigue cracks have been visually identified in the past. A crack on the mid-span of the 12th bridge girder (diaphragm #4) was selected for the bridge monitoring. The crack occurred at the welded joint between the bridge web and a stiffener plate and has three crack branches. The last reported crack size of the particular crack has been found to be 1 11/16 in. in length.

Four AE sensors were used to monitor the crack. The sensors are placed in a parallelogram pattern as seen in Figure 6, and 2-D source location technique was used to identify the source of the AE signals. The location technique helps in eliminating undesirable noises from sources outside of the sensitive region. A weldable strain gage was also used during the monitoring to measure the strain in the proximity of the crack. From the strain information one can estimate the relative loads that are being applied to the bridge and extract the bridge vibration information. Actual field monitoring was conducted continuously for 6 hours under regular traffic loads. Preliminary AE data indicated that there are active AE sources surrounding the visual crack profile, and the visually measured crack length is larger than the last reported length in 1993.

ACKNOWLEDGMENTS

The authors would like to thank the Federal Highway Administration for their support.

The authors also thank Shen-En Chen for his valuable input.

REFERENCES

Dunegan, H. L., D. O. Harris and C. A. Tatro. (1968). *Fracture Analysis by Use of Acoustic Emission*, Engineering Fracture Mechanics, vol. 1, pp.105.

Dunegan, H. L. and D. O. Harris. (1973). *Acoustic Emission Techniques*, Experimental Techniques in Fracture Mechanics, Society for Experimental Stress Analysis Monograph No. 1, Kobayashi, A. S. ed., pp.38-72.

Gong, Z., E. O. Nyborg and G. Oommen. (1992). *Acoustic Emission Monitoring of Steel Railroad Bridges*, Materials Evaluation, v. 50, no. 7, pg. 883.

Miller, R. K. and P. McIntire. edit (1987). *Acoustic Emission Testing*, Nondestructive Testing Handbook, Vol. 5, American Society NT.

Vannoy, D. W. and Azmi, M. (Apr. 1991), *Acoustic Emission Detection and Monitoring of Highway Bridge Components*, Final Rept., Maryland Univ., College Park. Dept. of Civil Engineering, Maryland State Highway Administration, Baltimore.

WVDOH (1991), Bridge Inspection Report on the Uffington Bridge (No. 31-79-149.29), WVDOH.

WVDOH (1993), Bridge Inspection Report on the Uffington Bridge (No. 31-79-149.29), WVDOH.

Acoustic Emission Detection, Characterization, and Monitoring of Steel Bridge Members

M. G. LOZEV, G. G. CLEMEÑA, J. C. DUKE, JR. AND M. F. SISON, JR.

ABSTRACT

This paper describes an acoustic emission (AE) monitoring technique to identify active fatigue cracks in steel bridge members. The current approach for AE field monitoring was modified, using guard sensors, electronic filtering, and post-monitoring analysis, to obtain and evaluate AE data in the time and frequency domains. A data base of AE characteristics of various structural components during field applications was developed to improve the reliability of the AE technique.

INTRODUCTION

Past studies (Miller,1987; Vannoy and Azmi, 1991) of AE have not fully developed the engineering application of AE to bridge monitoring. In particular, unwanted noises associated with bolt fretting and rubbing or traffic have not distinguished systematically from sounds associated with crack initiation or growth, during monitoring of critical structure regions. Continued advances in electronics, such as faster microprocessors, provide testing capabilities that were not possible a few years ago, since AE relies heavily on instrumentation. Recommendations for a successful AE test program are given in recent studies (Ghorvanpoor, 1994; Clemeña et al., 1995). In this study three methods of separating unwanted noise from relevant AE were explored: spatial discrimination, load discrimination, and signal discrimination.

TESTING PROCEDURES

An 8-channel Spartan AE data acquisition system manufactured by PAC was used to test five bridges. Six of the 8 channels were used for signal measurement. The other two channels digitized and stored AE waveforms using TRA-212 Transient Recorder Analyzer. One parametric input was used to record load data from general-purpose strain gage, designed for strain averaging

M. G. Lozev, G. G. Clemeña, VTRC, 530 Edgemont Rd, Charlottesville, VA 22903
J. C. Duke, Jr., M. F.Sison, Jr, VPI & SU, Blacksburg, VA 24061

measurements on large specimens. Most of the tests were configured for both planar (zonal) and linear location.

Two types of piezoelectric transducers, the R30I resonant and the WD wideband, and a 40 dB preamplifier with a highpass frequency filter of 20 KHz, were used. With all sensors in place, the traditional pencil lead break test was performed for each sensor and source location sensor array. The transducers were in good acoustic contact with the part monitored if the breaks registered amplitudes of at least 80 dB for a reference voltage of 1 mV and a total system gain of 80 dB. The breaks also checked the accuracy of the source location setup, indirectly determining the actual value of the acoustic wavespeed using the differences in the time of arrival of lead break signals at two separate transducers. A gain of 40 dB and a floating threshold of 25 dB were used.

RESULTS AND DISCUSSION

The results of AE tests performed on five bridges in Virginia are summarized in Table 1.

TABLE I. SUMMARY OF ACOUSTIC EMISSION BRIDGE TESTS

Bridge	Detail; Monitoring	Problem	AE Test Results
Rte. 460 New River, Glenlyn VA	Pin and hanger; 120 min	Ultrasonically-detected crack on pin	No crack activity
Rte. 29 Staunton River, Altavista VA	Girder web; 45, 60 min	a) Web crack retro-fitted with splice plate b) Web crack arrested with stop drill holes	No crack activity
Rte. 671 Moormans River, Albemarle County VA	Diagonal counter; 90 min	Visible transverse crack on diagonal counter	No crack activity
I-66 south exit over Rte. 29, Gainesville VA	a) Coverplate weld b) Lower web-to-flange weld; 85 min	New repair welds	No crack activity
Rte. 29 Robinson River, Madison County VA	Pin and hanger; 40, 60, 95 min	Four visible cracks on 2 hangers	Crack activity detected from 3 cracks

Two pin-and-hanger connections on the western end of the west-bound lanes on the Rte. 460 bridge over the New River were chosen for AE monitoring. One had a crack; the other was newly installed. The live loading of the bridge was solely the normal passing traffic. However, there was no obvious clustering of AE events at the location, where ultrasonic inspection had indicated a crack 102 mm from AE sensor.

The sensor array attached to the members at different locations on the Rte. 29 bridge over the Staunton River, the Rte. 671 bridge over the Moormans River and I-66 south exit over Rte.29 detected no events coming from the crack tip and repair welds. The results showed that cracks were inactive.

The Rte. 29 northbound bridge over the Robinson River was monitored more extensively than any other bridge in this study. During a regular inspection in October 1992, cracks were found on two of the hangers. The east exterior hanger had a 9.4 mm crack. Three similarly located cracks were found on the west interior hanger. The longest was 36 mm; the shorter upper crack was 6 mm long. The third crack was not a through-part crack and was visible only at the surface of the internal slot. All the cracks initiated at hanger's internal cavity at practically the same location. Bolted catch plates were installed on both hangers to prevent collapse in the event of a sudden failure of the hangers.

Both cracked hangers were monitored during the first AE test. Sensor placement was not originally intended to perform source location. The bridge was loaded by normal noon-time traffic. Analysis of the collected AE data made it apparent that spatial discrimination using source location was necessary to distinguish relevant AE signals from noise.

On the second AE test, only the cracked west inner hanger was monitored. The WD sensor used for recording waveforms was mounted between the R30I sensors close to the crack (Figure 1). Guard sensors were positioned to eliminate rubbing noise from the top and lower pin, and noise from the girder itself. The AE monitoring setup was further improved by attaching a strain gage to the left side of the link. The bridge was loaded by normal traffic.

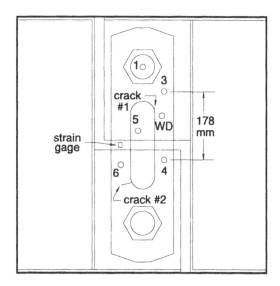

Figure 1. Location of sensors on cracked west inner hanger of Robinson River bridge.

The third AE test on the Robinson River bridge gathered more AE and waveform data from crack #1, and monitored the longer crack #2. The threshold setting for R30I sensor was decreased from 30 dB to 25 dB, because the first test showed that some crack-related signals had peak amplitudes less than 30 dB. TRA thresholds were likewise adjusted so weak, long-duration noise signals would not be detected. The digitization rate was kept at 5 MHZ while waveform record lengths were decreased to 1.17 msec for the WD sensor and 0.778 msec for the R30I sensor on the basis of results from the previous test. Monitoring times were not continuous; on occasion, data were only collected when large vehicles such as tractor trailer trucks passed over the bridge.

Spatial Discrimination

The first test indicated a strong AE source close to crack #1, although failure to arrange the sensors for source location made it impossible to determine whether the signals were crack-related or were fretting noise from the pin. A linear source location detected only 83 events during the second test; but even with so few events, a cluster of events can be recognized where the crack was located. The effectiveness of the guard sensors was shown by the removal of most of the signals that appeared close to sensor 4. Most of the signals eliminated by the guard sensors came from a source closest to sensor 6, which detected the greatest number of signals of all the sensors.

The array configured for linear source location on crack #2, detected 247 events, all of which were detected when a single tractor-trailer truck passed over the bridge. Crack #2 was obviously much more active than the other cracks. These events were more spread out than expected. The accuracy of the source location function of the data acquisition system was undoubtedly a factor in this spread. However, another explanation is that many of the events were due to crack face rubbing which may have occurred anywhere along the 36 mm crack faces. A longer crack surface, with more rubbing surfaces, can also partly explain the larger number of detected events compared to the shorter cracks #1 and #3.

Waveform and Frequency Analysis

Waveforms of 11 crack-related events from crack #1 were stored from the second test. Figure 2 shows a representative waveform as detected by the wideband transducer, together with its normalized frequency spectrum, and a typical noise waveform and a frequency spectrum as detected by the WD sensor. All of the crack-related AE signals had nearly the same waveform envelope shapes. However, the most notable feature distinguishing crack-related AE from noise was the peak frequency. All the crack signals had peak frequencies of 275 KHz, while all other waveforms, presumably noise, peaked no higher than 200 KHz, mostly less than 100 KHz. This, it should be noted, applied only to signals from the WD wideband transducer. Figure 2 shows the same crack and noise emissions, as detected by R30I sensor. Both waveforms have frequency contents that peak at about 325 KHz, which is close to the transducer resonant frequency. This illustrates how the R30I resonant sensor modifies the AE signals it detects with its own characteristics, thereby masking the inherent differences between crack-related AE and noise.

Continuing work in this study includes load discrimination and further analysis of the waveforms collected from crack #2 to identify features which distinguish crack growth signals from crack-face rubbing.

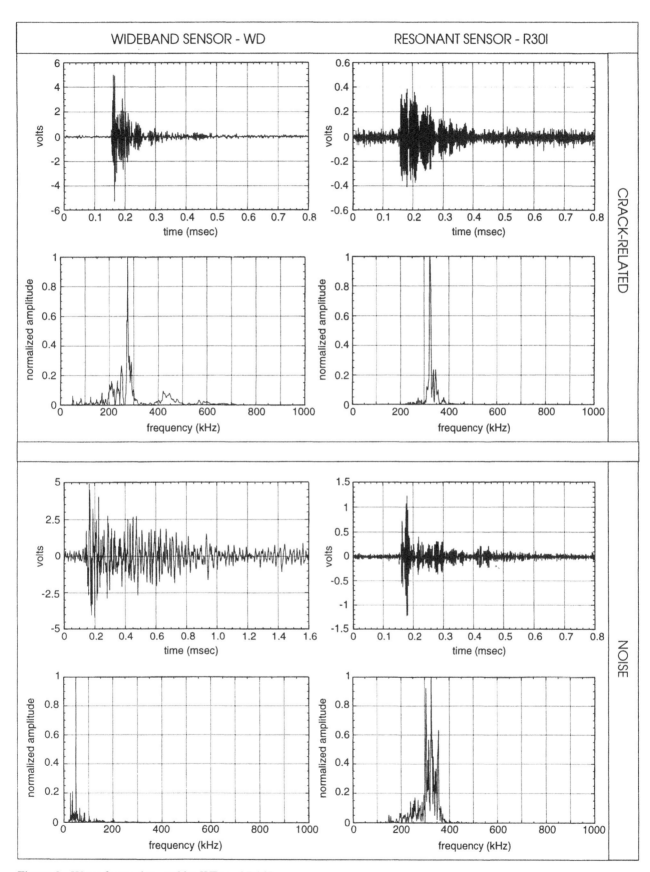

Figure 2. Wave forms detected by WD and R30I sensors.

CONCLUSIONS

Source location and spectrum analysis are highly effective for filtering noise sources that are spatially and frequency distinct and separable from monitored flaws. Guard sensors are necessary when noise activity outside the zone being monitored is high.

ACKNOWLEDGMENTS

The authors thank D. B. Sprinkel, R. Truxell, W. L. Sellars, and J. W. Lillard of VDOT for their assistance; A. R. Zadeh of VPI & SU, C. Apusen and A. French of VTRC for technical support; R. Combs for graphics; and C. Napier of FHWA, E. G. Henneke II of VPI & SU, R. Truxell of VDOT, and M. M. Sprinkel of VTRC for peer review. G. Mawyer edited the paper. The authors would like to express their appreciation to Dr. G. R. Allen of VTRC for his continuing support of NDE research.

REFERENCES

Clemeña, G. G., Lozev, M. G., Duke, J. C, Jr., and Sison., M. F., Jr. 1995. "Acoustic Emission Monitoring of Steel Bridge Members", FHWA/VTRC 95-IR1. Charlottesville: Virginia Transportation Research Council.

Ghorbanpoor, A. 1994. "An Assessment of the Current Acoustic Emission Evaluation of Steel Bridges", Proceedings of the Structural Materials Technology-An NDT Conference, Technomic, Lancaster, PA.

Miller, R. K. (ed.). 1987. "ASNT Nondestructive Testing Handbook", vol. 5, ASNT, Inc.

Vannoy, D. W., and Azmi, M. 1991. "Acoustic Emission Detection and Monitoring of Highway Bridge Components. Report No. FHWA-MD-89-10. College Park, MD: University of Maryland.

Application of Acoustic Emission, Strain Gage, and Optical Sensors to Moveable Bridges

D. W. PRINE

ABSTRACT

Recently, acoustic emission and strain gage and laser displacement gage monitoring technology have been successfully applied to the nondestructive evaluation of several lift bridges. The applications cover a wide range of problems ranging from machinery diagnostics to crack characterization. This research is being sponsored by Northwestern University's Infrastructure Technology Institute and the states of Wisconsin and Oregon. This paper will discuss test procedures and results on two of these bridges. One of the bridges is a rolling bascule type while the other is a vertical lift bridge. The technology was used to assess the functionality of the segmental casting attachments on the rolling bascule structure and to confirm fatigue crack growth in the trunnion shaft of the vertical lift bridge.

Wisconsin DOT Structure B-47-40, Prescott, WI

Wisconsin DOT structure B-47-40 carries east and westbound traffic on U.S. highway 10 over the St. Croix River in the town of Prescott in Pierce County, Wisconsin. The bridge consists of 5 spans and has an overall length of 682.7 feet. The center span (span 3) is a two leaf, rolling bascule lift bridge with an overall length of 205.5 feet. The bridge deck is 66 feet in width and has four lanes of vehicular traffic and a pedestrian sidewalk.

The three piece segmental castings that the bridge rolls on are attached to the bascule girders with high strength friction bolts. This is a relatively recent design modification (8 to 10 years ago) and replaces the traditional turned bolts or rivets that were commonly used in this application. This design change is basically a cost cutting measure. Wisconsin DOT bridge inspectors have observed cases of bolt failure and casting slippage on similarly constructed bridges. These occurrences can lead to dangerous operating conditions or structural failure.

Prior to this test, concerns for safe bridge operation by Wisconsin DOT bridge inspection personnel led them to call upon BIRL engineers to apply advanced NDE technology to the Tayco Street lift bridge (WI-DOT structure B70-97-93) in Menasha, WI in

David W. Prine, Northwestern University, BIRL Industrial Research Laboratory, 1801 Maple Avenue, Evanston, IL 60201

September of 1993. This bridge utilized the same design modification as the Prescott bridge. Bridge operation and inspection personnel had observed loud audible impact noises during bridge operation.   BIRL engineers applied acoustic emission monitoring techniques to determine that the source of the impact noises was the high strength bolts. Strain gages were applied to the casting/girder interface and large permanent displacements of the casting with respect to the girder were observed

The loud impact noises that were observed on the Tayco Street bridge were not observed during operation of the Prescott bridge. However, continuing concerns on the part of WI-DOT inspection personnel over the performance of the friction bolts led to an agreement with BIRL to perform the type of testing on this structure that was previously applied to the Tayco St. bridge.

Tests utilizing acoustic emission and strain gage monitoring were performed. Additional experiments were performed using a displacement sensor based on laser triangulation. The tests were completed on November 8 and 9, 1994.

Acoustic emission (AE) testing was performed on each of the four casting assemblies using a field portable AE monitoring system.   AE sensors were attached to each of the three casting segments. An additional sensor was attached in the vicinity of the pinion gear on the upper part of the bascule girder. The sensors were 175 KHz resonant piezoelectric devices. Silicone grease was used for acoustic couplant and magnetic hold downs were used to clamp the sensors to the structure.   AE data was recorded on disk files using a portable PC attached to the AE monitor via an RS232C serial port. The recorded data was analyzed post test. The casting mounted sensors were used to detect any impact or fretting events related to the casting attachments. The pinion mounted sensor was utilized to intercept drive gear and deck related AE signals and to act as a guard for the casting mounted sensors.

Examination of the three casting mounted sensor's hits and energy shows that the Prescott bridge had both lower hit counts and much lower energy than the Tayco St. tests. The difference in AE test data between the two bridges is even more apparent if we look at the product of event counts and their average energy which is a relative measure of the total detected AE activity resulting from a complete bridge operating cycle. Table 1 shows this result for the two bridges. Clearly, the Prescott bridge has less overall AE activity.

AE Sensor	Tayco St.	Prescott NE	Prescott NW	Prescott SE	Prescott SW
1	0	2,325	646	53	417
2	111,544	178	544	504	0
3	0	4,734	0	0	4,730
4	1,776	0	136	8,640	5,150

Table 1 Summary of AE Activity

Strain Gage Testing

A weldable foil strain gage was mounted diagonally across the center casting to bascule girder mating surface on both the north and south corners at the east end of the bridge. This application is a non-standard approach to using strain gages. The purpose is to attempt to observe displacements between the casting segment and the girder flange. The observations are more qualitative than quantitative because the gage is not subjected to a uniform strain field across the gage width. This mounting approach was developed during experiments performed on the Tayco St. Bridge in Menasha, WI.

The strain gage data taken on the Prescott bridge showed three significant departures from similar tests on the Tayco St. bridge. The peak strain on Prescott was 50 to 75 micro-inches per inch while Tayco St. was 600 to 1200 micro-inches per inch. Secondly, the Prescott strain gage data showed even symmetry and good repeatability on the strain wave forms recorded during raising and lowering while poor odd symmetry and no repeatability was observed in the Tayco St. data. The Prescott strain gages returned to their original zero within the quantization error of the monitoring system (approx. 10 micro-inches per inch) while the Tayco St. strain data was offset by as much as 150 micro-inches per inch following a complete bridge cycle.

Laser Displacement Gage Testing

The laser displacement gage (Aeromat LM300) had source and target mounted on opposite sides of the casting to girder interface and aligned parallel to this interface. It allowed us to observe the elastic deformation of the casting under dynamic loading conditions during bridge opening and closure and was easily capable of detecting slippage of the casting. The shape and symmetry of both the laser gage and the strain gage data agreed well and the laser gage showed no appreciable slippage between the girder flange and the casting following a complete bridge operating cycle. A typical laser gage plot is shown in Figure 1. The laser gage typically showed some offset when the bridge was at the maximum opening point (mid-range). The gage was mounted on the center casting in all tests. We believe the offset observed at the bridge up position is the result of load sharing between the castings that is fostered by the wedges applied between castings.

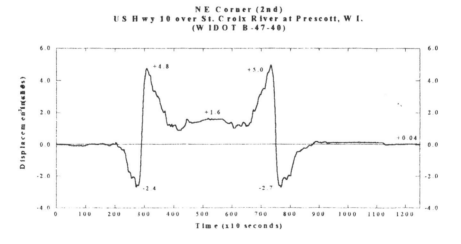

Figure 1 Typical Laser Gage Plot

These test results indicate that under current operating conditions the segmental casting attachments on the Prescott bridge (B-47-40) exhibit no abnormal behavior as evidenced by AE, strain gage, and laser displacement testing. However, future re-testing would be prudent based on past experience with this design.

Oregon DOT Bridge # 1377A

The Interstate Bridge carries I-5 traffic over the Columbia River between Portland Oregon and Vancouver Washington. The bridge consists of two separate bridges each carrying three lanes of traffic. The east bridge was built in 1917 with rehabilitation work done in 1960. This bridge carries the northbound traffic. The west bridge was built in 1958 and carries the southbound traffic. Each bridge has a span drive vertical lift span to permit large vessels to pass through. The lift span trusses are 272 feet long between the live load supports. Vertical clearance at low water is 39.86 feet and the maximum vertical lift is 139 feet. Oregon DOT records indicate that the bridge openings average 400 to 500 per year and the average daily vehicular traffic volume is 87,000.

According to information furnished by Oregon DOT, the trunnion shafts on the east bridge were modified during the 1960 rehabilitation to allow application of tapered roller bearings in place of the plain sleeve bearings that were originally installed. The drawings indicate that prior to the taper machining operation on the shafts to accommodate the roller bearings, the longitudinal grease grooves were to be filled with weld metal.

Laboratory testing performed on the shafts in 1987 under the supervision of Sverdrup Corp. determined that the forged steel had carbon content ranging from 0.50% to 0.87%. Carbon levels this high have a severe adverse effect on the weldability of the material. Charpy V-notch tests on samples taken from the bore area indicated low impact properties with a ductile to brittle transition temperature of 70° F or higher. Field Ultrasonic tests (UT) performed during the 1987 Sverdrup study and again in 1993 by Oregon DOT indicated the presence of a crack indication in the vicinity of the shoulder at the outboard end of the north east trunnion shaft. The presence of the roller bearings makes access to the shaft surface impossible without major costly disassembly thus precluding direct examination of the area containing the UT indication. The high cyclic loads and questionable material properties in conjunction with the UT indications led Oregon DOT to consider the use of AE to aid in their understanding of the trunnion shaft condition.

AE Testing

During the period of October 3 through 5, 1994, BIRL engineers performed AE tests on each of the four trunnion shafts of the East Interstate Bridge in Portland Oregon. The AE monitoring was done using a field portable AE system that has 6 input channels and is computer based.

A total of five AE sensors were used to acquire data during the opening and closing of the bridge. The AE sensors used for these tests were piezoelectric resonant devices with center frequencies of approximately 175 KHz. Line driving pre-amplifiers were used to eliminate any cable loading effects on the sensors and to reduce vulnerability to electrical interference. Over-all system gains of 80 db were used for these tests. Sensitivity was checked before and after each test with both a pulser and pencil lead breaks.

A sensor was coupled to each end of the shaft using silicone grease as an acoustic couplant and magnetic clamps to hold the sensors in place. This pair was used to perform linear AE source location on the shaft. On the N.E. shaft, these sensors were lined up with the UT crack indications probable position. The remaining three sensors were mounted on the outer portion of the sheave and spaced at 120° intervals to act as guards to intercept AE signals produced by cable slippage and other potential noise producing mechanisms. AE sources that are located off of the line joining the pair being used for linear location will not locate properly and so they must be rejected from the data set which is the function of the guard sensors

Source location tests were run with simulated AE sources. Both pencil lead breaks and a sensor driven by an electronic pulser were used. These tests showed that the predominant acoustic propagation mode was the bulk longitudinal wave. This results from the shaft dimensions (19 inches in diameter and 67 inches long). This mode of acoustic propagation has a higher velocity than the normal plate or surface wave mode that is typically encountered in the thin plates (0.5 inches to 3 inches) that are common in bridge structural members. Plate or surface waves typically have velocities near the shear wave velocity which is the velocity that our AE monitor's internal software uses. The ratio between shear and longitudinal wave velocity for steel is .552. The measured length of the trunnion shaft was 67 inches. Our monitor which uses shear wave velocity (stored in firmware) should give a shaft length (measured by the internal calibration routine) of 37 inches. We actually measured 38 inches so our agreement with the theoretical shear to longitudinal velocity is within 2.7%. The presence of longitudinal wave propagation forced us to analyze all of the data off-line and thus not be able to make use of the powerful real-time noise rejection algorithm that the AE monitor has in its operating software. The time and location clustering that is the primary noise rejection tool can be done off-line on recorded data but it is a time consuming process.

A total of ten data files were recorded on the nights of October 3 and 4, 1994. The first four tests were recorded on the north east shaft and the remaining six runs were done two each on the remaining three shafts. The recorded data was filtered to remove low level background noise and source location was plotted for each of the runs. The first AE test run was recorded with only two channels to determine if guard sensors were needed. Examination of the recorded data for run one (no guards) and run two (guards in place) in the field confirmed that large numbers of AE bursts were being generated outside the locating array thus making the use of guards mandatory. Table 1 below summarizes the tests.

Table 1 AE Test Summary

Test Run	Location	Results	Comments
1	N.E. Shaft	NA	No Guards
2	N.E. Shaft	Location & Time Clustering in Crack Area	High Event Count
3	N.E. Shaft	Location & Time Clustering in Crack Area	High Event Count
4	N.E. Shaft	Location & Time Clustering in Crack Area	High Event Count
5	N.W. Shaft	No Clustering in Shoulder Areas	Low Event Count
6	N.W. Shaft	No Clustering in Shoulder Areas	Low Event Count
7	S.E. Shaft	No Clustering in Shoulder Areas	Moderate Event Count
8	S.E. Shaft	Location Clustering only in Shoulder Area	Moderate Event Count
9	S.W. Shaft	No Clustering in Shoulder Areas	Moderate Event Count
10	S.W. Shaft	No Clustering in Shoulder Areas	Moderate Event Count

The clustering analysis used on these data was developed to allow acoustic emission to reliably detect flaw growth in welds during the welding process. It is based on observations of acoustic emission signal characteristics that accompany known crack growth under controlled laboratory conditions. Crack growth typically produces AE event rates of the order of several per second (typically 3 to 5). Cracks are also localized sources of acoustic emission. In other words, the events produced by a crack have tightly clustered source locations. This algorithm has been successfully tested over the past 15 years in a wide variety of acoustic emission applications and has allowed AE to be utilized for applications that have large amounts of background noise. It is particularly useful in cases where noise sources are coincident with cracks such as cracks growing out of fastener holes. The data recorded during these tests had high background noise and the clustering algorithm allows us to see differences in the data for each of the shafts that probably would have been over looked in simpler analysis. To perform the analysis, we filtered the recorded data to remove low level noise and then computed source locations using the correct acoustic velocity as determined by our calibration process. The application of clustering analysis to source location allows us to identify the areas of the shaft that produce significant AE signals. If we examine the source location clustering data for the shoulder region of the shaft (15 inches) with a window of plus or minus 2 inches, we see locational clustering only on the N.E. shaft outboard shoulder region (runs 2, 3, and 4) between 13 and 17 inches, and the outboard shoulder region of the S.E. shaft (run 8). These results are shown below in Figure 2. Close examination of the recorded data for these clusters shows that only the N.E. shaft indications exhibit time clustering. The three event cluster on the S.E. shaft at 15 inches has a minimum time interval between events of over 3 seconds thus failing the time clustering test. This analysis shows that only the N.E. shaft produces AE events in the shoulder area of the shaft that satisfy our location and time clustering criteria. This is the location that includes the ultrasonic crack indication so AE adds additional credibility to the UT findings with an indication of possible crack growth.

These tests point out the usefulness of applying multiple NDE methods to a difficult inspection problem. Either UT or AE by itself does not provide data that is positively conclusive. However, when we combine the two and the results correlate as they did in these tests, confidence in the test results is considerably improved.

Figure 2 Showing AE Source Location Clusters in the Crack or Shoulder Area

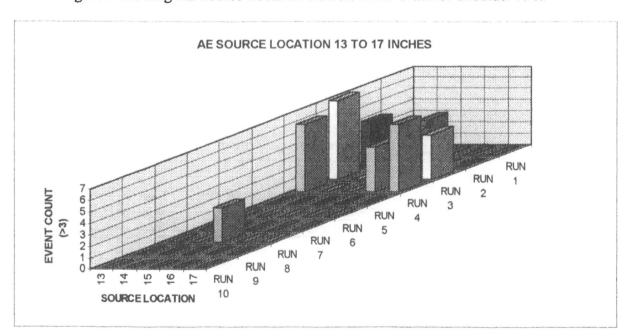

Use of Plate Wave Analysis in Acoustic Emission Testing to Detect and Measure Crack Growth in Noisy Environments

H. L. DUNEGAN

ABSTRACT

One of the major problems encountered in using Acoustic Emission(AE) techniques to monitor structures in the field is the difficulty in separating AE signals from crack growth , from signals due to extraneous noise sources. These extraneous noise sources can be created by frictional rubbing, impact of particles on the structure being monitored and leaks by pressurized components. Most extraneous noise sources of this type are out-of-plane (OOP sources) and although they can have very high frequency components in an undamped structure, most of the energy in the stress waves created in most structures constructed from plates, can be found at frequencies below 100KHz. This energy is carried by a low frequency flexure wave in the plate. Since most field tests in the past have been conducted with high pass filters at 100KHz and above using resonant transducers, few practitioners were even aware that these waves were present prior to recent work by (Gorman and Prosser 1990). AE signals generated by crack growth are in-plane sources(IP sources) and most of the energy in the stress wave is carried by high frequency extensional and shear waves. This report shows how special transducer and instrumentation techniques can be used to recognize the type of wave predominate in a plate to allow filters to be constructed in the instrumentation to not only eliminate extraneous noise sources from the AE data, but also have the potential of measuring the depth of a growing crack in a plate.

INTRODUCTION

Acoustic Emission techniques have used in the field for over 25 years for the testing of metal and composite pressure vessels and piping. It has also found wide application in the testing of composite man lift booms. It is used primarily for locating cracks and potential problem areas in metal pressure boundary applications and other types of Nondestructive techniques are used to provide acceptance or rejection criteria. The technology has not achieved the acceptance of other Nondestructive techniques for the testing of bridges, and other components of the infrastructure for two primary reasons: One, the difficulty in

Harold L. Dunegan, Dunegan Engineering Consultants Inc. (DECI), P.O. Box 1749, San Juan Capistrano, CA 92693

BACKGROUND

We have recently reported results (Dunegan 1995) of an experimental program conducted to study and analyze the types of wave modes present in plates. Our results have shown that a small aperture mass loaded transducer is the best type of transducer to used for accurately defining the displacement and frequency content of OOP sources. This type of transducer was found to be very insensitive to IP sources which one associates with crack growth. On the other hand it was found that a large aperture transducer without mass loading was very sensitive to IP signals and less sensitive to OOP signals. These results lead to the design of a "false" aperture transducer which comprises a piezoelectric crystal that is mass loaded over a small area of the crystal in the center. We have found that by adjusting the size of the mass and the relative area of the crystal covered by the mass that its sensitivity to OOP and IP sources of AE signals can be adjusted so that it has equal sensitivity for both types of signals. Examples will be given for two types of transducers, one for the testing of small specimens and the other for the testing of larger components.

PROCEDURE

Published literature to date dealing with plate wave analysis of AE signals has been on thin metal and composite plates. We were interested in applying the techniques to bridges and other thicker plate structures and therefore starting our experiments on steel bars of 1/4 and 1/2 inch thickness.. Experiments were also conducted on a 3/4 inch thick compact tension fracture toughness specimen (CT). Figure 1 shows the experimental setup on the CT specimen. The larger transducer on the edge of the plate was used to provide a trigger signal for the digital voltmeter when 0.3mm pentil pencil lead breaks were made at different depths in the specimen shown by the parallel bars on the edge of the specimen. The pencil lead breaks were made at each level on the vertical line drawn through the different levels. The small transducer shown was used to detect the AE data from the specimen for each lead

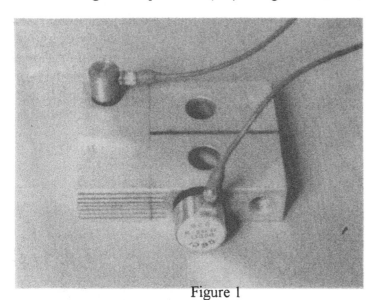

Figure 1

break. The trigger transducer was used so that velocity of sound for the different waves could be measured and phase shift of the signals could be observed.

This transducer was designed with a 1/8 inch aperture and the piezoelectric crystal was mass loaded. The aperture and size of the mass was adjusted so that the transducer was equally sensitive to OOP and IP signals in the specimen. The out-of-plane (OOP) signal was generated by breaking the pencil lead on the top or bottom surface in the vicinity of the vertical line on the specimen. The transducers were attached to the CT specimen with hot glue.

The standard method used for years to calibrate an AE system in the field has been to place a pulser on the structure or break pencil leads to provide a calibration signal. We have found that both of these methods act as out-of-plane (OOP) sources, resulting in most of the energy being dispersed in the form of a flexure wave, with very little of the energy propagating in the extensional and shear modes associated with crack growth. (Prosser and Gorman) found that breaking a pencil lead on the edge of a small plate coupled to a structure provided a source of extensional waves in the structure. We were interested in finding a method of field calibration, and therefore experimented with coupling the CT specimen shown in figure 1 to a steel bar to see how effective the specimen would "clone" itself to the bar. Figure 2 is a photograph showing this experimental setup.

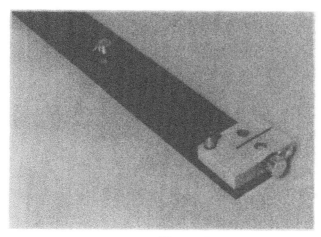

Vaseline was used to couple the CT specimen to the end of the 1/2 X 3 X 42 inch long steel bar . The "false aperture" transducer described previously was then coupled to the bar with Vaseline at a distance of 24 inches from the end of the bar. 0.3mm pencil leads were broken on the CT specimen at different depths signified by the horizontal bars on the specimen as in the previous experiment (figure 1).

Figure 2

In order to provide a calibration of structure in this manner, one needs to be assured that the data recorded is representative of what one would obtain on the structure alone. With this in mind we then removed the CT specimen, coupled our trigger transducer to the end of the bar with hot glue and broke pencil leads at the same percentage depth in the bar as was used for the CT specimen with the data transducer at 24 inches from the end of the bar.

Figure 3 is a block diagram of the instrumentation used to detect and record the signals from each setup. The trigger transducer starts the sweep of the digital voltmeter. The signal from the data transducer is split and passed through a 100KHz hipass filter and a 20-80 KHz bandpass filter. The digital voltmeter captures the transient signals and passes them to the computer through a fiber optic cable. The signals are then printed. When a pencil lead is broken at a given depth, both the high frequency and low frequency signals are captured and printed.

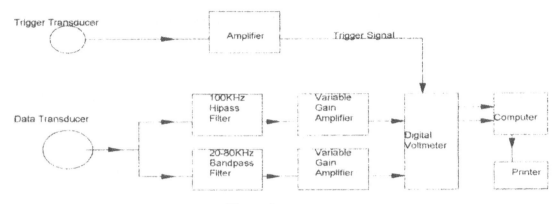

Figure 3

EXPERIMENTAL RESULTS

Figure 4 shows the low frequency data from the compact tension (CT) specimen for an OOP

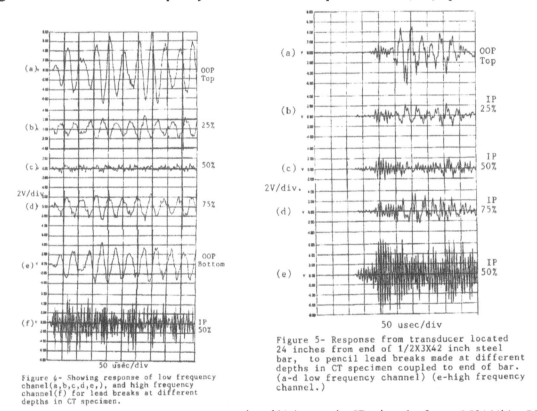

Figure 4- Showing response of low frequency channel(a,b,c,d,e,), and high frequency channel(f) for lead breaks at different depths in CT specimen.

Figure 5- Response from transducer located 24 inches from end of 1/2X3X42 inch steel bar, to pencil lead breaks made at different depths in CT specimen coupled to end of bar. (a-d low frequency channel) (e-high frequency channel.)

signal(4a),   and   IP signals from   25%(4b),   50%(4c) and 75%(4d) depth in the specimen. An OOP signal from the bottom surface of the specimen is also shown(4e) The high frequency signals change very little as a function of depth , so only the signal from the 50% (4f) depth is shown.

Figure 5 shows the data obtained from the transducer located at 24 inches on the steel bar when the CT specimen is coupled to the end of the bar and the same procedure used for figure 4 is repeated.   The high frequency signal shown was for the 50% depth(5e). The gain of the amplifiers were increased approximately 6 dB in order to obtain signals of approximately the same voltage level as figure 4. It was surprising how well the specimen "cloned" its response to the bar, with only 6dB of signal attenuation.

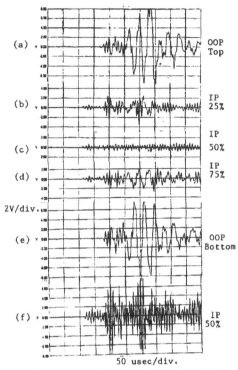

Figure 6-Response from transducer located at 24 inches from the end of a 1/2X3X42 inch steel bar. Lead breaks were made at the end edge of the bar at different depths, and on the top and bottom surface. (a-e low frequency channel). (f- high frequency channel)

Figure 6 shows the data from the bar alone for lead breaks at the same percentage depth used in the CT specimen. The gain increase was reduced to the same level used in the CT experiments in order to keep the signal levels at approximately the same as shown in figure 4 for the CT specimen. The vertical scale on each of these figures is 2V/division. Approximately 70dB of total gain was used for figures 4 and 6, and 76dB of gain was used for figure 5.

ANALYSIS AND DISCUSSION OF RESULTS

The breaking of pencil leads at different depths in the CT specimen (figure 4) produce dramatic changes in signal level of the low frequency channel. This is an expected result in thin plates, but was not expected in such a thick un-plate like structure. As can be seen in 4a and 4e, an OOP source results in a large low frequency flexure type wave being set up in the specimen. Lead breaks at different depths on the edge of the specimen result is very little low frequency signal at the point of symmetry at the center of the specimen (4c) with increasing amounts of low frequency signal at 25%(4b) and 75%(4d) for unsymmetrical input of the lead break. Note the phase shift in the signals between 4a and 4e, also between 4b and 4d.

The data in figure 5 demonstrates that the response of the CT specimen to lead breaks at different depths can be "cloned" into the bar. This is illustrated by the large flexure wave signal generated by the lead break on the top surface, compared to the very small signal observed from the lead break made at the center of the CT specimen. Note that the signals arriving at approximately 120 microseconds correlate well with the extensional wave velocity of 200,000 in/sec in steel. The first large signal seen in 5d has an arrival time corresponding to the shear velocity of approximately 130,000 in/sec.

The data were analyzed from each of the three combinations by dividing the peak amplitude of the high frequency signal by the peak amplitude of the low frequency signal. The results of this analysis is shown in figure 7. Each data point represents an average of 5 pencil lead breaks. It is very difficult to always put the same signal into the bar with a pencil lead break. Also one must be very careful to hold the pencil parallel to the plane of the bar when breaking the leads. Any force perpinduclar to the plane of the bar will tend to induce flexure waves in the bar. Since we are taking a ratio of two components of the same signal, the ratio is not highly dependent on the absolute amplitude of the signal. It is very dependent on the location where the lead is broken and the pencil orientation in respect to the plane of the bar.

Figure 7-High frequency ratio/Low frequency ratio as a function of depth for all three conditions

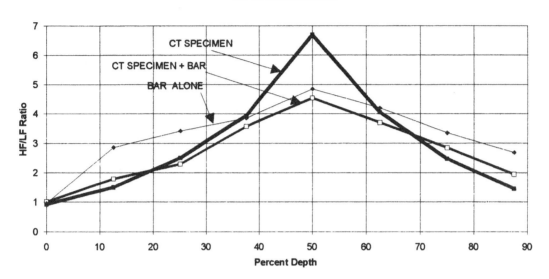

One can observe from the data in figure 7 that out-of-plane (OOP) sources for all three situations represented by figure 7 have ratios of less than one. A filter can therefore be constructed for data of this type by setting up the instrumentation and software to only accept as valid signals those signals having a high frequency to low frequency ratio greater than 1. Since extraneous noise sources are primarily OOP sources, they can be eliminated from the data set early by simple analog front end filtering as used in these experiments coupled with a simple software calculation and algorythm.

The excellent correlation of the data between the CT specimen coupled to the bar and a data from the bar alone suggest that structural calibration in the field can be accomplished using this technique in order to measure crack depth. A hypothetical example is given to show how this might be accomplished:

Assume that a crack is present in a bridge structure and one wishes to monitor the crack to see if it is growing, and if growing at what depth the crack is in the bridge plate structure.
In this example, one would couple the CT specimen to the plate in the vecinity of the crack and place a trigger transducer on the CT specimen and a data transducer, say at 24 inches away. Pencil leads are broken at different depths in the CT specimen and signals are recorded by the data transducer. A calibration curve similar to figure 7 is constructed in order to correlate frequency ratios to crack depth, and separate OOP from IP sources. Once calibrated the CT specimen is removed and a trigger transducer put in its place. Continuous monitoring of the crack proceeds and ratios are calculated for all received signals, as well as time of flight between the trigger transducer and data transducer (this time is used to construct a time related filter). The ratio of each signal is calculated and signals rejected that come from OOP sources. If a signal passes the ratio test the time of flight between trigger and data transducer is observed. If the time does not fall within a predetermined window the data is rejected. It the signal passes

the time test its ratio is compared to the calibration curve to estimate the depth of the crack responsible for the signal. As the crack continues to grow it is observed whether on not the ratio is increasing or decreasing. If increasing the crack is less than half way through the plate, if decreasing the crack has passed the mid point of the plate (figure 7). The trigger transducer will allow the change in phase of the low frequency signal to be observed. This information can also be used to determine crack depth information.

CONCLUSIONS

The data in this report presents a new technique for using Acoustic Emission data to eliminate extraneous noise sources from a data base, and measure the depth of a growing crack in metallic structures. This can have tremendous impact on use of the technology for monitoring of structures in the field. Only crack depth information can be measured, but this is the most important parameter effecting structural integrity for surface cracks. Some assumptions concerning the crack shape and physical measurement of the crack length at the surface, might allow one to determine the profile of the crack tip for a crack that has not penetrated through the thickness. The CT specimen used in these experiments has the capability of being bolt loaded. Future studies will involve loading the specimen after hydrogen charging to create crack growth, and attempting to correlate the AE data with actual crack growth with a more automated computer based system.

REFERENCES

M.R. Gorman and W.H. Prosser, Journal of Acoustic Emission 9(4), 283-288 (1990)

H.L. Dunegan,"The Effects of Aperture Size and Mass Loading on the Response of AE Transducers to Plate Waves.", 38th AEWG conference, NASA Langley VA 1995.

W.H. Prosser and M.R. Gorman, Patent disclosure, NASA Langley 1993.

Electrical Resistance Tomography for Imaging the Spatial Distribution of Moisture in Pavement Sections

M. BUETTNER, A. RAMIREZ AND W. DAILY

ABSTRACT

Electrical Resistance Tomography (ERT) was used to image spatial moisture distribution and movement in pavement sections during an infiltration test. ERT is a technique for determining the electrical resistivity distribution within a volume from measurement of injected currents and the resulting electrical potential distribution on the surface. The transfer resistance (ratio of potential to injected current) data are inverted using an algorithm based on a finite element forward solution which is iteratively adjusted in a least squares sense until the measured and calculated transfer resistances agree to within some predetermined value.

Four arrays of ERT electrodes were installed in vertical drill holes 1.22 m (4 ft) placed at the corners of a square 61 cm (2 ft) on a side into a pavement section which is used for a truck scale ramp on U.S. Highway 99 just north of Sacramento, CA. Water was introduced slowly into the pavement through a shallow hole in the center of this pattern and ERT data were collected in various planes as the water infiltrated into the pavement and subgrade materials over a period of several hours.

The ERT data were inverted, and the resulting images show 1) the basic structure of the pavement section and 2) the movement of water through the image planes as a function of time during infiltration. An interesting result is that the water does not appear to drain from the section toward the shoulder as had been expected based on the design.

CONCEPT OF ERT

ERT is a method for determining the electrical resistivity distribution in a volume based on discrete measurements of current and voltage on the boundary. Resistivity data can be taken in a variety of configurations, including borehole-to-borehole, borehole-to-surface, or pure surface. Detailed technical concepts and theory used for ERT can be found elsewhere (Daily and Owen, 1991; Daily et al., 1992).

In order to obtain an image of a body from surface measurements, a number of electrodes are placed on the surface of the body and in electrical contact with it. A pair of adjacent electrodes are driven by a known current, and the resulting voltage difference is measured between other pairs of electrodes. Then, the known current is applied to another

Michael Buettner, Lawrence Livermore National Laboratory, 7000 East Avenue, Livermore, CA 94550
Abe Ramirez, 7000 East Avenue, Livermore, CA 94550
William Daily, 7000 East Avenue, Livermore, CA 94550

pair of electrodes and the voltage is again measured between other pairs. This procedure is repeated until current has been applied to all pairs of electrodes. The ratio of a voltage at one pair of terminals to the current causing it is a transfer resistance. For n electrodes there are n(n-3)/2 independent transfer resistances.

The next step is to calculate the distribution of resistivity in the volume given the measured transfer resistances and to construct an image. However, the calculation for the distribution of resistivity is highly nonlinear because the currents flow along the paths of least resistance and are dependent on the resistivity distribution. Finite element algorithms and least square methods are used to invert the transfer resistances.

Image construction can be on the same finite element mesh used to calculate the measurements or on a different array. An absolute image shows a resistivity structure whereas a comparison image shows changes in resistivity. Comparison images can be used to study dynamic processes in structures, such as imbibition, by comparing data taken at different times. This paper presents both absolute images and comparison images.

A more detailed description of the theory can be found in a companion paper entitled "Electrical Resistance Tomography for Imaging Concrete Structures" (Buettner, Ramirez, and Daily) from this conference.

INFILTRATION EXPERIMENT

A set of ERT electrodes was installed in September, 1992 on the truck scale pull-off ramp just north of Riego Road on Highway 99 near the Sacramento, CA Metro Airport. The pavement structural section has open graded asphalt concrete (OGAC, 1.8 cm), dense graded AC (DGAC, 19.8 cm), asphalt treated permeable base (ATPB, 7.6 cm), aggregate base (AB, 13.7 cm) and lime treated subgrade (LTS, 30.5 cm) over native soil as illustrated in Figure 1.

Figure 2 shows a plan view of four, 10.2 cm diameter boreholes where electrode arrays were inserted. Four image planes were defined by four linear arrays of electrodes placed into holes at corners of a 61 cm square. The electrode arrays extended to 122 cm below the pavement surface. There were 12 electrodes in each hole and 5 surface electrodes on each edge of the square. Surface electrodes were placed into 1.3 cm diameter holes drilled into the bottom of sawcuts on the pavement surface. These holes extended about 5.1 cm into the DGAC layer. The sawcut was approximately 5.1 cm wide and 5.1 cm deep.

Electrodes in boreholes and at the surface were spaced at every 10.2 cm. One 10.2 cm diameter hole was drilled at the center of the square (see Figure 2) down to the bottom of ATPB layer for a water infiltration experiment. The water infiltration experiment was planned to better understand water flow (vertical and horizontal direction) in this drained pavement section.

Two kinds of backfill material were tried in this experiment: cement mortar and excavated pavement materials. Cement mortar with 1 to 3 ratio (by weight) was used with a water-cement ratio of 0.5. Boreholes No. 1 and 2 (see Figure 2) were backfilled with the excavated native soil, LTS and AB materials which were removed from the holes during drilling. The space of ATPB, DGAC and OGAC was filled with cold AC patching material. Boreholes No. 3 and 4 were filled with cement mortar up to the top of the aggregate base and cold AC patching material filled the remaining space to the surface. A steel rod was used to compact the backfill materials to eliminate air pockets around the electrodes and to make good contact between the backfill materials and electrodes. A commercial concrete patching and cold AC patching materials were used to fill the sawcut trench. Before introducing water into the center hole, ERT data were collected for baseline images. Baseline images of the 1-2, 2-3 and 3-4 planes are shown in Figure 3 and the images are matched along their common edges. The images clearly show the resistivity structure with high resistivity in upper layers and low resistivity in the LTS and native soil. Also, the OGAC layer and the interface between OGAC and DGAC show low resistivity in all three planes. This effect is

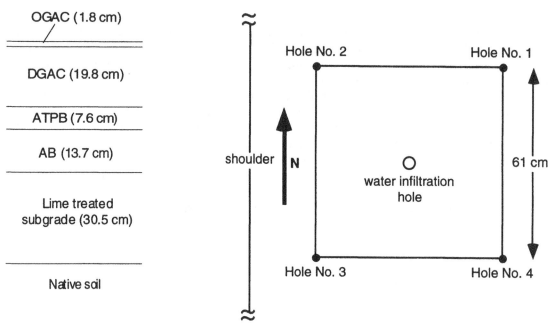

OGAC (1.8 cm)

DGAC (19.8 cm)

ATPB (7.6 cm)

AB (13.7 cm)

Lime treated
subgrade (30.5 cm)

Native soil

Figure 1. Pavement structural section
for infiltration test

Hole No. 2   Hole No. 1

shoulder   N

water infiltration
hole

61 cm

Hole No. 3   Hole No. 4

Figure 2. Layout of ERT boreholes for the
infiltration experiment

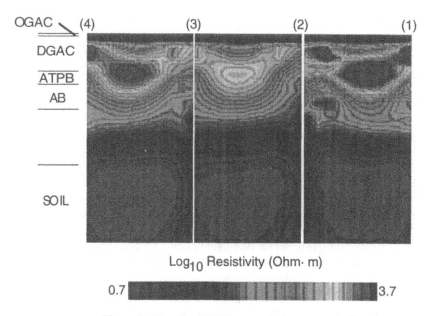

OGAC (4)   (3)   (2)   (1)

DGAC

ATPB

AB

SOIL

Log_{10} Resistivity (Ohm· m)

0.7   3.7

Figure 3. Baseline ERT images before water infiltration

probably not real and is thought to be caused by the low-resistivity AC patching material used to backfill the surface sawcuts.

The infiltration experiment was conducted on April 20, 1993. During a 6.5 hour time period, approximately 5.5 gallons of water was introduced uniformly and slowly into the pavement through the center hole. Data were collected at several times during the infiltration period. The change in resistivity associated with moisture movement was imaged as a function of time. Four complete data sets were collected at 0914, 1017, 1231 and 1443 hours for the 2 - 3 plane because the researchers expected most of the water to drain through this plane. One data set was collected for the 3 - 4 plane at 1336 and one for the 1 - 2 plane at 1125. Because each data set required about one hour to collect, the images

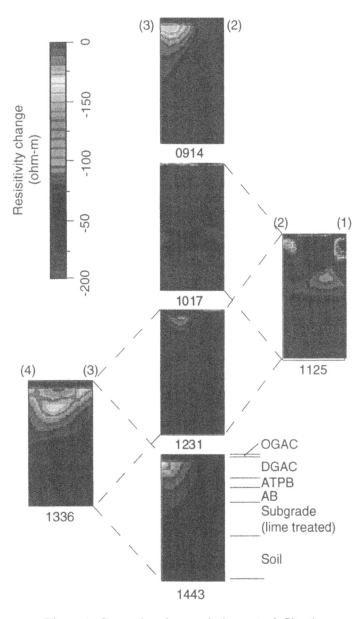

Figure 4. Comparison images during water infiltration

represent a time average. Figure 4 shows a sequence of comparison images which show differences from the baseline conditions. That is, if there were no changes in resistivity because of water infiltration, the images would be a uniform single color. Progressive decreases in resistivity (i.e., increase in moisture) appear as progressively lighter as shown on the grayscale.

DISCUSSION AND CONCLUSIONS

The sequence of images for the 2 - 3 plane shows that at 0914 there is a decrease in resistivity (interpreted as water flow) in the pavement structural section around the No. 3 borehole. At 1017 this feature has disappeared but there is perhaps some ponding of water across the section and in the LTS. By 1231 this ponding disappeared and there is more moisture movement around the borehole No. 3. Finally at 1443 there appears to be more moisture movement around the borehole No. 3 much like at 0914. The sequence of four ERT images does not show what could be termed a steady state flow phenomenon which is expected during slow, constant infiltration.

The single image in the 3 - 4 plane at 1336 shows moisture movement through the pavement structural section. The image of this plane matches well along its borehole No. 3 edge with the 2 - 3 plane image at 1443. The other single image of the 1 - 2 plane at 1125 shows moisture movement, though not as pronounced as the one for 3 - 4 plane at 1336. The images in Figure 9 do not appear to show that infiltrated water drained preferentially to the shoulder through the 2 - 3 plane.

There are several factors which complicate the interpretation of the images from Fig. 4. Among them are 1) possible leaks or water movement between layers via boreholes, 2) the image reconstructions assume 2-dimensional water flow when the flow is really 3-dimensional, and 3) the data acquisition time is probably too long (about 1 hour) compared to the time scale of changes in the flow pattern in the pavement structure. Because of 3) each image must be thought of as a time-average of the processes involved.

In spite of the difficulties, is appears that ERT can be used to delineate moisture movement in pavement structures. Future work should seek to ameliorate the above problems by 1) improving ERT electrode placement in boreholes, 2) inverting the data with a true 3-dimensional approach, and 3) by speeding up the data acquisition process.

APPLICATIONS

ERT imaging has potential applications relating to structures, soils, pavements, and environmental remediation and monitoring. ERT should be applicable for detecting cracking in structures through the changes in resistivity that occur when water infiltrates cracks. In addition, ERT should be capable of detecting rebar location and possibly the state of corrosion. Applications of ERT for soils include monitoring slope stability and the stability of footings. In pavements, ERT has already been used to monitor moisture content and movement (see companion paper). Finally, ERT has been used to detect and monitor leaks from storage tanks, for monitoring thermal processes during environmental remediation, and for detecting and monitoring contaminants in soil and groundwater.

ACKNOWLEDGMENT

John Carbino provided technical support for this project. This work was performed under the auspices of the U.S. Department of Energy by Lawrence Livermore National Laboratory under contract no. W-7405-Eng-48.

REFERENCES

Daily, W. D., and E. Owen. 1991. "Cross-borehole Resistivity Tomography," *GEOPHYSICS*, 56(8): 1228-1235.

Daily, W. D., A. Ramirez, D. LaBrecque and J. Nitao. 1992. "Electrical Resistivity Tomography of Vadose Water Movement," *Water Resources Research*, 28(5):1429-1442.

Determining Concrete Highway Pavement Thickness Using Wave Speed Measurements and the Impact-Echo Method

M. SANSALONE, J.-M. LIN AND W. B. STREETT

ABSTRACT

Engineers involved in quality control of new concrete pavements must determine the thickness of new pavements to ascertain whether the actual thickness meets the specification before payment to the contractor is made. If the actual thickness is less than that specified, payment is reduced. Currently pavement thickness is determined by taking cores. This method is time consuming and expensive, and it leaves holes in the new pavement, which not only need to be repaired, but remain problem spots during the life of the pavement.

This paper describes the use of the impact-echo method in conjunction with a wave-speed measurement technique which can nondestructively and accurately determine the thickness of new concrete pavements. Portable instrumentation (computer, data-acquisition hardware, transducers, and impactor) that can be used to perform both wave speed and impact-echo measurements accurately and reliability has been assembled. Results of preliminary field studies carried out on a Strategic Highway Research Program (SHRP) Test Section in Arizona are reported, including results obtained from pavements having different thicknesses, different strengths, and placed on different types of subgrades.

Ongoing improvements in data-acquisition equipment which take advantage of new hardware and software are also described briefly. Comprehensive field studies to test the new hardware and software are planned for Fall 1995 on SHRP test section in Delaware and Ohio. The ultimate goal is to develop an ASTM Standard Test Method for determining the thickness of concrete pavements.

INTRODUCTION

The research that led to this development of the pavement thickness measurement technique based on the impact-echo method was performed in response to the frequent requests received from Department of Transportation engineers about whether the impact-echo method could be used to determine the thickness of new pavements. What was needed to make this application feasible

Mary Sansalone, Jiunn-Ming Lin and W. B. Streett, Cornell University, Ithaca, New York 14853

was an independent means to determine the P-wave speed in the pavement, so that the known P-wave speed could then be used in an impact-echo test to determine thickness. Work carried out by Dr. G. Clemena [1] of the Virginia State DOT showed that the impact-echo method worked well for determining pavement thickness if the P-wave speed was known .

The first two authors had developed a method based on the use of Rayleigh-wave speed for estimating P-wave speed when testing concrete mine shaft liners [2]. However this approach relies on knowing the Poisson's ratio of the concrete if the results are to be very accurate [3]. Because of the difficulty in measuring Poisson's ratio and the variability of the results, a different approach based on the use of direct P-waves was investigated and found to be very accurate and reliable.

This paper begins with a brief explanation of the impact-echo and wave speed measurement techniques. Next results of preliminary field trials are presented. The paper concludes with a discussion of ongoing work on the development of an ASTM standard test method. Readers interested in obtaining more information should refer to References [3,4] and/or contact the authors.

THE IMPACT-ECHO AND WAVE SPEED MEASUREMENT TECHNIQUES

Impact-echo is a technique based on the use of transient stress (sound) waves for nondestructive testing [5]. The method is based on using a short-duration mechanical impact (produced by tapping a small steel ball against a concrete surface) to generate low frequency stress waves that propagate into the structure and are reflected by flaws and external surfaces. Surface displacements caused by the reflections of these waves are recorded by a transducer located adjacent to the impact, as shown in Figure 1(a). This signal is sent to a portable computer which contains a high-speed data-acquisition card and signal processing and display software. The recorded displacement versus time signals are then transformed into the frequency domain. Multiple reflections of waves between the impact surface, flaws, and/or other external surfaces give rise to transient resonance conditions which can be identified and used to determine structure thickness or the location of flaws, if the wave speed in the concrete is known. When testing on a plate-like structure, such as a highway pavement, the pattern in the spectrum (see Figure 1(b)) consists of a single, high amplitude peak at a frequency, f, which corresponds to the multiple reflections of waves between the top and bottom surfaces of the pavement. The following equation gives the relationship between pavement thickness, T, and this measured frequency:

$$T = C_p / 2f \qquad\qquad Eq\,(1)$$

where C_p is the P-wave speed in the concrete.

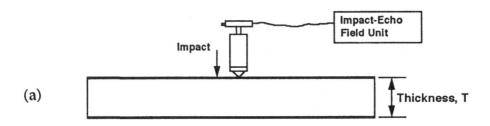

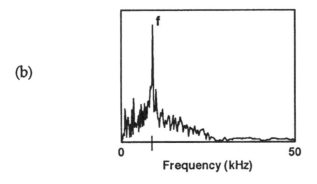

Figure 1.   (a) Schematic representation of an impact-echo test, (b) spectrum
             obtained from an impact-echo test on a highway pavement.

To perform the calculation for thickness, T, in Equation (1), the P-wave
speed, Cp, must be known. In many applications, the wave speed in the concrete
is determined from an impact-echo test on an area of known thickness.   This
wave speed is then used to test other areas of the structure.   However, because
the wave speed in concrete often varies in different parts of a structure, or in
some cases, such as in new highway pavements the thickness is the unknown,
an independent means of determining wave speed must be used.

A schematic representation of a wave speed measurement is shown in
Figure 2(a).   The technique involves measuring the time it takes for a wave
traveling along the surface to travel a known distance L between two
transducers. The transducers used are the same conical displacement sensors
used in impact-echo testing. The distance can be divided by the measured time
to obtain the speed of the P-wave.   The arrival of the P-wave at the transducers
can be identified from the first displacements that occur in the displacement
versus time graphs (waveforms).   Figure 2(b) shows a typical set of waveforms
obtained from the set-up shown in Figure 2(a); the arrow indicates the arrival of
the P-wave in each waveform.

(a)

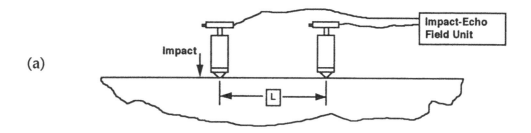

(b)

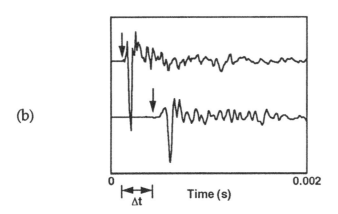

Figure 2.   (a) Schematic representation of a P-wave speed measurement, (b) results of measurement. The location of the P-wave arrival in each waveform is shown by an arrow.

This technique allows the wave speed to be determined easily, quickly, and accurately at any point along a pavement.   Subsequently an impact-echo test can be performed at the same point using the known wave speed, and the thickness of the pavement can be determined.   The accuracy with which both the wave speed in the concrete and the thickness of the pavement can be determined is discussed in detail in References [3,4].

FIELD STUDY

Table 1 summarizes results obtained from a SHRP concrete pavement test section in Arizona in May of 1995.   Eight sections of pavement were selected to provide a range of combinations from those available in the three-mile stretch of test site.   Test points at locations having two different nominal thicknesses, two different concrete flexural strengths, and placed on three types of sub-bases were selected. Sub-base types included a lean concrete base (LCB), a permeable bituminous treated base (PBTB), and a dense graded aggregate base (DGAB).   The surface of the pavement was grooved, and wave speed tests were carried out parallel to the grooves [3,4].

Table 1.   Field Test Results from SHRP Test Section in Arizona

Test Point	Sub-base Type	Nominal Thickness (in/mm)	Flexural Strength (psi)	Core Length (mm)	P-wave speed (m/s)	Calculated Thickness (mm)	Error (mm/%)
1	LCB	8 / 203	900	205	4255	203	-2 / 1%
2	LCB	12 / 305	900	294	4255	291	-3 / 1%
3	PBTB	8	900	209	4080	212	+3/ 1.4%
4	PBTB	12	900	294	4255	300	+6 / 2%
5	DGAB	8	900	212	4080	209	-3 / 1.4%
6	DGAB	8	550	197	3920	195	-2 / 1%
7	DGAB	12	550	288	4080	279	-9 / 3.1%
8	DGAB	12	900	287	4080	279	-8 / 2.8%

The data-acquisition equipment used in this field trial included a portable, rugged PC-based battery operated system.   Two channels of data in the wave speed measurements were sampled at a rate of 500 kHz per channel.   The spacing between transducers was 0.2 m.   Impact-echo data was sampled for 4096 microseconds, resulting in a resolution of 0.24 kHz in spectra.   With this equipment (DOCter impact-echo device sold by Germann Instruments), accuracies to within several millimeters (less than 0.1 inch) can be obtained when testing pavements on lean concrete sub-bases, within about 0.25 inches on permeable bituminous treated sub-bases and within about 0.35 inches on dense graded aggregate sub-bases.   It is interesting to observe that the magnitudes of these "errors" generally reflect the variations in actual thickness around the circumference of cores taken from pavements on each type of sub-grade.

FUTURE WORK

The accuracy of the current system is limited in part by the rate at which the data-acquisition hardware can record digital waveforms [3,4].   Recently, higher-speed data-acquisition hardware for use in portable PC and notebook-based field systems has become available.   Using this new hardware it will be possible to improve the accuracy of the wave speed measurements, and thereby obtain more accurate thickness measurements in many cases.   In the new system, wave speed data will be sampled at 1 MHz per channel improving the accuracy of the wave speed measurements. The spacing between transducers is being increased to 0.3 m which will also reduce the possible error inherent in the wave speed measurement. With these changes, the accuracy of the thickness measurement is also expected to improve.

In addition, a new Window's based software program which features automated determination of P-wave speed is being developed. The software is designed to recognize the arrive of the P-wave at teach transducer, and to use the arrival times and the distance between the transducers to calculate and display a wave speed. No interpretation of waveforms will be required by the operator.

This new hardware and software system will be tested by the authors in conjunction with the Office of Technology Applications of the FHWA and State DOTs in upcoming field trials in Delaware and Ohio on SHRP test sections in fall of 1995. The goal is to make the new system commercially available in 1996. When field trials on SHRP test sections in Delaware and Ohio are completed, work will begin on developing an ASTM test method for determining pavement thickness using P-wave speed measurement technique and the impact-echo method. The authors will be working with Dr. Nicholas J. Carino of the National Institute of Standards and Technology to write this standard method. It is hoped that a draft standard will be ready by summer of 1996, although it will take several years of round-robin testing and committee work to actually establish an ASTM standard.

REFERENCES

1. Clemena, G. "Use of the Impact-Echo Method in Nondestructive Measurements of the Thickness of New Concrete Pavements," Final Report No. FHWA/VA-95-R10, Virginia Transportation Research Council, March 1995.
2. Lin, J.M., and Sansalone, M., "Impact-Echo Response of Hollow Cylindrical Structures Surrounded by Soil and Rock: Part II - Experimental Studies," Geotechnical Testing Journal, June 1994, pp. 220-226.
3. Lin, J.M. and Sansalone, M., "A Procedure for Determining the P-Wave Speed in Concrete Using Rayleigh Wave Speed Measurements" to be published in <u>Innovations in Nondestructive Testing,</u> a Special Publication of the American Concrete Institute, 1996.
4. Sansalone, M., Lin, J.M., and Streett, W.B., "A New Method for Determining Highway Pavement Thickness Using Direct P-Wave Speed Measurements and the Impact-Echo Method, " Department of Structural Engineering Report, Cornell University, December 1995.
5. Sansalone, M., and Carino, N.J., "Impact-Echo: A Method for Flaw Detection in Concrete Using Transient Stress Waves," NBSIR 86-3452, National Bureau of Standards, Gaithersburg, Maryland, Sept., 1986, 222 pp.

Instrumentation of Geosynthetically Stabilized Flexible Pavement

I. L. AL-QADI, T. L. BRANDON, B. A. LACINA AND S. A. BHUTTA

ABSTRACT

Nine instrumented flexible pavement test sections were constructed in a rural secondary road in southwest Virginia. The 15.25 m long test sections were built to examine the effects of geogrid and geotextile stabilization. Three test sections were constructed using a geogrid, three with a geotextile, and three were non-stabilized. The test section base course thicknesses ranged from 10.2 cm to 20.3 cm and the hot-mix asphalt (HMA) thickness averaged 8.9 cm. Geosynthetic stabilization was placed on top of the subgrade layer. An extensive instrumentation infrastructure was constructed to place all instrumentation, cabling, and data acquisition facilities underground.

The pavement test sections were heavily instrumented with two types of pressure cells, soil and HMA strain gages, thermocouples, and soil moisture blocks. In addition, strain gages were installed directly on the geogrid and geotextile. Instrument survivability has ranged from 6% for the strain gages mounted on the geotextile to 100% for the soil moisture blocks after eight months of operation. The majority of instrument failures occurred either during construction or in the first few weeks of operation.

The data acquisition system is triggered by traffic passing over piezoelectric sensors and operates remotely. The data collected is transferred via modem to Virginia Tech for processing. It is planned to monitor the performance of the pavement test sections for a minimum of three years.

INTRODUCTION

The use of geosynthetics in civil engineering construction has increased dramatically over the past 20 years. These uses have included stabilization of weak road subgrades, increasing the height and slope angles of embankments, and providing hydraulic barrier and drainage layers for landfill construction. Of particular interest in civil engineering is to

Imad L. Al-Qadi and Thomas L. Brandon, Associate Professors of Civil Engineering, The Via Department of Civil Engineering, Virginia Polytechnic Institute and State University, Blacksburg, VA 24061.
Bruce A. Lacina and Salman A Bhutta, Graduate Research Assistants, The Via Department of Civil Engineering, Virginia Polytechnic Institute and State University, Blacksburg, VA 24061.

employ certain geosynthetics in secondary road construction (Barksdale, Brown, and Chan, 1989; Hass, Walls and Carroll, 1988). Their implementation can be advantageous in that the pavement service life can be increased and costs can be reduced if geosynthetics can be used in lieu of more expensive natural materials.

Considerable progress has been made in recent years towards the development of geosynthetics for pavement stabilization. In particular, many laboratory studies have been conducted to evaluate geosynthetic performance (Barksdale, Brown and Chan, 1989; Hass, Walls and Carroll, 1988; Valentine et al., 1993).

Recently a research effort at Virginia Tech has concentrated on examining the use of geotextiles and geogrids in pavement sections (Valentine, et al., 1993). This research has involved constructing scaled model pavement sections and loading them cyclically by applying a 40.0 kN load through a 30.5 cm diameter rigid plate until a predetermined displacement is reached. Pavement sections were constructed with different types of geosynthetic materials, thicknesses of aggregate layers, and subgrade bearing capacities. The study concluded that geosynthetic stabilization can significantly increase pavement resistance to rutting as compared to non-stabilized pavement over a weak subgrade. Geotextiles provide different mechanisms of stabilization as compared to geogrids, and may be best suited for pavement applications. In addition, the separation mechanism provided by a geotextile appears to be more significant than what has been reported in the literature.

While the previous research program at Virginia Tech has shed light on many facets of using geosynthetics for stabilized pavements, several limitations remain. These include differences in the laboratory test section size and loading procedure compared to actual field pavement systems.

To provide a mechanistic design methodology for stabilized pavements, nine instrumented stabilized and non-stabilized flexible pavement sections were built to validate the laboratory results. These test sections incorporated either geotextile, geogrid, or no stabilization. The main difference between the series of test sections is the thickness of the base course layer. Thus, the long term performance of geosynthetics can be examined under normal loading and environmental conditions.

SITE SELECTION

The chosen test section is on Routes 757 and 616 located in Bedford County, Virginia. The road has an estimated annual average daily traffic (AADT) of approximately 550 vehicles per day (1988) with 5% trucks. The average California Bearing Ratio (CBR) of the subgrade soil was measured to be about 2%. The road is a realignment of an existing road. This would facilitate the installation of the instrumentation and reduce disturbances to traffic. The road was to be constructed on approximately 3 m of cut. The subgrade consisted of a residual soil. The proposed alignment has a constant grade (4%) and radius of curvature. Thus the loading conditions of each of the test sections would be nearly identical.

Nine instrumented pavement sections were constructed. Each test section is 15 m in length. The main variables to be studied in the field pavement sections are the influence of the stabilization type and base course thickness. The hot-mix asphalt (HMA) and aggregate thicknesses, and the type of stabilization for each test section are given in Table I. The thicknesses of the different layers were determined by surveying and core sampling after construction. Figure 1 is a plan view showing the location of each test section.

TABLE I DETAILS OF INSTRUMENTED TEST SECTIONS

Section No.	HMA Thickness (cm)	Base Course Thickness (cm)	Reinforcement Type
1	7.8	10.2	None
2	8.8	10.2	Geotextile
3	9.9	10.2	Geogrid
4	9.8	15.2	None
5	9.1	15.2	Geotextile
6	9.0	15.2	Geogrid
7	8.9	20.3	None
8	8.7	20.3	Geotextile
9	8.6	20.3	Geogrid

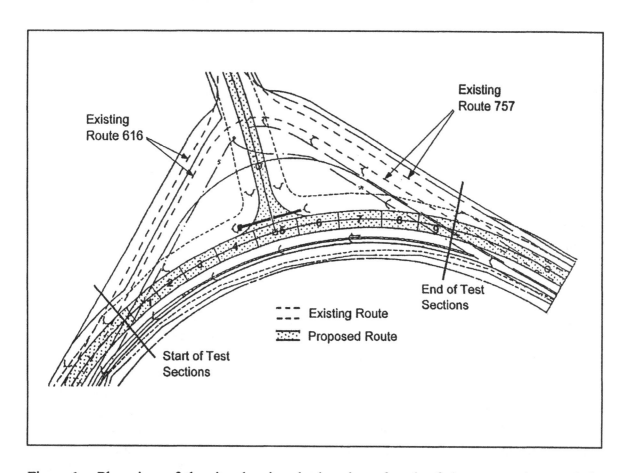

Figure 1. Plan view of the site showing the location of each of the test sections and the related instrumentation infrastructure.

INSTRUMENTATION

Five basic types of instruments were used in this research: earth pressure cells, strain gages, temperature sensors, soil moisture sensors, and piezoelectric polymer traffic sensors. The majority of the instruments were placed in the right side wheel path of the inside lane of the test sections.

Two different types of earth pressure cells were installed in the pavement sections: Kulite type 0234 and Carlson type TP-101. Their primary function was to measure the stresses imposed by the dynamic traffic loading. The Kulite type 0234 earth pressure cells are designed to operate within a vertical pressure range of 0 to 690 kPa and are 5.40 cm diameter, with a thickness of 1.43 cm. The Kulite 0234 contains a diaphragm that excites a silicon strain gage bridge upon diaphragm deformation. The Carlson type TP-101 earth pressure cells also have a pressure range of 0 to 690 kPa. The TP-101's stainless steel pressure head is 11.4 cm diameter and 0.64 cm thick, and is welded to a 1.6 cm outside diameter stainless steel tube that is attached to a silicon strain gage pressure transducer. The Carlson TP-101 was modified from its as-manufactured condition to accommodate the large pressure transducer housing; the stainless steel tube was bent to accommodate the proposed installation provisions.

Carlson JO-1 concrete joint meters were installed to measure the strain in the top of the subgrade layer of the non-stabilized sections. The JO-1 is a displacement type strain gage, measuring the ratio of resistance between two coils of steel wire. The JO-1 was modified by the manufacturer, RST Instruments, to include 7.6 cm diameter end plates (increased from 3.8 cm to accommodate the large displacements that may be realized in soil).

The construction of the temperature sensors consisted of a twisted, welded pair of T type thermocouple wire. T type thermocouples are composed of a constant and copper wire pair. After the wire pair were welded, the exposed end was surrounded by 0.64 cm inside diameter copper tubing.

Boyoucous gypsum blocks were utilized for moisture measurement in the subgrade and base course layers. The soil moisture block is equipped with two stainless steel electrodes, separated by nylon impregnated gypsum. As the gypsum is exposed to moisture, the resistance between the two electrodes decreases, indicating an increase in the moisture content of the soil.

Kyowa KM type embedded HMA strain gages were used to monitor the strain at the bottom of the HMA wearing surface layer. The gage construction consisted of a foil type strain gage, 10 cm in length, sandwiched between two sheets of composite fiberglass with a 3-conductor lead cable. The Kyowa HMA strain gages were modified from their original condition to include aluminum end bars. The end bars were fabricated from 7.6 cm long aluminum bar stocks with a 1.0 by 1.0 cm cross-section, and fastened to the Kyowa gage using a two-part epoxy adhesive.

Measurements Group N2A 06 40 CBY 120 foil type strain gages were used to monitor the changes in horizontal strain at the bottom of the geotextile. The 10.2 cm long foil strain gages were attached to the underside of the geotextile along the gage lengths using Measurements Group M-Bond epoxy type adhesive. A layer of Teflon tape was secured over the gage. The gages were then coated with Measurements Group J-Bond adhesive and a layer of RTV silicone over the Teflon tape to protect the gage assembly from environmental effects.

Texas Measurements FLK-6-1L foil type strain gages were used to monitor the changes in horizontal strain at the bottom of the geogrid. Each strain gage consists of a 2.5 cm long foil type strain gage with three lead wires. The foil strain gages were protected in the same manner as the geotextile strain gages.

The AMP Sensors Inc. Roadtrax® Series P Traffic Sensor is a permanent, in-the-road, class II traffic sensor, used for vehicle classification and counting. The sensor consists of an aluminum channel, the sensor element, and a polyurethane elastomer to protect the sensor. The function of a sensor is to convert mechanical energy into an electrical charge when a stress is imparted to the piezoelectric element. Each sensor is 1.8 m long, 2.5 cm wide, and approximately 2.5 cm thick.

LAYER CONSTRUCTION AND INSTRUMENTATION INSTALLATION

During the construction of the test section, the project personnel moved to the site in early July 1994, and lived in tents for about six weeks. Completion of the test sections did not occur until mid-August.

Instrumentation of Subgrade

After the native soil was excavated to the proper elevation, the subgrade instrumentation was installed. Instruments located in the subgrade included Kulite and Carlson earth pressure cells, soil strain gages, thermocouples, and gypsum blocks. After the subgrade was swept clear of debris, the locations of the gages were determined.

The subgrade was excavated using small hand tools for placement of the pressure cells. The pressure cells were installed about 2.5 cm below the subgrade surface. The pressure cells and the wires were then covered with a layer of fine sand to protect them from damage. Thermocouples and gypsum blocks were installed in the pavement subgrade layer at a depth of 15.2 cm below the surface; gypsum blocks were also installed at 0.6 m below the subgrade layer. The soil strain gages were installed with their upper surface lying 5.1 cm below the surface of the non-stabilized sections' subgrade. Small trenches were excavated to accommodate the wires from instruments to the shoulder where conduit pipes were used. The wires placed in the trenches were covered with a thin layer of sand backfill to protect them from damage.

Concurrent with the instrument installation, nuclear density tests were conducted on the subgrade soil using a Troxler nuclear density gage Model No. 3440. Measured moisture contents, using backscatter, ranged from 29% - 36% and averaged 32%. Dry densities measured using the backscatter procedure ranged from 11.2 - 13.9 kN/m³, while dry densities using direct transmission ranged from 13.4 - 14.4 kN/m³; the average dry densities were 12.3 and 14.0 kN/m³, respectively.

Installation of Geosynthetics

After the subgrade instrumentation was installed, the subgrade surface was swept clean of debris before installing the geosynthetics. The geosynthetics had been cut to the proper size and instrumented at Virginia Tech prior to being transported to the site. Sections 1, 4, and 7 were control sections with no geosynthetic stabilization. Geotextiles were installed on sections 2, 5, and 8. Sections 3, 6, and 9 were stabilized with geogrid. The strain gages on the bottom of the geotextile and geogrid were protected from heavy equipment by constructing a cushion beneath each strain gage. The cushion was constructed by digging a hole in the subgrade, 1.3 cm deep and approximately twice the size of the gage, and filled with fine sand. The cushion was used to reduce the risk of damage to the gages by heavy equipment rolling over them when placing the base course aggregate layer.

Placement and Instrumentation of the Base Course Layer

The base course material was a Virginia Department of Transportation (VDOT) class 21-B limestone aggregate. The aggregate was first end-dumped on the side of a non-stabilized test section not having any instrumentation. The aggregate was then carefully pushed over the test sections with a front-end loader and bladed over the instrumentation laterally using a motor-grader to avoid damage to the instruments and compacted using a vibratory roller. Sections 1-3 received 10.2 cm of compacted aggregate. Sections 4-6 and sections 7-9 received 15.2 cm and 20.3 cm of compacted aggregate, respectively. The base course moisture contents ranged from 2.0% - 3.2%, with an average of 2.6%. The average dry density of the base course layer was 22.4 kN/m^3.

The base course layer was instrumented in the same manner as the subgrade. Pressure cells were installed 2.5 cm below the compacted surface of the base course, and gypsum blocks and thermocouples were installed 5.1 cm below the base course layer surface. The thermocouples and gypsum blocks were installed in the same manner as the ones installed in the subgrade layer with the exception that poorly graded sand and fine aggregate were used to surround and backfill the instruments to avoid damage from large angular aggregate.

The HMA layer thermocouples were placed on the base course, and were exposed 2.5 cm above the surface. AC-30 based emulsion (CRS-2) mixed with a poorly graded sand at a ratio of approximately 1 to 1 was used to coat the exposed thermocouple prior to placement of the chip seal and wearing surface.

Construction and Instrumentation of HMA Wearing Surface

A chip seal layer was placed prior to the HMA wearing surface and was compacted using a vibratory roller. The maximum aggregate size of the chip seal was 1.3 cm. The HMA strain gages were placed on the chip seal just prior to placement of two 4.4 cm class SM-2A HMA layers. The gages were coated with an emulsion mixture to protect them during paving.

The roadway was surveyed after the construction of each layer to determine its thickness. Core samples were taken by VDOT from six of the sections which allowed the thickness of the HMA layer to be determined. Table I shows the approximate thicknesses of the HMA layers from the core sample measurements. The thicknesses given are representative of the wheel path where the instruments are installed.

Four piezoelectric traffic sensors were installed. Two sensors act as triggers for the data acquisition system and the other two are used as weigh-in-motion sensors.

The positions of each of the four sensors were decided based on the average traffic speeds and driving habits on the test sections. The primary data acquisition trigger sensor was installed approximately 15.2 m before the instruments in test section 1 and it acts as a trigger for the data acquisition for all nine test sections. The first of the two weigh-in-motion sensors was installed between test sections 3 and 4. The second of the two weigh-in-motion sensors was installed between sections 5 and 6. The secondary data acquisition trigger was installed before section 7, and it acts as a trigger for the data acquisition for sections 7 through 9. Sensors were protected with bituthane tape.

DATA ACQUISITION INFRASTRUCTURE

The instruments and conduits were all buried and all connections were either buried or locked in junction boxes. At the midpoint of each section, wires from the instruments were run through flexible PVC pipes up the hillside into plastic junction boxes. All junction boxes were interconnected with PVC pipe to the middle of section 5. At the middle of section 5, a large reinforced concrete "bunker" was installed to conceal and protect the data acquisition equipment.

Data Acquisition System

The data acquisition system for this project consists of an external backplane unit that communicates with a personal computer by means of an internal card. The data acquisition hardware was manufactured by Keithley Metrabyte Instruments, Inc. Ten different modules can be installed in the unit to accommodate different types of input, such as analog, digital, strain gage bridge, or thermocouples. For this application, 1 strain gage input, 5 analog input, and 1 thermocouple input modules were installed. The analog input of data acquisition system modules has 16 bit resolution.

The software for the test section operation was written by research personnel in QuickBASIC™ using subroutines and libraries supplied by the data acquisition system manufacturer. The output from each instrument is sampled continuously at a frequency of 200 Hz. The length of data acquisition for all nine test sections is 12 seconds while 6 seconds is used for sections 6 through 9. The thermocouples are sampled by the system only once, immediately after a data acquisition system trigger is activated. The soil moisture cells are read manually using an impedance bridge on a weekly to bimonthly basis. The on-site computer system is accessible for data transfer by modem to the research personnel.

PERFORMANCE OF THE TEST SECTION INSTRUMENTATION

The test sections are to be monitored for a minimum of three years. As with other instrumented road test sections, the instrument survivability proved to be a reason for concern. Approximately eight months after the initial construction, the instruments have had the survivability shown in Table II. Most of the instruments that failed did so during pavement construction or within a few weeks after construction. Some instruments that had ceased to function immediately after construction came "back to life" in the following months. The failure rate after the first month has been very low.

TABLE II INSTRUMENT SURVIVABILITY AFTER 8 MONTHS

Gage Type	Number Installed	Number Survived	Percent Survived
Kulite earth pressure cells	6	3	50%
Carlson earth pressure cells	21	16	76%
HMA strain gages	35	26	74%
Geotextile strain gages	18	1	6%
Geogrid strain gages	18	5	28%
Soil strain gages	6	5	83%
Thermocouples	17	15	88%
Gypsum blocks	18	18	100%

During the first year falling weight deflectometer (FWD) was used three times to back-calculate the resilient modulus of the layers of the pavement system. In addition, a truck, operated with different axle loads, tire pressures and speeds, were used to calibrate the instrument. A typical results from the process is shown in Figure 2.

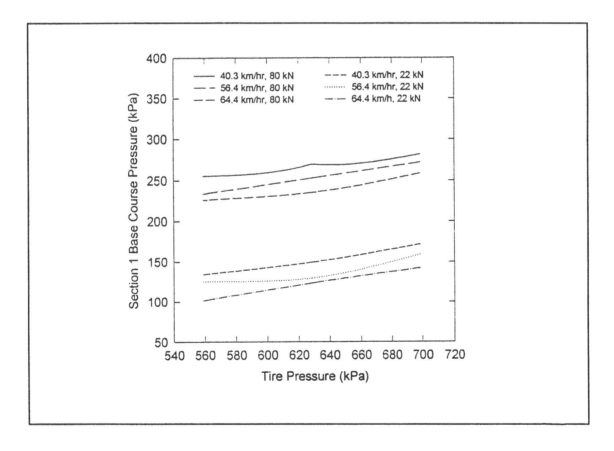

Figure 2. Base course pressure cell response for different vehicle speeds as a function of tire pressure.

SUMMARY

Nine instrumented test sections were constructed in a rural secondary road in southwest Virginia. These 15 m test sections were built to examine the effects of geogrid and geotextile stabilization. Three test sections were constructed using a geogrid, three with a geotextile, and three with no stabilization. The test sections differed in the thickness of the base course, which ranged from 10.2 to 20.3 cm. The geosynthetic stabilization was placed between the subgrade and base course layer.

The test sections were heavily instrumented with two types of pressure cells, soil and HMA strain gages, thermocouples, and soil moisture cells. In addition, strain gages were installed directly on the geogrid and geotextile. After eight months of operation, the instrument survivability has ranged from 6% for the strain gages mounted on the geotextile to 100% for the soil moisture blocks. Most of the instrument failures occurred either during construction or in the first few weeks of operation.

Piezoelectric sensors allow the data acquisition system to be triggered by passing traffic. The data acquisition system operates remotely, and the data are transferred via modem to Virginia Tech for data processing. It is planned to monitor the test section for a minimum of three years.

REFERENCES

Barksdale, R. D., S. F. Brown, and F. Chan 1989. *Potential Benefits of Geosynthetics in Flexible Pavement Systems*, Report No. 315, NCHRP, Transportation Research Board, Washington, D.C.

Hass, R., J. Walls, and R. G. Carroll 1988. Geogrid Reinforcement of Granular Bases in Flexible Pavements. *Transportation Research Record 1188*, TRB, National Research Council, Washington, D.C., 19-27.

Valentine, R. J., T. L. Brandon, I. L. Al-Qadi, and L. van't Hoog 1993. *Laboratory Performance of Geosynthetic Reinforced Pavement Sections*, Proceedings of the Thirty-Eighth Annual Conference of the Canadian Technical Asphalt Association, Polyscience Publications, Inc., 111-122.

Yang, C. and Yu, X 1989. *Mud-Pumping Prevention of Railway Subgrade Soil Using Geotextiles*, Proceedings of the Twelfth International Conference on Soil Mechanics and Foundation Engineering, Rio de Janeiro, Brazil, 3:1693-1696.

ACKNOWLEDGMENTS

This research is sponsored by the Civil Engineering Division of Amoco Fabrics and Fibers Co. and the Virginia Center for Innovative Technology. VDOT cooperation and assistance are greatly appreciated.

New Wave Propagation Devices for Nondestructive Testing of Pavements and Bridge Decks

S. NAZARIAN AND M. BAKER

ABSTRACT

The Seismic Pavement Analyzer (patent pending) is a new nondestructive testing device that can help the maintenance and design engineers in pavement evaluation. The device was developed under the Strategic Highway Research Program.

Several seismic testing techniques are combined: 1) Ultrasonic Body Wave, 2) Ultrasonic Surface Wave, 3) Impulse Response, 4) Spectral Analysis of Surface Waves (SASW), and 5) Impact Echo. With these methods eight pavement parameters are measured (including moduli of each layer). The operation of the system has been automated to make it quite simple. Most of the data reduction is done rapidly in the field, and the results are saved in a database for later analysis.

A portable version of the SPA has also been developed. The main tests used are: the Impact Echo for finding the thickness of the slab, the Ultrasonic Body Wave for estimating the modulus, and the Ultrasonic Surface Wave for also detecting the modulus. The device is attractive for detecting the onset of delamination of slabs and bridge decks, and quality control of pavement layers.

Extensive field testing on many types of base and subgrade, and bridge decks, the techniques document the suitability of the two devices, and the techniques in general, for many quality control and diagnostic projects.

INTRODUCTION

In recent years, the focus of pavement engineering has shifted from design and construction to preventive maintenance and rehabilitation of the existing highways. A highway maintenance program is usually based on a visual condition survey and, to a lesser extent, on appropriate in situ tests. By the time symptoms of deterioration are visible, major rehabilitation or reconstruction is often required. If the onset of deterioration can be measured accurately in the early stages, the problem can often be resolved or stabilized through preventive maintenance.

Soheil Nazarian, Center for Geotechnical & Highway Materials Research, The University of Texas at El Paso, El Paso, TX 79968, 915-747-6911

Mark Baker, Geomedia Research and Development, 6040 S. Strahan Road, El Paso, TX 79932, 915-877-2777

Another important issue is control of the quality of the material being placed. Both the contractor and the agency involved in the construction can benefit if any problems with the materials can be detected shortly after placement. Finally, the existing nondestructive testing devices are not adequately sensitive to the properties of the pavement layers, which can cause some uncertainties in the design and evaluation of pavements.

Two relatively inexpensive and precise device for project-level measurements has been developed. The Seismic Pavement Analyzer (SPA, patent pending) and a portable version of it, can be of help in addressing these types of problems.

OVERVIEW OF DEVICES

SEISMIC PAVEMENT ANALYZER

To diagnose a pavement, mechanical properties of each of the pavement system layers should be determined. The Seismic Pavement Analyzer (SPA) lowers transducers and sources to the pavement. The SPA then digitally records surface deformations induced by a large pneumatic hammer that generates low-frequency vibrations, and a small pneumatic hammer that generates high-frequency vibrations (see Figure 1).

This transducer frame is mounted on a trailer that can be towed behind a vehicle. The SPA is controlled by an operator at a computer connected to the trailer by a cable. The computer may be run from the cab of the truck towing the SPA or from various locations around the SPA.

All measurements are spot measurements; that is, the device has to be towed and situated at a specific point before measurements can be made. A complete testing cycle at one point takes less than one minute. A complete testing cycle includes situating at the site, lowering the sources and receivers, making measurements, and withdrawing the equipment. During this one minute, most of the data reduction is also executed.

The SPA collects two levels of data. The first level is raw data. These are the waveforms generated by hammer impacts and collected by the transducers. The second level is processed data. These are pavement layer properties derived from the raw data through theoretical models.

Some unique and significant practical features of the system are:

1. Almost all the data reduction is carried out in the field; little data reductions is required in the office.
2. The flexible structure of the software allows easy upgrade of the data reduction or interpretive capabilities of the device by either the manufacturer or the owner.
3. Rapid graphical presentation of all the pavement parameters allows the testing program to be modified or extended in the field.
4. All the measured pavement parameters are archived in a format that can be exported to a commercially available database or spreadsheet software; therefore, reports can be quickly produced.
5. The SPA can be diagnosed, and recommendations for its repair can be given, over any cellular or public phone, reducing the down time of the equipment. The data can be transmitted over a cellular telephone to an experienced engineer in the headquarters, while a technician collects data in the field.

PORTABLE SEISMIC PAVEMENT ANALYZER (LUNCH BOX)

A picture of the Lunch Box is shown in Figure 2. The device is operable from a computer located in a vehicle. The computer is tethered to the hand-carried transducer unit through a cable.

Two accelerometers and a high-frequency source are used. The receivers are connected to a data acquisition system. The rise-time of the computer-controlled source is about 10 μsec. The accelerometers adequately respond up to a frequency of 50 KHz. The data acquisition system can acquire data at a rate of 500 Ksample per channel. The collection and reduction of data at one point take less than 15 sec when an i-386-20 MHz IBM-PC compatible is used.

The device has many benefits of the SPA, except that it provides information only about the top layer of the pavement. Several advantages over other ultrasonic devices are:

1. The device is specifically designed to be functional on grooved pavements.
2. The device simultaneously measures the modulus and thickness of the pavement layer, which is imperative for accurate thickness measurement.
3. The device can average (stack) signals for higher quality data.
4. The source and receivers are controlled by the computer, therefore the results are more repeatable.
5. The ultrasonic surface wave method, the most robust way of determining modulus, can be automatically performed.

DESCRIPTION OF MEASUREMENT TECHNOLOGIES

IMPULSE-RESPONSE (IR) METHOD

Two parameters are obtained with the IR method—the shear modulus of subgrade and the damping ratio of the system. These two parameters characterize the existence of several distress precursors. In general, the modulus of subgrade can be used to delineate between good and poor support. The damping ratio can distinguish between the loss of support or weak support. The two parameters are extracted from the flexibility spectrum measured in the field. An extensive theoretical and field study (Nazarian et al, 1993) shows that except thin layers (less than 75 mm) and soft paving layers (i.e., flexible pavements), the modulus obtained by the IR method is a good representation of the shear modulus of subgrade. In other cases, the properties of the pavement layers (AC and base) affect the outcome so that the modulus obtained from the IR test should be considered an overall modulus.

SPECTRAL-ANALYSIS-OF-SURFACE-WAVES (SASW) METHOD

The Spectral-Analysis-of-Surface-Waves (SASW) method is a seismic method that can determine shear modulus profiles of pavement sections nondestructively.

The set-up used for the SASW tests is depicted in Figure 1. All accelerometers and geophones are active. The transfer function and coherence function between pairs of receivers are determined during the data collection.

A computer algorithm uses the phase information of the transfer function and the coherence functions from several receiver spacings to find a representative dispersion curve in an automated fashion (Nazarian and Desai, 1993).

The last step is to estimate the elastic modulus of different layers, given the dispersion curve. A recently developed automated inversion process (Yuan and Nazarian, 1993) determines the stiffness profile of the pavement section.

ULTRASONIC-SURFACE-WAVE METHOD

The ultrasonic-surface-wave method is an offshoot of the SASW method. The major distinction between these two methods is that in the ultrasonic-surface-wave method the properties of the top paving layer can be easily and directly found without a complex inversion algorithm.

ULTRASONIC COMPRESSION WAVE VELOCITY MEASUREMENT

Once the compression wave velocity of a material is known, its Young's modulus can be readily determined. The same set-up used to perform the SASW tests can be used to measure compression wave velocity of the upper layer of the pavement.

An automated technique for detecting the arrival of compression waves has been developed. Times of first arrival of compression waves are measured by triggering on an amplitude range within a time window (Willis and Toksoz, 1983).

IMPACT-ECHO METHOD

The impact-echo method can effectively locate defects, voids, cracks, and zones of deterioration within concrete. Once the compression wave velocity of concrete, V_p, is known, the depth-to-reflector, T, can be found from:

$$T = V_p / 2f \qquad (1)$$

where f is the resonant (return) frequency obtained by transforming the deformation record into the frequency domain.

CASE STUDIES

Due to space limitation only general descriptions of some projects where the SPA or the Lunch Box has been involved are included herein. The SPA has shown distinct advantage over other existing devices in projects such as:

1.   Pavement evaluation in a major metropolitan where the thickness and the nature of the pavement were highly irregular. Due to many overlays and utility repairs, the nature of the pavement could not even be determined by extensive coring.

2. Comprehensive pavement diagnostics in terms of the quality of concrete, potential for delamination, existence of voids beneath the slab and the modulus of subgrade reaction.
3. Determining the effects of moisture penetration on the subgrade and base moduli under a flexible pavement, where other devices could not yield results because of the presence of dipping bedrock.
4. Quality control/quality assurance of concrete pavement layers during construction.
5. Bridge deck evaluation on original and overlaid decks to detect the existence of delamination, to determine the quality of concrete, and to optimize the repair plans.

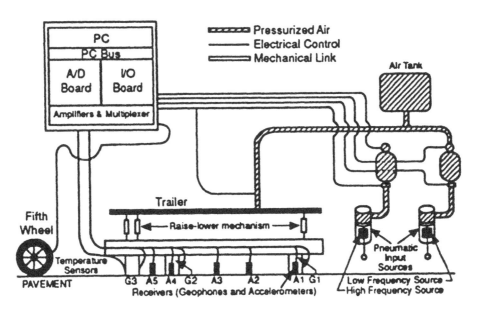

Figure 1 - Schematic of Seismic Pavement Analyzer

REFERENCES

Nazarian S. and M. Desai. 1993. "Automated Surface Wave Testing: Field Testing," *Journal of Geotechnical Engineering* (American Society of Civil Engineers, New York) 119, no. GT7:1094-112.

Nazarian, S., R.M. Baker and K. Crain. 1993. "Development and Testing of a Seismic Pavement Analyzer," Research Report SHRP-H-375, Strategic Highway Research Program, National Research Council, Washington, D.C.

Willis, M. E., and M. N. Toksoz. 1983. "Automatic P and S Velocity Determination from Full Wave form Digital Acoustic Logs." *Geophysics* 48, no. 12: 1631-44.

Yuan, D., and S. Nazarian. 1993. "Automated Surface Wave Testing: Inversion Technique." *Journal of Geotechnical Engineering* (American Society of Civil Engineers, New York) 119, no. GT7:1112-26.

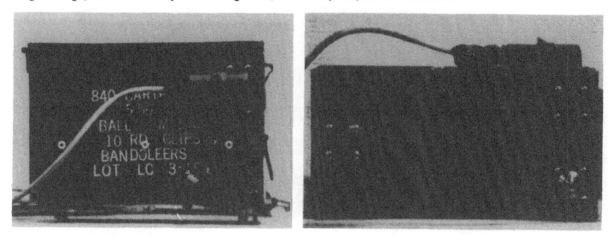

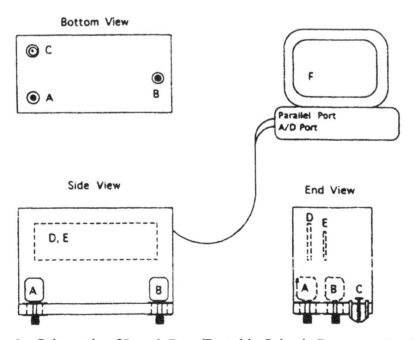

Figure 2 - Schematic of Lunch Box (Portable Seismic Pavement Analyzer)

Use of NDE Methods in Early Detection of Deterioration in Concrete Pavements

G. CLEMEÑA, T. FREEMAN, S. LANE AND M. LOZEV

ABSTRACT

This paper describes the results of recent assessments of several potentially useful nondestructive evaluation (NDE) techniques for their effectiveness in the detection of concrete pavement deterioration, at its earliest stages of development.

INTRODUCTION

Many sections of continuously reinforced concrete pavements (CRCP) in Virginia are showing signs of progressive damage. The most significant patterns of distress observed have been map cracking (with a strong longitudinal trend) and the development of potholes, cracking and punchouts at closely spaced tranverse cracks, and delamination of the slab at the level of the steel. The major mechanisms that have been associated with these patterns of distress are: excessive shrinkage and alkali silica reaction (ASR) in the concrete, corrosion of rebars, and poor drainage.

It is suspected that many newer sections of CRCP will likely suffer similar deterioration in the future. Because of the high cost of rehabilitating damaged concrete pavements, this situation presents a major challenge to pavement maintenance engineers to develop an effective maintenance strategy. The development of such strategy would, however, require the ability to detect and quantify certain types of distresses at their early stages before they become obvious on the surface of the concrete and likely more expensive to repair. Unfortunately, there isn't any proven inspection method available for doing this, as previous attempts to detect these types of damage by using the falling-weight deflectometry, and the simple method of sounding concrete with chain-drags and other similar tools have not been successful.

In response to this need, several potentially useful NDE techniques or methods were assessed.

Gerardo Clemeña, Thomas Freeman, Stephen Lane, and Margarit Lozev, Virginia Transportation Research Council, 530 Edgemont Road, Charlottesville, VA 22903

EXPERIMENTAL APPROACH

NDE METHODS

Three NDE methods, including the surface mode of ultrasonic pulse velosity (UPV) measurement, the impact-echo (IE) method, and the seismic pavement analyzer (SPA) were tested. These NDE methods all involved propagation of elastic and/or sonic waves, which are very sensitive to the presence of defects in concrete. In addition, these methods were either already sufficiently developed, or appeared to have the potential to be developed, for fulfulling the need.

Surface Measurement of UPV

Using the direct transmission mode of measurement, wherein the concrete is sandwiched between an ultrasonic transmitter and a receiver, Bungey observed a reduction in UPV (as indicated by increase in transit time) in concrete specimens affected by ASR (Bungey, 1991). This measurement mode, however, is not possible on pavements, where only the surface of the pavements is accessible to any probes or transducers being utilized. This constraint can be overcome by the use of the surface mode, although likely sacrificing some sensitivity. In this mode of measurement, both transducers would be coupled to the surface.

Impact-Echo Method

Another technique tested is the recently developed impact-echo method, which is also based on the propagation of stress waves in the concrete member being tested (Carino, Sansalome and Hsu, 1986). Briefly, the method involves introducing a transient stress pulse into the concrete by mechanically impacting its surface with a spherical steel ball (of a size appropriate for the thickness of the slab). The reflections of this stress pulse from defects in the concrete (and its boundaries) can be picked up by a piezoelectric transducer coupled to the surface of the concrete and then analyzed to determine their depths.

Seismic Pavement Analyzer

A mechanized version of the above two methods, the sonic /ultrasonic wave measurement system, was originally included in this investigation. Developed by the former Weston Geophysics, this system utilizes high-velocity, low-mass projectiles as impactors and a series of three equally spaced receivers, which are set on rolling wheels for mobile inspection. By observing the arrival times of various direct and refracted waves, the compressional and shear wave velocities of the concrete can be estimated (from known thickness), which can, in turn, serve as indirect indicators of the presence of flaws in the concrete and also as indirect measures of its quality.

That system was replaced by the seismic pavement analyzer (SPA), which measures the thickness of the concrete slab and detects some pavement conditions by estimating the compression and shear wave velocities in the pavement, base, and subgrade through techniques that included: impact-echo, impulse response, ultrasonic surface waves, and ultrasonic body waves (Nazarian and Desai, 1993). This system uses high- and low-frequency pneumatic hammers to introduce waves into the pavement and then records its response with five accelerometers and three geophones.

TEST SITES AND PROCEDURES

Six different CRCP sections were used in the investigation. The cross-sections of the pavements at these sites were similar and consisted of 200 mm-thick concrete slab, with longitudinal rebars at about 130 mm deep, over a 150-mm thick base (stabilized with either cement or asphalt), over subgrade. The concrete slabs at three of the sites (1, 3, and 5) showed closely spaced transverse cracks at many locations, while those at the other sites (2, 6, and 7) showed various degrees of fine longitudinal cracks and streaks of wet-look. Several 60-cm wide by 90-cm long test areas, longitudinal with respect to traffic, were selected from each site for a total of 47.

Measurements by all of the above methods were made in each area. For surface-mode of UPV measurements at each test area, the transit time was measured with the transmitter at the edge of the area and the receiver at either 5 or 30 cm away. Then the measurement was repeated by moving the receiver away, in the increment of either initial transducer spacing, along a straight line. At each of the test area, one series of these measurements was made -- with the transducers alignment at the longitudinal center line of the test area. For sites 2, 6, and 7, where fine longitudinal cracks were present, a second series of measurements was made with the transducer alignment perpendicular to each area. Theoretically, each series of measurements should exhibit a linear relationship between the distance of the receiver (from the transmitter) and the corresponding transit time, if the concrete slab is free of damage. Otherwise, a deviation from this linearity in transit time may be expected.

Three to four impact-echo measurements were made in each test area along the longitudinal center line, while the SPA was used along another parallel line about 15 cm off.

The surface condition of the concrete in each test area was recorded with a digital camera and a 35-mm camera. From the resulting imageries, the total number and length of cracks were estimated for each area. Cores were also extracted from all the test areas for examination.

RESULTS AND DISCUSSION

Among the 47 test areas, the concrete slabs at 29 areas were considered defective in various degrees because of presence of any sign of distress, such as transverse or longitudinal cracks, delamination, and ASR products around aggregates. As suspected from visual inspection of the general surface condition of the concrete slabs in sites 2, 6, and 7, some of the cores taken from these sites confirmed that ASR was a major mechanism of distress by the presence of varied amount of the white ASR product around some aggregates.

As the following tabulation shows, when the predictions or results of the different NDE methods were compared with the actual slab condition at the test areas of sites 1, 3, and 5, where the distress was mostly in the form of transverse cracks, it appeared that impact-echo method made the best diagnosis -- 71% correct -- when the concrete was fine. However, when the concrete slab had some transverse cracks or delamination, the impact-echo method and the UPV measurements made the best diagnosis -- 100% and 93% correct, respectively. Overall, the impact-echo appeared to provide the best diagnosis.

Method	Sites 1,3,5		Sites 2,6,7	
	Concrete Sound	Concrete Defective	Concrete Sound	Concrete Defective
UPV	29%	93%	36%	100%
IE	71%	100%	64%	100%
SPA	0%	47%	45%	86%

For sites 2, 6, and 7, when the concrete was fine, impact-echo method again performed the best, although its accuracy was far from being satisfactory. And, when there was problem with the concrete, both UPV measurements and the impact-echo method performed extremely well, with SPA not far behind.

Overall, i.e., for both groups of pavements tested, the impact-echo method appeared to have provided the best diagnosis of the condition of the concrete slabs tested.

In addition to predicting the condition of the concrete slabs, the PSA also provided estimates of their thickness, which was compared with the actual length of the extracted cores. It appeared that the estimates of PSA were off by +1 to -20 mm, excluding three test areas (7-5, 7-6, and 7-7) in site 7 where the SPA was off by -40 to -109 mm -- these areas happened to be approach slabs to a bridge.

REFERENCES

Bungey, J. H. 1991. "Ultrasonic Testing to Identify Alkali-Silica Reaction in Concrete." *British Journal of NDT*, 33(5): 227.

Carino, N. J., M. Sansalome and N. N. Hsu. 1986. "Flaw Detection in Concrete by Frequency Spectrum Analysis of Impact-Echo Waveforms." International Advances in Nondestructive Testing, 12th Edition, Gordon & Breach Science Publishers, New York, N.Y.

Nazarian, S., and M,. Desai. 1993. "Automated Surface Wave Testing: Field Testing." *Journal of Geotechnical Engineering*, 119 (GT7): 1094-1112.

Laboratory Radar Evaluation of Concrete Decks and Pavements with and without Asphalt Overlay

U. B. HALABE, H. L. (ROGER) CHEN, M. K. ALLU AND L. PEI

ABSTRACT

This paper presents the preliminary results of an ongoing study using Ground Penetrating Radar (GPR) for locating subsurface anomalies in concrete bridge decks and pavements with and without asphalt overlay. Specimens were cast in the laboratory with varying internal conditions such as with or without reinforcement and with air- or water-filled cracks. These specimens simulate the conditions of field bridge decks and pavements. The radar waveform obtained from these specimens were studied to determine the effect of subsurface cracks on the radar waveform and to assess the sensitivity of the technique.

INTRODUCTION

Bridge deck and pavement deterioration is one of the leading problems faced by many state highway departments. Assessment of the condition of concrete is critical especially when it is overlaid with asphalt. The overlay covers the deterioration until it is far advanced. Therefore, in order to prioritize program rehabilitation activities and estimate service life, significant efforts have been focused on developing methods for rapid determination of deterioration in bridge decks and pavements.

The early stages of deterioration in asphalt covered concrete bridge decks and pavements is not detectable by the traditional methods such as chain drag or resistivity method. Moreover, they are laborious and less reliable. Core sampling, which gives fair results, causes localized damage to the deck or pavement and hinders traffic. Hence nondestructive techniques such as ground penetrating radar (GPR), which offers rapid and non-contact measurement, are becoming increasingly popular.

Considerable research has been done to evaluate the asphalt layer thickness in overlaid concrete decks and pavements using GPR. Maser (1994) used radar for asphalt layer thickness

Udaya B. Halabe, Assistant Professor, H. L. (Roger) Chen, Associate Professor, Mani K. Allu and Lianfeng Pei, Graduate Research Assistants, Department of Civil and Environmental Engineering, Constructed Facilities Center, West Virginia University, Morgantown, WV 26506-6103

evaluation and achieved an accuracy within 5-10%. Scullion et al. (1994) showed that the GPR system developed by Texas Transportation Institute is capable of estimating asphalt layer thickness up to 600 mm within 5% error. GPR has also been used to detect subsurface cracks and delaminations in field concrete decks (e.g., Maser, 1991). Also, a study was previously conducted at WVU (Chen et al., 1994; Halabe et al., 1996) to assess the effectiveness of GPR technique in detecting subsurface cracks in 6" thick concrete deck specimens under controlled laboratory conditions. This previous study showed that it is possible to detect 0.12" cracks in concrete decks, especially when they are water-filled. The objective of the present research is to determine the detectability of subsurface cracks in concrete pavement specimens. Also, the effect of asphalt overlay on the detectability of subsurface cracks will be investigated at a later date.

TEST SPECIMENS

The laboratory test specimens include the following: (a) eight concrete pavement specimens of size 24" x 24" x 10" and two specimens of size 24" x 24" x 12" (b) fourteen concrete bridge deck specimens of size 24" x 24" x 6". The above specimens were cast to simulate varying internal conditions such as with and without reinforcement and with air and water filled cracks. The design concrete compressive strength is 4300 psi, with a water/cement ratio of 0.5, and the ratio of cement:sand:aggregates as 1:2:3. The cracks in the 10" thick specimens were located at approximately 5" depth. For the 6" thick concrete specimens, the cracks were located are about 1.5" depth.

COMPARISON OF WAVEFORMS

The waveforms obtained from the laboratory specimens contain noise due to internal reflections within the antenna. Analysis of these waveforms has been conducted after subtracting the antenna noise. The radar waveforms from different specimens were compared to assess the effect of subsurface cracks on the waveform. Figure 1 shows comparison of radar waveforms from 10" thick plain concrete specimen and specimens with embedded plexiglas and plain water crack. The plexiglas simulates an air-filled crack. A small reflection is observed due to the water-filled crack at the location of the crack (about halfway deep into the specimen). The reflection from the plexiglas (air crack) is also observable, but to a lesser extent compared to the water-filled crack. Figure 2 shows comparison of radar waveforms from 10" thick concrete specimens without crack, and with plain and saline water-filled cracks. Both cracks show similar reflections, and it appears that the locations of the cracks are slightly different due to casting irregularities. The differences between waveforms from cracked and uncracked specimens are rather small due to high wave attenuation within the concrete. On the other hand, the cracks located at lower depths (about 1.5") in 6" thick concrete specimens (not shown in the figure) have been found to be more prominent (Halabe et al., 1996; Chen et al., 1994).

CONCLUSIONS

This study has attempted to detect cracks within concrete deck and pavement specimens in the laboratory. The cracks have been detected by comparing the individual waveforms from specimens with air- or water-filled cracks with the waveform from sound concrete specimen. While a previous study indicated significant reflections due to cracks located within concrete specimens at depths of about 1.5", the current study shows that the reflections from cracks located at about 5" depth are significantly lower. This is due to the high radar wave attenuation within concrete. The detectability of these subsurface cracks for asphalt overlaid specimens is currently being investigated, and the results will be reported in a later paper.

ACKNOWLEDGEMENTS

The authors appreciate the financial support provided by the West Virginia Division of Highways (WVDOH) for this research. The authors especially thank the project monitor, Mr. Don Lipscomb, for his assistance and valuable comments.

REFERENCES

Chen, H. L., U. B. Halabe, Z. Sami, and V. Bhandarker. 1994. "Impulse Radar Reflection Waveforms of Simulated Reinforced Concrete Bridge Decks," *Materials Evaluation*, 52(12):1382-1388.

Halabe, U. B., H. L. Roger Chen, V. Bhandarker, Z. Sami. 1996. "Detection of Subsurface Anomalies in Concrete Bridge Decks Using Ground Penetrating Radar," *American Concrete Institute (ACI) Materials Journal*, to be published.

Maser, K. R. 1991. "Bridge Deck Condition Surveys Using Radar: Case Studies of 28 New England Decks," *Transportation Research Record 1304*, Washington, D. C., 94-102.

Maser, K. R. 1994. "Highway Speed Radar for Pavement Thickness Evaluation," *Proceedings of Fifth International Conference on Ground Penetrating Radar*, Kitchener, Ontario, Canada, June 12-16, 423-431.

Scullion, T., C. L. Lau and Y. Chen. 1994. "Pavement Evaluation Using Ground Penetrating Radar in Texas," *Proceedings of Fifth International Conference on Ground Penetrating Radar*, Kitchener, Ontario, Canada, 449-463. June 12-16.

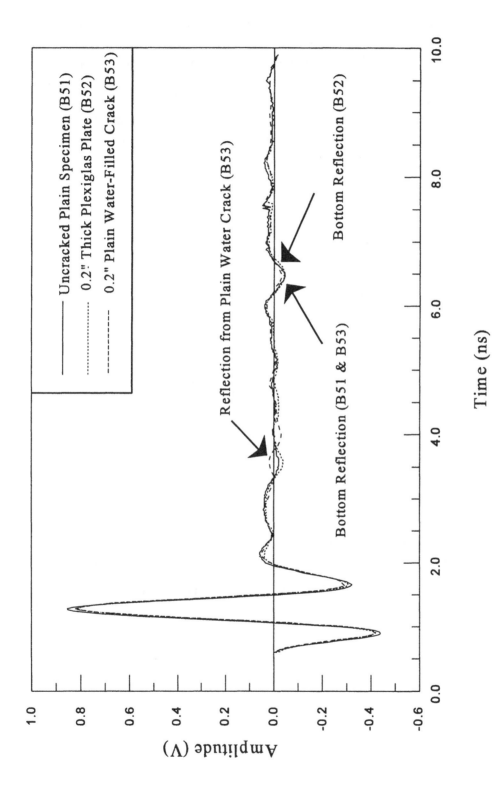

Figure 1. Comparison of Plain (B51), Plexiglas (B52), and Water-Filled Crack (B53) Specimens

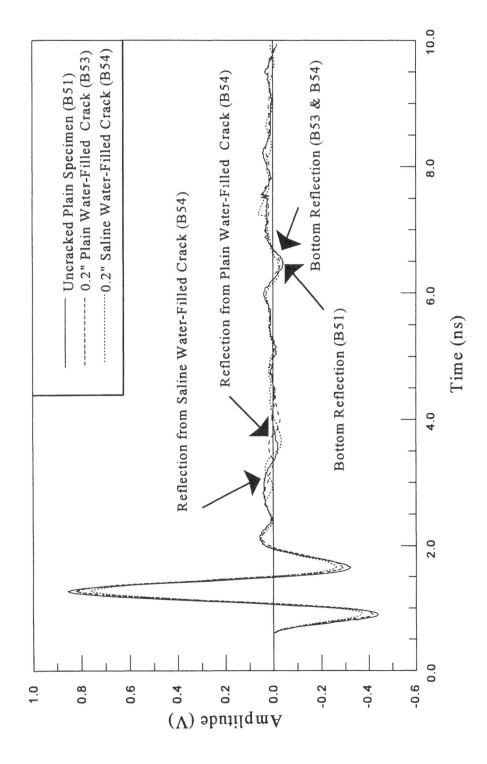

Figure 2. Comparison of Plain (B51), Plain Water-Filled Crack (B53), and 8% Saline Water-Filled Crack (B54) Specimens

377

Current NDT Procedures for Pavement Structural Evaluation

N. F. COETZEE, R. C. BRIGGS, W. A. NOKES AND P. J. STOLARSKI

ABSTRACT

This paper provides an overview of typical non-destructive test (NDT) procedures currently used for pavement structural evaluation applications. In particular, surface deflection measurements, spectral analysis of surface waves (SASW) and ground penetrating radar (GPR) approaches are described and discussed. Data interpretation and application, including problems encountered with each type of measurement, are also covered. Potential future developments in this field are discussed.

INTRODUCTION

Evaluation of pavement condition is desirable from both a network level, for pavement management system purposes, and a project level for specific rehabilitation design purposes. Testing techniques for evaluation can generally consist of destructive measures, usually involving removal of material for laboratory testing purposes, or in-situ non-destructive testing (NDT) approaches. This paper provides an overview of the current pavement NDT procedures, primarily from the structural response perspective. Other NDT techniques are available for consideration of layer thicknesses, shape and roughness, surface distress and surface texture (1). A brief discussion of the most common NDT approach for layer thickness determination, involving the use of ground-penetrating radar (GPR) is also included.

The most common NDT approach for pavement structural evaluation involves consideration of surface deflections under an applied load using various types of equipment, ranging from the Benkelman beam, originally used in 1956 (2), to the current equipment of choice viz. the Falling Weight Deflectometer (FWD). Other approaches, such as spectral analysis of surface waves (SASW) (3,4) and laser measurements of deflections under moving loads (5) are at various stages of development and are not in common usage at this point. However, rapid developments are occurring in these applications (6,7) and it is expected that use will increase as the technology and ease of use improves.

N.F. Coetzee, Dynatest Consulting, Inc., P.O. Box 71, Ojai, CA 93024; R.C. Briggs, Dynatest Consulting, Inc., P.O. Box 337, Starke, FL 32091; W.A. Nokes, California Dept. of Transportation, 650 Howe Ave., Ste. 400, Sacramento, CA 95825; P.J. Stolarski, California Dept. of Transportation, 5900 Folsom Blvd., Sacramento, CA 95819

TYPES OF NDT DEFLECTION MEASUREMENTS

The commonly used NDT equipment for deflection testing can be categorized according to the nature of the load applied to the pavement surface. Static or slow moving load deflection measurements represent the first generation approach, which basically originated with the development of the Benkelman beam at the WASHO Road Test in the early 50's. The next generation involved application of a dynamic vibratory load, exemplified by the Dynaflect and Road Rater. These pieces of equipment are more mobile and productive than the static equipment, and led to deflection measurements becoming a routine pavement condition survey task. Falling weight deflectometers can be considered third generation deflection equipment, and measure deflections resulting from a dynamic impulse load which attempts to simulate the effect of a moving wheel load. Current development is directed at equipment that will measure deflections caused by a loaded wheel moving at highway speeds.

DEFLECTION USES

Early use of deflection data typically involved consideration of maximum deflection directly under the load, relative to empirical standards. Usually some statistical measure of deflections on a pavement section is compared with a "tolerable" deflection level for that section under the expected traffic. If the measured value exceeds the tolerable deflection then an empirical procedure determines the corrective measure required, usually an overlay, to reduce the measured deflections to the tolerable level. Examples of this approach include The Asphalt Institute's MS-17 (8) and CalTrans' Test Method 356 (9). In some states maximum deflections are monitored during spring thaw and load restrictions are placed when the thawing pavement's deflection reaches a certain level. Empirical use of deflection basin data usually involves one of the "basin parameters" which combine some or all of the measured basin deflections into a single number.

With a trend towards mechanistic pavement analysis and design, which is based on fundamental engineering principles, the use of deflection data has become more sophisticated. Complete deflection basins are used, in a procedure known as backcalculation, to estimate in-situ elastic moduli for each pavement layer. Knowledge of the existing layer thicknesses are typically necessary for this procedure. The backcalculated moduli themselves provide an indication of layer condition. They are also used in an elastic layer or finite element program to calculate stresses and strains resulting from applied loads. These stresses and strains are used with fatigue or distress relationships to evaluate damage accumulation under traffic and predict pavement failure. They can also be used to evaluate corrective measures such as overlays, rehabilitation or reconstruction. It is these mechanistic analyses of pavement deflection that this paper is intended to address. Briefly, the backcalculation procedure involves calculation of theoretical deflections under the applied load using assumed pavement layer moduli. These theoretical deflections are compared with measured deflections and the assumed moduli are then adjusted in an iterative procedure until theoretical and measured deflection basins match acceptably well. The moduli derived in this way are considered representative of the pavement response to load, and can be used to calculate stresses or strains in the pavement structure for analysis purposes.

Calculation of theoretical deflections, and the subsequent stress or strain calculations, currently typically involve linear elastic theory. Application of elastic theory may be through the use of:

(i)   Traditional layered elastic programs based on numerical integration procedures such as ELSYM5, CHEVRON (various versions), BISAR and WESLEA.

(ii)   The Odemark-Boussinesq transformed section approach rather than numerical integration.

(iii)   Finite element programs, either those that have been specifically oriented towards pavement analysis, such as ILLI-PAVE or MICHPAVE, or general structural analysis programs such as SAP (various versions), ANSYS, ABACUS, ADINA, etc.

(iv)   Plate theory such as the Westergaard solutions for PCC pavements.

(v)   neural networks trained to reproduce results that emulate one of the above applications (10,11).

BACKCALCULATION

Lytton (12) provides an in-depth summary of the historical developments of NDT, backcalculation and theoretical considerations as well as associated technology in his state-of-the-art presentation in 1988. He also illustrates some of the concerns regarding the differences between backcalculated results using different backcalculation programs on the same deflection data. These are typically technical problems but they are exacerbated by the continuing development of similar backcalculation programs. In many cases, new programs have little to differentiate them from existing software other than a name.

Most of the programs rely on linear elastic layered theory, or a variation thereof, for the basic structural model. In comparing results from these programs, the primary criterion used for evaluation of accuracy is based on the goodness of fit of computed deflections to measured deflections. As computing power has increased, so has the ability to improve the goodness of fit. An important fact to keep in mind is that, in many cases, improving the goodness of fit does not necessarily mean that the theoretical model better represents actual pavement response. If an existing pavement structure is in such a condition that it clearly violates some of the fundamental assumptions of elastic theory, then a good fit between measured and calculated deflections should *not* be expected, and goodness-of-fit should *not* be the determining factor for deciding if a solution is realistic or not. This point is also made by Lytton (12) who discusses the need for experience in analysis, with materials and with deflections to ensure that the backcalculation process yields the most acceptable set of moduli for a given deflection basin. It should be noted that essentially all pavements violate the fundamental assumptions of linear elastic theory, albeit to differing degrees.

Also important, and related to the issue discussed above, is the fact that backcalculation provides a modulus value that is a layer parameter and not necessarily the layer material modulus, which can be measured using laboratory tests on a sample of the layer material. This is due to the geometry of typical deflection basin measurements, which is typically on the order of a 1.8 meter (6 ft.) length, so that the effect of horizontal layer and material variability over that dimension is included in the backcalculated moduli. This variability includes damage such as cracking, both on the macro- and micro-structural level. Simply stated, the problem lies with the fact that the in-situ modulus is not known, so that backcalculated values cannot be validated directly.

Specific problems encountered during backcalculation which relate primarily to the solution procedures or to violation of the fundamental assumptions of elasticity, continuity, homogeneity and isotropy are discussed in detail elsewhere (13, 14).

SASW APPROACHES

Spectral analysis of surface waves (SASW) approaches evaluate pavements using Rayleigh wave measurements involving low strain levels. Until very recently, both data acquisition and analysis was cumbersome and time consuming. Recent equipment and software development under the SHRP-IDEA program appears to have made significant advances in the application of SASW techniques (6). Hiltunen (14) describes the basic approach as follows:, "Current practice calls for locating two (or more) vertical receivers on the pavement surface a known distance apart and a transient wave containing a large range of frequencies is generated in the pavement by means of a hammer. The surface waves are detected by the receivers and are recorded using a Fourier spectrum analyzer. The analyzer is used to transform the waveforms from the time to the frequency domain and then to perform spectral analyses on them. The spectral analysis functions of interest here are the phase information of the cross power spectrum and the coherence function. Knowing the distance and the relative phase shift between the receivers for each frequency, the velocity of the surface wave (phase velocity) and wavelength associated with that frequency are calculated. The final step is application of an inversion process that constructs the shear wave velocity profile from the phase velocity-wavelength information (dispersion curve)."

The SASW approach has been used for determination of the pavement layer modulus profile, as well as layer thickness, depth to stiff layer, modular ratios between layers and detection of voids or delamination problems. To date, the most successful application appears to be layer thickness determination, which suggests that the SASW approach could be used as a complement to FWD testing. Layer moduli determined from SASW applications do not appear to be representative of the pavement response exhibited by a moving wheel load, probably due to the fact that SASW approaches involve high frequency, low strain measurements. Wheel loads typically result in higher stresses and strains, and the apparent layer modulus is often a function of applied stress levels due to material non-linear response (14,15).

GPR MEASUREMENTS

Ground Penetrating Radar (GPR) is currently used fairly extensively for pavement layer thickness determination. Pavement layer thickness is a critical input parameter to the backcalculation process for structural evaluation (14,16) and has traditionally been inferred from destructive coring at limited locations on a given project. GPR allows continuous high-speed measurement of layer thickness, as well as providing a measure of subgrade moisture variation (17,18). Harris (17) provides the following brief description of the procedure:

"GPR directs pulses of electromagnetic radiation into the ground or pavement structure. A portion of this energy is reflected back to the surface, and picked up by the GPR receiver, at each location in the pavement structure where a significant difference in electrical properties of the materials occur. The electrical property of interest is the material's dielectric constant. GPR is effective for pavement evaluations as long as there is sufficient contrast in the dielectric constant of the paving materials. Additionally, the dielectric constant is frequency dependent. The following dielectric ranges are typical for paving materials at a frequency of approximately 1 GHz:

- Air   1
- Asphalt Surface / Black Base   5 to 6
- Concrete / Cement Stabilized Base   8 to 9
- Flexible Base   10 to 11 (highly moisture dependent)
- Water   80
- Steel   81

From the list above, it is evident that pavement layers composed of materials having significantly different dielectric constants can be identified. Pavement structures having multiple layers with similar dielectric constants are more difficult to evaluate, and it may not be possible to identify each individual layer and measure its thickness. Additionally, city streets often have utility patches and maintenance practices that can confuse data reduction.

The wavelength of a 1-gigahertz GPR system is approximately three inches. Layer thicknesses of approximately one-quarter of the radar wavelength or greater can be resolved. Consequently, GPR systems cannot resolve pavement layers which are less than one inch in thickness. The 1-gigahertz system has a depth of penetration of approximately 24 inches. The penetration depth is a function of the overall dielectric constant of the pavement structure. Materials possessing a high dielectric constant tend to attenuate the radar signal, thereby decreasing its effective depth of penetration.

A 500-megahertz GPR system has a wavelength of approximately six inches. Since the signal has a longer wavelength, it can penetrate deeper into the pavement structure. The 500 MHz system is capable of measuring to depths of four to five feet, depending on the dielectric of the material. The trade-off is that there is less thickness measurement capabilities with the 500 MHz system when compared to the 1 GHz system."

SUMMARY AND POSSIBLE FUTURE DEVELOPMENTS

(i) Direct measurement of pavement deflection response under load remains a convenient method for assessing structural capacity, and future developments are focusing on performing these measurements at highway traffic speeds.

(ii) Combinations of current measurement techniques such as FWD and SASW or GPR are likely to provide more reliable backcalculated moduli since this approach can reduce the effect of layer thickness variation on the analysis techniques.

(iii) The analytical techniques typically used for the basis of backcalculation programs are limited in their ability to realistically model actual pavement material response. It is likely that finite-element based approaches, which are more capable of realistic modelling, will become more widely used as computing power increases.

(iv) Neural network approaches for modelling pavement response are likely to become more widespread. Training of such networks requires realistic modelling of pavement response, which is related to (iii) above.

(v) Validation of pavement analysis results, including those related to NDT measurements, is likely to become more common. The current widespread interest in full-scale and accelerated pavement testing (APT) generally involves extensive and sophisticated pavement instrumentation which will provide some means of validation.

REFERENCES

(1) Epps J.A. and Monismith C.L., *"Equipment for Obtaining Pavement Condition and Traffic Loading Data"*, NCHRP Synthesis of Highway Practice No. 126, Transportation Research Board, Washington D.C., 1986.

(2) Hudson W.R., Elkins G.E., Eden W. and Reilley, K.T., *"Evaluation of Pavement Deflection Measuring Equipment"*, USDOT, Federal Highway Administration, Report No. FHWA-TS-87-208, Washington D.C., 1987.

(3) Heisey J.S., Stokoe K.H., Hudson W.R. and Meyer A.H., *"Determination of In Situ Shear Wave Velocities from Spectral Analysis of Surface Waves"*, Center for Transportation Research, Research Report 256-2, The University of Texas at Austin, Austin, TX, 1982.

(4) Nazarian S. and Stokoe K.H., *In Situ Determination of Elastic Moduli of Pavement Systems by Spectral-Analysis-of-Surface-Waves Method (Theoretical Aspects)"*, Research Report 437-2, Center for Transportation Research, University of Texas at Austin, Austin, TX, 1987.

(5) Elton D.J. and Harr M.E., *New Nondestructive Pavement Evaluation Method"*, ASCE, Journal of Transportation Engineering, Vol. 114 No. 1, New York, NY, 1988.

(6) Nazarian S. and Baker M., *"A New NDT Device Under Development For Preventative Pavement Maintenance"*, Technical Quarterly TQ7-3, Sept. 1993.

(7) —— Rolling Weight Deflectometer Product Brochure, Quest Integrated, Inc., Kent, WA, 1995.

(8) —— *"Asphalt Overlays for Highway and Street Rehabilitation"*, The Asphalt Institute, Manual Series No. 17 (MS17), June 1983.

(9) —— Asphalt Concrete Overlay Design Manual, California Department of Transportation, 1979.

(10) Meier R.W. and Rix G.J., *"Backcalculation of Flexible Pavement Moduli Using Artificial Neural Networks"*, Presented at the 73rd Transportation Research Board Annual Meeting, Washington D.C., 1994.

(11) Meier R.W. and Rix G.J., *"Backcalculation of Flexible Pavement Moduli from Dynamic Deflection Basins Using Artificial Neural Networks"*, Presented at the 74th Annual Meeting, Transportation Research Board, Washington D.C., 1995.

(12) Lytton R.L., *"Backcalculation of Pavement Layer Properties"*, ASTM STP 1026, 1988.

(13) Ullidtz P. and Coetzee N.F., *"Analytical Procedures in NDT Pavement Evaluation"*, Presented at the 74th Annual Meeting, Transportation Research Board, Washington D.C., 1995.

(14) —— *"Pavement Deflection Analysis — Participant Workbook"*, prepared by Dynatest Consulting, Inc. and Soils and Materials, Inc. for the FHWA National Highway Institute, 1993.

(15) Stubstad R.N., Mahoney J.P. and Coetzee N.F., *"Effect of Material Stress Sensitivity on Backcalculation and Pavement Evaluation"*, <u>Nondestructive Testing of Pavements and Backcalculation of Moduli (Second Volume)</u>, American Society for Testing and Materials, Philadelphia, PA, 1994.

(16) Briggs R.C., Scullion T. and Maser K.R., *"Asphalt Thickness Variation on Texas Strategic Highway Research Program Sections and Effect on Backcalculated Moduli"*, Transportation Research Board, TRR #1377.

(17) Harris R.L., Personal Communication, 1995.

(18) Lau C.L., Scullion T. and Chan P., *"Modelling of Ground Penetrating Radar Wave Propagation in Pavement Systems"*, Presented at the 71st Annual Meeting, Transportation Research Board, Washington D.C., 1992.

Author Index

Printed and bound by CPI Group (UK) Ltd, Croydon, CR0 4YY

23/10/2024

01777686-0018